Additive Manufacturing of Ceramics

Additive Manufacturing of Ceramics

Fundamentals, Techniques, and Applications

Edited by Zhangwei Chen and Paolo Colombo

WILEY VCH

Editors

Prof. *Zhangwei Chen*
Additive Manufacturing Institute
Shenzhen University, Shenzhen
China

Prof. *Paolo Colombo*
Department of Industrial Engineering
University of Padova, Padova
Italy

Cover Design: Wiley
Cover Image: Zhangwei Chen

Typesetting Straive, Chennai, India

Library of Congress Card No.: applied for

British Library Cataloguing-in-Publication Data A catalogue record for this book is available from the British Library.

Bibliographic information published by the Deutsche Nationalbibliothek
The Deutsche Nationalbibliothek lists this publication in the Deutsche Nationalbibliografie; detailed bibliographic data are available on the Internet at <http://dnb.d-nb.de>.

Print ISBN: 978-3-527-35502-0
ePDF ISBN: 978-3-527-85174-4
ePub ISBN: 978-3-527-85173-7
oBook ISBN: 978-3-527-85175-1

Printed and bound by
CPI Group (UK) Ltd, Croydon, CR0 4YY

C9783527355020_030626

Contents

About the Editors *xi*
Preface *xiii*

1 **Introduction** *1*
Zhangwei Chen and Paolo Colombo
1.1 Most Important Properties and Applications of Ceramics *2*
1.1.1 Most Important Properties of Ceramics *2*
1.1.2 Main Applications of Ceramics in Advanced Science and Technology *3*
1.2 Typical Processing Routes and Manufacturing Techniques for Ceramics *7*
1.2.1 Typical Processing Routes for the Fabrication of Ceramic Components *7*
1.2.2 General Manufacturing Techniques for Ceramics *8*
1.3 Challenges and the Need for Additive Manufacturing *9*
1.4 Brief Introduction to Ceramic Additive Manufacturing Techniques *10*
1.5 Summary of the Book *14*
1.6 Conclusions *18*
References *19*

2 **Vat Photopolymerization (VPP)** *25*
Vincent Pateloup and Gweni Alonso Aruffo
2.1 Introduction *25*
2.2 Photopolymerization 3D Printing Processes for Ceramic Parts *26*
2.2.1 Stereolithography Process *27*
2.2.2 Digital Light Processing Process *27*
2.2.3 Top-Down and Bottom-Up Approaches *28*
2.2.4 Feedstock Formulations for VAT Photopolymerization Processes *30*
2.2.5 Thermal Treatment *32*
2.2.6 Summary of Both Processes *33*
2.3 Process Modeling and Simulation *34*
2.3.1 Context *34*

2.3.2 Mechanical Modeling 35
2.3.3 Limits of Actual Modeling Approaches 37
2.4 Advantages and Limits of VAT Processes 37
2.5 AM Hybridization Opportunities 39
2.6 Next Generation Ceramic Printing: Two-Photon Photopolymerization and Volumetric Additive Manufacturing Methods 41
2.6.1 Two-Photon Polymerization 42
2.6.2 Volumetric Additive Manufacturing 43
2.7 Conclusion 45
Acknowledgments 46
References 46

3 Binder Jetting (BJT) and Material Jetting (MJT) 51
Andrea Zocca, Jens Günster, Anna-Katharina Hofer, and Stefan Waldner
3.1 Inkjet Printing 51
3.1.1 Continuous Inkjet Printing 52
3.1.2 Drop-on-Demand Inkjet Printing (DOD) 53
3.1.2.1 Piezoelectric DOD Inkjet Printing (PIJ) 54
3.1.2.2 Thermal DOD Inkjet Printing (TIJ) 57
3.1.2.3 Other Approaches of DOD Inkjet Printing 58
3.1.3 Theory of Ink Printability 59
3.2 Binder Jetting (BJT) 62
3.2.1 Definition of the Technology (ISO/ASTM) and Working Principle 62
3.2.2 Historical Development and Applications 63
3.2.2.1 BJT of Gypsum Plaster Models 64
3.2.2.2 BJT of Sand Casting Molds and Cores 64
3.2.2.3 Large-Scale BJT (Concrete and Construction Materials) 65
3.2.2.4 BJT of Ceramics for Technical Applications 70
3.2.3 Feedstock Properties and Process Parameters 77
3.2.3.1 Powder Feedstock Properties 77
3.2.3.2 Layer Recoating 80
3.2.3.3 Process Parameters and Binder Saturation 82
3.3 Material Jetting 84
3.3.1 Definition and Classification of Material Jetting Technologies 84
3.3.2 Historical Development 86
3.3.2.1 Material Jetting of Ceramics 86
3.3.2.2 Nano-Particle Jetting (NPJ)TM 87
3.3.2.3 Multi-Material Jetting (MMJ) 87
3.3.2.4 Micro Particle Jetting^{TM} (MPJ) 88
3.3.3 Process Variations 88
3.3.4 Challenges in Material Jetting 90
3.3.4.1 Nozzle Failures 90
3.3.4.2 Droplet Ejection, Impingement, and Fusion 92
3.3.4.3 Leveling 93

3.3.4.4 Curing and Crack Formation *93*
3.3.5 Applications *94*
3.3.6 Potential of Material Jetting *96*
3.4 Post-Processes *98*
References *100*

4 Material Extrusion (MEX) *109*
Frank Clemens and Esther García-Tuñón
4.1 Introduction *109*
4.2 Direct Ink Writing *109*
4.2.1 DIW Systems *110*
4.2.2 Feedstock Preparation: Additives and Processes *111*
4.2.3 Feedstock Characterization *114*
4.2.4 Printing Parameters *121*
4.3 Fused Filament Fabrication *124*
4.3.1 Printhead Systems *124*
4.3.2 Feedstock Preparation: Additives and Processes *126*
4.3.3 Feedstock Characterization *128*
4.3.4 Printing Parameters *129*
4.4 Heat Treatment and Post-Processing *131*
4.4.1 Post-Processing for DIW-Printed Ceramic Parts *132*
4.4.2 Post-Processing for FFF-Printed Ceramic Parts *132*
4.5 AM of Fiber-Reinforced Composites *135*
4.5.1 Fiber-Reinforced Composites by DIW Printing *135*
4.5.2 Fiber-Reinforced Composites by FFF Printing *138*
4.6 Summary *139*
Acknowledgments *140*
References *140*

5 Powder-Bed Fusion (PBF) *153*
Enrique Juste and Fabrice Petit
5.1 Short Historical Background of PBF *154*
5.2 Terminology Focus *155*
5.3 Indirect – Powder-Bed-Selective Laser Processing (PBSLP) *157*
5.4 Direct – Powder-Bed-Selective Laser Processing (PBSLP) *161*
5.4.1 Electron Beam Melting (EBM) *161*
5.4.2 Powder-Bed-Selective Laser Processing (PBSLP) *164*
5.5 Process Parameters PBF *171*
5.5.1 Powder-Bed Features *171*
5.5.1.1 Powder Properties *172*
5.5.2 Hardware Parameters *174*
5.6 Heat Treatment and Post-Processing *182*
5.7 Conclusion *186*
References *187*

6 Directed Energy Deposition (DED) of Ceramics *195*
Amit Bandyopadhyay, Nathaniel W. Zuckschwerdt, and Susmita Bose
6.1 Introduction *195*
6.2 Advantages and Limitations of DED *197*
6.3 Powder Characteristics *199*
6.4 DED Processing-Related Challenges *200*
6.5 Processing via DED *201*
6.6 Monolithic Ceramics via DED *201*
6.7 Metal-Matrix Composites via DED *203*
6.8 Current Challenges and Future Directions *206*
6.9 Summary *207*
Acknowledgments *207*
References *207*

7 Hybrid AM of Ceramics and AM of Ceramic-Based Multi-Material Components *211*
Giorgia Franchin, Uwe Scheithauer, and Tassilo Moritz
7.1 Hybrid Additive Manufacturing of Ceramics *211*
7.1.1 Concurrent Hybrid Processes *212*
7.1.1.1 UV-Assisted MEX *212*
7.1.1.2 Laser-Assisted MEX *214*
7.1.2 Sequential Hybrid Processes *214*
7.1.2.1 Additive/Additive *214*
7.1.2.2 Additive/Subtractive *215*
7.1.3 Conclusions for Hybrid Manufacturing *216*
7.2 Additive Manufacturing of Ceramic-Based Multi-Material Components *217*
7.2.1 Introduction Concerning AM of Multi-Material Ceramic Components *217*
7.2.2 Multi-Material AM Methods for Ceramics *219*
7.2.2.1 Process Category Vat Photopolymerization *219*
7.2.2.2 Process Category Material Extrusion *221*
7.2.2.3 Process Category Material Jetting *225*
7.3 Applications *227*
7.4 Conclusion and Outlook for AM of Ceramic Multi-Material Components *231*
References *232*

8 Design and Characterization of Additively Manufactured Ceramics *243*
Riccardo Balzarotti, Marco Pelanconi, Giovanni Bianchi, and Alberto Ortona
8.1 Introduction *243*
8.2 Model Design – General Rules for Ceramic AM *243*
8.3 Topological Optimization and Numerical Simulation *244*
8.4 Characterization of the Feedstocks Prior to Printing *245*
8.4.1 Powder Based *245*

8.4.2 Liquid/Slurry Based *250*
8.4.3 Bulk Solid Based *251*
8.5 Characterization of the Parts Before and After Heat Processing *252*
8.5.1 Mechanical Properties *252*
8.5.2 Composition and Microstructure Characterization, from the Green to the Consolidated Part *255*
8.5.3 Surface Roughness *256*
8.5.4 Dimensional and Geometrical Stability *256*
8.5.5 Density and Porosity *257*
8.6 Summary *259*
References *260*

9 Industrial VPP Equipment for AM of Functional Ceramic Composites *269*
Martin Schwentenwein, Serkan Nohut, and Johannes Homa
9.1 Introduction *269*
9.2 Materials and Rheology *270*
9.3 Processing *270*
9.4 Reproducibility *271*
9.5 Quality Management, Compliance, and Performance Metrics *271*
9.6 Economic and Regulatory Considerations *273*
9.7 Industrial Applications of Lithography-Based Ceramic AM *273*
9.7.1 Biomedical Engineering *273*
9.7.2 Energy and Aerospace Engineering *274*
9.7.3 Chemical and Process Engineering *274*
9.7.4 Electronics and Micro-Systems *274*
9.8 Market Considerations and Selection Guidance *274*
9.9 Multi-Material Additive Manufacturing *275*
9.10 Multi-Material AM of Ceramic Composites with LCM *276*
9.10.1 Lithography-Based Ceramic Manufacturing (LCM) *276*
9.10.2 LCM Multi-Material Printing *277*
9.11 Applications for LCM Multi-Material Printing *279*
9.11.1 Alumina/Zirconia-Toughened Alumina *279*
9.11.2 Porous/Dense Alumina *282*
9.12 Future Research *286*
9.13 Summary *289*
Acknowledgments *289*
References *290*

10 Applications and Impacts of AM of Ceramics *295*
Kareen Malsallez, Maxence Bourjol, Natalia Kostretsova, Marc Torrell, Albert Tarancón, Malte Hartmann, Johannes Noll, Georgia Astbury, Matthew Drane, and Elsa Antunes
10.1 Aerospace and Defense Applications *296*
10.1.1 Aerospace Applications: Large-Scale, Lightweight, and Integrated Structures *296*

10.1.2 Space Communication: Ceramic Antennas and NewSpace Innovation *297*
10.1.3 Defense: High-Performance, Tailored Components *297*
10.2 Chemical Applications *298*
10.3 Electronic and Semiconductor Applications *299*
10.3.1 Downholders for Welding *299*
10.3.2 Semiconductor Industry: From Prototyping to Industrial-Scale Production *301*
10.4 Biomedical Applications *301*
10.4.1 Personalized Healthcare: The Expanding Role of Ceramic 3D Printing in Biomedicines *301*
10.4.2 Orthopedic and Dental Solutions *302*
10.4.3 Next-Generation Surgical Tools and Devices *303*
10.5 Detailed Case Studies of Ceramic AM Applications *306*
10.5.1 Solid Oxide Cells for High-Efficiency Hydrogen Production *306*
10.5.2 Ecological Restoration for Environmental Protection *311*
10.6 Cross-Domain Value and Systemic Impact of Ceramic Additive Manufacturing *315*
10.7 Conclusion *317*
References *318*

11 New Developments and Perspectives *321*
Fiona Spirrett and Soshu Kirihara
11.1 Introduction *321*
11.2 Green/Smart Design and Manufacturing *322*
11.2.1 Sustainable Feedstocks and Material Circularity *323*
11.2.2 Environmental Impact of Ceramic Sintering *325*
11.2.3 Design for Sustainability – Computational Modeling *326*
11.2.4 Challenges and Barriers to Green Design and Manufacturing *328*
11.3 4D/5D Printing *328*
11.3.1 4D Ceramic Printing *329*
11.3.2 5D Ceramic Printing *333*
11.4 Potentials and Challenges *336*
References *338*

Index *341*

About the Editors

Professor Zhangwei Chen

Zhangwei Chen is currently a tenured professor and the director of Additive Manufacturing Institute at Shenzhen University, China. He also serves as the Deputy Director of the Key Laboratory of Electromagnetic Control and Intelligent Robotics of Guangdong Province and the Group Leader of High-performance AM Innovation Team, Department of Education of Guangdong Province. He was awarded Distinguished Scholar by Shenzhen University in 2021. He holds a PhD degree from Imperial College London, with his thesis awarded John Kilner Prize for Energy Materials. He has received the Second Prize of Chinese Innovation Association Award, First Prize of Shenzhen Science and Technology Award, and Second Prize of Guangdong Science and Technology Award. He is a Fellow of the International Association of Advanced Materials (IAAM) and Stanford University Top 2% Global Scientist for the year 2021–2025. He has also received the Top Young Investigator Award of the year 2020 from the Division of Advanced Ceramics of the Chinese Ceramic Society. He works on the fields of preparation and manufacturing of high-performance materials and components, involving materials processing, additive manufacturing, and characterizations. He has spent over 18 years, particularly on the investigation of additive manufacturing of ceramics, including ceramic powder-based slurries, polymer-derived ceramics and composite materials, and the property–structure relationships. He has published over 160 peer-reviewed articles, with a citation number over 9000 and an H-index of 43, in prestigious journals including *Progress in Materials Science*, *Materials Science & Engineering: Reports*, *Advanced Functional Materials*, *Nano Energy*, *International Journal of Extreme Manufacturing*, *Additive Manufacturing*, *Journal of Advanced Ceramics*, *Virtual and Physical Prototyping*, *Journal of the European Ceramic Society*, etc. He has been invited to be co-organizer/chairman in a number of national and international conferences/symposia on ceramic additive manufacturing, such as ICACC, PACRIM, ECerS, and CICC. In particular, he has initiated and organized the Forum on Additive Manufacturing of Ceramics (FAME2021) in 2021. He currently also serves as editor of *Journal of Advanced Ceramics*, *Materials Science in Additive Manufacturing*, *Journal of Inorganic Materials*, and *Journal of Materials Engineering*. He is also a panel referee of funding agents at an international level, including European Commission Fund, Canada EPSRC, Singapore A STAR Fund, New Zealand Marsden Fund, NSF Switzerland, NSF China, and so on.

Professor Paolo Colombo

Paolo Colombo is a professor of Materials Science and Technology at the Department of Industrial Engineering, University of Padova, Padova, Italy. He graduated from the University of Padova with a degree in chemical engineering in 1985 and a diploma in Glass Engineering in 1988. He was an assistant professor at the University of Padova from 1990 to 1998 and then an associate professor at the University of Bologna until 2005. He is also Adjunct Professor of Materials Science and Engineering at the Pennsylvania State University, Honorary Professor in the Department of Mechanical Engineering of University College London, UK, Honorary Researcher at the Henan Academy of Sciences, Zhengzhou, Henan, CN and Honorary Professor at Xi'an Jiaotong University, Xi'an, Shaanxi, CN. He was a Foreign Scientist at INSA, Lyon, France, a DGF Mercator Professor at the Technical University Bergakademie Freiberg, Germany, an Academic Advisor at the Institute for Advanced Ceramics, School of Materials Science and Engineering, Harbin Institute of Technology, China, and a Visiting Professor at the University of São Paulo, Brazil.

He was elected Academician of the World Academy of Ceramics, the European Academy of Sciences, Academia Europaea, and the Italian Academy of Engineering and Technology. He is Fellow of the American Ceramic Society, the Institute of Materials, Minerals and Mining, the European Ceramic Society, and the International Core Academy of Sciences and Humanities. He was awarded a Fulbright Scholarship for the Pennsylvania State University in 1991 and has received numerous awards. He is Past President (2022–2024) of the International Ceramic Federation (ICF) and Vice President of the Italian Ceramic Society (2024–2026), and a Stanford University Top 2% Global Scientists for the years 2019–2025.

He published about 400 papers in peer-reviewed journals, with a citation number over 27,200 and an H-index of 82, in prestigious journals including *Progress in Materials Science*, *Materials Science & Engineering: Reports*, *Science*, *Nature Materials*, *Advanced Materials*, *Advanced Science*, *Small*, *Matter*, *Additive Manufacturing*, *Journal of the European Ceramic Society*, etc. He is co-editor of a book on cellular ceramics, a book on polymer-derived-ceramics, 11 proceedings books, and guest editor of 10 Special issues in international scientific journals. He holds 20 international patents.

He is Editor-in-Chief of Open Ceramics (Elsevier) and is on the editorial board of 11 international scientific journals. He has co-organized several international conferences and has given numerous invited talks, keynote, and plenary in the field of porous ceramics, polymer-derived ceramics, and additive manufacturing of ceramics. He was chair of the XVI conference of the European Ceramic Society (Turin, Italy, 16–20/6/2019) and of the Shaping 10 conference (Padova, 20–23/9/2026).

Paolo Colombo's research interests include novel processing routes to architected glasses and ceramics (currently focusing mainly on Additive Manufacturing, using different technologies), the development of ceramic components from preceramic polymers and geopolymers, and the upcycling of industrial and natural waste.

Preface

The field of additive manufacturing (AM), commonly known as 3D printing, has revolutionized the way we design and fabricate objects across industries. From rapid prototyping to the production of end-use parts, AM technologies have unlocked unprecedented geometric freedom, enabling the creation of complex, lightweight, and customized structures that were previously impossible or prohibitively expensive to manufacture. While the adoption of AM began with polymers and has expanded significantly into metals, 3D printing of ceramics has presented a unique and compelling frontier.

Ceramics, in fact, are notably challenging to process. Their inherent brittleness, high melting point, and sensitivity to thermal gradients have long confined their manufacturing to traditional techniques like pressing, casting, and machining, which often involve complex multistep processes, expensive tooling, and limitations in geometric complexity. The advent of AM for ceramics is dismantling these barriers, offering a pathway to fabricate intricate, high-performance ceramic components in a more flexible, more reproducible, more customizable, and digitally driven manner.

This book, *Additive Manufacturing of Ceramics: Fundamentals, Techniques and Applications*, is conceived as a comprehensive guide to this rapidly evolving field. It is designed to serve a broad audience, including materials scientists, mechanical and manufacturing engineers, researchers, students, and industry professionals seeking to understand and leverage the power of AM for ceramic materials.

The structure of the book is organized to provide a logical progression from fundamental principles to specific processes and advanced applications. Initial chapters lay the groundwork by exploring the unique characteristics of ceramic materials that influence their printability and the essential prerequisites for feedstock development, such as powder morphology, rheology, and binder chemistry. The core of the book is dedicated to in-depth examinations of the major AM techniques employed with ceramics. This includes detailed discussions of Vat Photopolymerization (e.g. Stereolithography – SLA, Digital Light Processing – DLP), Powder-Bed Fusion (e.g. Selective Laser Sintering/Melting – SLS/SLM), Material Extrusion (e.g. Direct Ink Writing – DIW, Fused Filament Fabrication – FFF), Binder Jetting, and Directed Energy Deposition (DED). Each technique is analyzed critically, with a focus on its working principles, necessary material modifications, key process parameters, and the resulting microstructures and properties of the printed parts. The book also addresses the significant challenges associated with each method, such as achieving high density, minimizing defects, controlling shrinkage, and implementing necessary post-processing steps like debinding and sintering.

Finally, the text explores the emerging applications of additively manufactured ceramics across diverse sectors. Examples span from biomedical implants (e.g. customized bone scaffolds, dental restorations) and advanced electronics (e.g. heat sinks, insulators) to aerospace components (e.g. turbine blades, radomes) and industrial tools. The book concludes by looking toward the future, examining emerging trends like multi-material printing, hybrid additive manufacturing, the integration of artificial intelligence for process optimization, and the development of new ceramic compositions designed specifically for AM.

Our goal is that this book will become a valuable resource, providing both a solid foundational understanding and practical insights that will empower readers to push the boundaries of what is possible in the field of additive manufacturing of ceramics. We hope it inspires further innovation and contributes to the continued growth and maturation of this exciting technology.

We would like to take this opportunity to thank all the authors and relevant parties for their excellent contributions. Without their experience, time, and energy in preparing each of the chapters, this book would not have been possible. We are also thankful to the Wiley team for editing, formatting, pressing, and follow-ups in all technical aspects in publishing this book.

October 2025

Zhangwei Chen
Paolo Colombo

1

Introduction

Zhangwei Chen[1] *and Paolo Colombo*[2,3]

[1] *Shenzhen University, Additive Manufacturing Institute, No. 3688, Nanhai Avenue, 518060 Shenzhen, China*
[2] *University of Padova, Department of Industrial Engineering, via Marzolo 9, 35131 Padova, Italy*
[3] *Pennsylvania State University Department of Materials Science and Engineering, University Park, PA 16802, USA*

Ceramics are a class of inorganic, nonmetallic materials that are typically hard, brittle, and resistant to deformation and abrasion. They are one of the oldest and most diverse classes of materials known to humankind, ranging from traditional pottery to cutting-edge advanced ceramics [1]. Traditional ceramics, such as earthenware, stoneware, and porcelain, are crafted from natural materials like clay and are fired at high temperatures to create durable and aesthetically pleasing items for everyday use and art. Advanced ceramics, on the other hand, are a category of materials engineered for specific high-performance applications. Structural ceramics, such as oxide ceramics (e.g. alumina and zirconia) and non-oxide ceramics (e.g. silicon carbide and silicon nitride), are chosen for their exceptional mechanical properties, including high strength, hardness, and wear resistance, making them ideal for components in extreme environments like aerospace and automotive industries [2, 3]. Functional ceramics, which include piezoelectric and ferroelectric materials, are valued for their unique electrical properties [4]. These materials can convert mechanical stress into electrical signals, or vice versa, and are crucial in sensors, actuators, and electronic devices. Advanced ceramics also encompass bioceramics, used in medical implants due to their biocompatibility, and ceramic matrix composite (CMC) materials, which exploit the properties of composites for enhanced performance [5, 6]. These advanced ceramics are revolutionizing industries with their high performance and specialized functionalities. The categories of different ceramics are illustrated in Figure 1.1.

In the domain of advanced science and technology, ceramics play a crucial role in various applications due to their unique properties. This section will delve into the most important properties of ceramics, their typical processing and fabrication techniques, as well as the challenges and the need for additive manufacturing (AM) technologies.

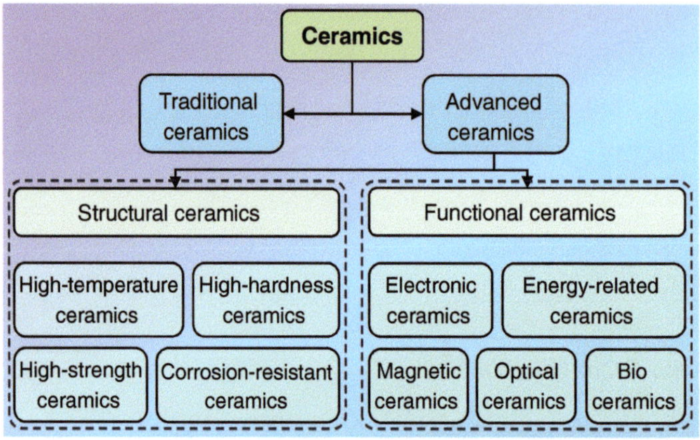

Figure 1.1 Categories of different ceramics, specifically of advanced ceramics.

1.1 Most Important Properties and Applications of Ceramics

Ceramics have been an integral part of human civilization for thousands of years, evolving from simple pottery to sophisticated materials used in advanced technologies. The unique properties of ceramics make them indispensable in various industries, from aerospace to medicine.

1.1.1 Most Important Properties of Ceramics

- **Chemical stability:** One of the most notable properties of ceramics is their chemical stability [7]. They are highly resistant to chemical attack and corrosion, which makes them suitable for applications in environments where metals would deteriorate. This resistance is particularly valuable in chemical processing plants, where ceramics can be used to line reactors and storage vessels, protecting them from corrosive substances [8].
- **Thermal stability:** Ceramics are known for their ability to withstand high temperatures without significant degradation [9]. This thermal stability is crucial in applications such as furnaces, kilns, and engine components, where materials must operate under extreme heat. For instance, ceramic tiles on spacecraft are designed to protect against the intense heat generated during re-entry into Earth's atmosphere [10].
- **Electrical insulation:** Most ceramics are excellent electrical insulators, a property that is vital in electronic devices and high-voltage applications [11]. Their insulating capabilities prevent the flow of electric current, making them ideal for use in insulators, capacitors, and other electronic components that require dielectric materials [12].
- **Mechanical strength:** Despite their brittleness, ceramics can exhibit high compressive strength, which makes them suitable for load-bearing applications [13].

This strength is leveraged in construction materials and in the manufacturing of components that must support heavy loads without significant deformation or fracture [14].

- **Hardness:** The hardness of ceramics makes them resistant to wear and abrasion, a characteristic that is beneficial in cutting tools and industrial machinery [15, 16]. Ceramic knives, for example, maintain their sharpness longer than their metal counterparts due to their hardness.
- **Optical properties:** Some ceramics possess unique optical properties, such as transparency, which is essential in applications like optical fibers and laser components [17]. The transparency of certain ceramics allows light to pass through with minimal scattering or absorption, making them ideal for fiber-optic communication systems [18].
- **Biocompatibility:** Certain types of ceramics are biocompatible, meaning they can be integrated with living tissue without causing adverse reactions [19]. This property makes them suitable for medical implants and prosthetics, such as hip and knee replacements, where the ceramic materials can mimic the physicochemical characteristics of bone [20, 21].

1.1.2 Main Applications of Ceramics in Advanced Science and Technology

The unique properties of ceramics – chemical stability, thermal stability, electrical insulation, mechanical strength, hardness, optical properties, and biocompatibility – make them a versatile and essential material in a wide range of applications. Figure 1.2 illustrates the main cutting-edge applications of advanced ceramics.

Structural ceramics are engineered to perform under high mechanical, thermal, and tribological stresses, as well as in corrosive environments and high temperatures [22]. They are used across a variety of domains and products due to their superior strength, thermal stability, and wear resistance. Here are some specific examples of applications and fields where structural ceramics are utilized. They are used in precision cutting tools to extend tool life, as well as in high-performance bearings and seals, and wear-resistant machine components, due to their high strength and high hardness. They are used in aerospace industry as part of gas turbines, nozzles, and heat exchangers due to their ability to resist high temperatures, mechanical stresses, corrosion, and chemical degradation under critical environments characterized by extreme conditions [23]. They can also be employed in automotive industry as high-temperature engine components, brake system elements, and exhaust systems to improve performance and durability [24]. Moreover, various laser components like pump chambers, feedthroughs, waveguides, and ion tubes are often made of porous ceramics, especially alumina [25]. Moreover, CMCs combine the high-temperature resistance of ceramics with the toughness of composite materials [26]. They have been developed for use in advanced aircraft engines and other high-stress environments. CMCs offer superior mechanical properties such as high strength, stiffness, and toughness, making them suitable for structural components in energy storage systems and lightweight, durable enclosures [27].

Figure 1.2 Advanced ceramics for cutting-edge applications.

Oxide ceramics, such as alumina (Al_2O_3), zirconia (ZrO_2), and silica (SiO_2), are known for their versatility, low cost, and relatively abundant resources. These ceramics are widely used as structural ceramics, such as in aerospace applications, for their high-temperature stability, hardness, and corrosion resistance. For instance, they are used in the ceramic cores for casting turbine blades for aircraft engines [28], as well as in heat-resistant tiles for spacecrafts [29]. Non-oxide ceramics, including carbides, nitrides, and borides, are characterized by higher temperature resistance, increased strength, and fracture toughness compared to oxide ceramics [30]. Silicon carbide (SiC) and silicon nitride (Si_3N_4) are used in structural applications requiring high thermal conductivity and mechanical strength, such as in aerospace, semiconductor equipment, energy and environmental protection, and nuclear power, as advanced structural components [31–38]. Non-oxide ceramics like tungsten carbide (WC) and boron carbide (B_4C) are utilized in superhard abrasives and cutting tools due to their extreme hardness [39, 40]. Furthermore, non-oxide ceramics are used in rocket nozzles and electrodes due to their ability to withstand high temperatures and chemically aggressive environments [41]. The above-mentioned applications highlight the versatility and importance of structural ceramics in modern manufacturing and engineering, where their unique properties provide solutions for demanding performance requirements.

Advanced ceramics often also possess certain required functional properties, alongside their structural features. For example, optical ceramics not only need excellent optical properties, high transparency in particular, but also require high resistance to fracture and durability for long-term use. Bioceramics are used in wear surface applications such as femoral hip and acetabular structures, as well as in spinal applications and dental implants [42]. Besides structural features like load bearing and wear resistance, their biocompatibility also makes them preferred materials over metals, as they better mimic the properties of bone and are less likely to cause rejection by the body's immune system. Oxide structural and functional ceramics represented by alumina and zirconia are widely used in optical communication and semiconductors, electronic packaging substrates, sensors, medical devices, smart wear, electric vehicles, and other fields. For example, they are widely used as functional electronic ceramics in the electronics industry, particularly in multilayer ceramic capacitors (MLCCs) and resistors [43]. Semiconductor equipment such as photolithography machines, plasma etchers, ion implanters, heat treatment furnaces, and diffusion furnaces required for semiconductor wafer and chip manufacturing use a large number of high-purity alumina, aluminum nitride, yttrium oxide, silicon nitride, and silicon carbide ceramic parts (such as electrostatic and vacuum chucks, handling arms, wafer carriers, and etched rings) [44]. Aluminum nitride (AlN) with high thermal conductivity is not only widely used in electronic packaging substrates but is also now part of insulated-gate bipolar transistors (IGBTs) in electric vehicles and wind power generation systems [45]. Other functional ceramics include piezoelectric ceramics, which generate an electric charge in response to mechanical stress, making them invaluable in sensors, actuators, and energy harvesting devices [46–48]. Their ability to convert mechanical stress into electrical signals is crucial in various technological applications, from medical imaging to industrial automation. Energy-related ceramics are critical in advancing energy storage and conversion technologies, combining high efficiency, durability, and tailored electrical properties. In solid oxide fuel cells (SOFCs), ceramics like lanthanum strontium cobalt ferrite (LSCF) and yttria-stabilized zirconia (YSZ) enable high electronic and ionic conductivity [49], while transition metal oxide-based functional ceramics such as lithium iron phosphate ($LiFePO_4$) serve as the cathode materials in lithium-ion batteries [50]. Electromagnetic wave absorption has now become a demanding functional feature for ceramic composite components used in aerospace applications [51, 52].

Table 1.1 demonstrates the selected applications of ceramics in various domains and relevant properties required. Oxide ceramics are favored for their electrical and thermal insulation properties in electronics and aerospace, while non-oxide ceramics are chosen for their superior mechanical properties and high-temperature performance in cutting tools, electronics, and high-stress environments.

Ceramics have come a long way from their ancient origins to become a cornerstone in advanced science and technology. Their unique properties make them indispensable in a wide range of applications, from aerospace to medicine. As research continues to push the boundaries of what ceramics can do, their importance is only set to grow. The development of nanoceramics [53], self-healing ceramics [54], and ceramic-based quantum technologies [55] is just a few examples of how ceramics are shaping the future of technology. As we look to the future, ceramics will undoubtedly continue to play a vital role in the advancement of human knowledge and the development of new technologies.

Table 1.1 Applications of ceramics in various domains and relevant properties required.

Application field	Ceramic components used	Ceramic properties required	Typical relevant ceramics
Aerospace	• Heat shields • Heat exchangers • Turbine blades • Reflecting mirrors	• High-temperature resistance • High strength • Thermal stability • Low density	• Boron nitride • Carbo-silicon carbide • Silicon carbide • Titanium diboride
Electronics	• Capacitors • Resistors • Substrates • Sensors	• Dielectric properties • Thermal conductivity • Thermal stability • Electrical insulation	• Alumina • Barium titanate • Lead zirconate titanate • Silicon nitride
Biomedicine	• Artificial bone implants • Dental implants • Dental veneers	• Biocompatibility • Biodegradability • Porosities • Hardness • Aesthetics	• Alumina • Hydroxyapatite • Silicon nitride • Tricalcium phosphate • Zirconia
Energy	• Fuel cells • Batteries • Furnaces • Nuclear power	• High-temperature resistance • Corrosion resistance • Chemical stability • Mixed electron and ion conductivities	• Alumina • Zirconia • Lanthanum strontium cobalt ferrite • Lithium iron phosphate • Lithium orthosilicate • Yttria-stabilized zirconia
Automobile	• Automotive engines • Brake system elements • Bearings	• Heat resistance • Wear resistance • High strength • High hardness	• Alumina • Silicon carbide • Silicon nitride • Zirconia
Environmental protection	• Ceramic filters • Ceramic membranes • Catalytic converters	• High-temperature resistance • Corrosion resistance • Thermal stability • Porosities	• Alumina • Cordierite • Silicon carbide • Silicon nitride
Optics	• Lenses • Light guides • Optical windows • Beam splitters	• Transparency • High strength • Thermal stability	• Aluminum oxynitride • Yttrium aluminum garnet • Zirconium silicate
Others	• Jewelry • Commercial device casings	• Strength • Hardness • Surface finish • Chemical stability • Aesthetics	• Zirconia • Alumina • Cordierite

1.2 Typical Processing Routes and Manufacturing Techniques for Ceramics

The processing routes and fabrication techniques for ceramics have undergone significant advancements since their traditional use as pottery and bricks, particularly with the development of advanced ceramics for specialized uses in industries such as aerospace, electronics, and biomedical.

1.2.1 Typical Processing Routes for the Fabrication of Ceramic Components

Advanced ceramics require sophisticated processing techniques due to their specialized applications and the need for high purity and precise control over microstructure. The typical processing route for the fabrication of ceramic components includes the following steps, as also illustrated in Figure 1.3:

- **Powder synthesis:** High-purity powders are synthesized through various chemical processes, such as precipitation from solution, sol–gel processes, or vapor-phase synthesis.
- **Feedstock processing:** The synthesized powders are then mixed with organic solvents and additives to form pastes or slurries that can be further processed.
- **Green body formation:** The processed powder is shaped to form a green body, which is the unsintered, pre-fired body of the desired ceramic part. This can involve techniques such as dry pressing (if no solvent is added), plastic forming, and casting.
- **Drying and binder removal:** The green body is dried, and any binders or organic additives are removed through a process called debinding, which can be carried out thermally or using chemical solvents.

Figure 1.3 Typical processing routes for the fabrication of ceramic components.

- **Sintering:** The green body is sintered at high temperatures to obtain the final dense and strong ceramic component. Sintering can be carried out through various methods, including conventional sintering, hot pressing, and spark plasma sintering (SPS).
- **Post-sintering treatments:** Depending on the application, the sintered ceramic may undergo additional treatments such as machining, glazing, or polishing to achieve the desired properties or appearance. Complicated post-processing such as polymer infiltration and pyrolysis may also be used for further densification and performance enhancement purposes.

1.2.2 General Manufacturing Techniques for Ceramics

The shaping and manufacturing techniques for ceramics have evolved to accommodate the diverse range of properties and applications expected from the components. Some of the key techniques include:

- Forming:
  - **Dry pressing:** A forming technique suitable for simple shapes, where dry or semi-dry powders are compacted in a die under high pressure to form the desired shape.
  - **Cold/hot isostatic pressing:** Techniques used for advanced ceramics where a flexible mold containing the ceramic powder is subjected to high pressure from all sides under high or low temperature simultaneously, resulting in a highly homogeneous, dense, and strong final product.
- Molding:
  - **Extrusion molding:** A continuous process where a viscous mixture of ceramic particles and binders is forced through a die to form a shape, which is then cut into pieces and dried.
  - **Injection molding:** Similarly to polymer processing, a mixture of ceramic powder, plasticizer, and thermoplastic polymer is injected into a mold under high pressure and then sintered after the polymer is removed. This method can produce relatively complex shapes with tight tolerances.
  - **Rolling molding:** A process that shapes ceramics by passing them between rollers.
- Casting:
  - **Slip casting:** A traditional method where a slurry of ceramic powder is cast into a plaster mold. The water is drawn out by capillary action, leaving a layer of ceramic particles that form the shape of the mold.
  - **Tape casting:** Used for producing thin, flat ceramic sheets, this method involves spreading a slurry of ceramic powder, binders, and plasticizers onto a carrier film, then drying it to form a tape.
  - **Gel casting:** A forming technique where a ceramic slurry is mixed with a monomer, poured into a mold, and then polymerized to form a green body.

As mentioned earlier, sintering at high temperature is needed after the binder removal process, where the ceramic body is further heated to a temperature below the melting point of the ceramic material to allow particle bonding and densification, together with coarsening. The densification can be further improved by high-pressure sintering to achieve better mechanical properties of the final ceramic parts. There are also precision machining processes involved for further refinement of the shaped/manufactured ceramic parts before and after sintering. They include abrasive grinding and polishing, as well as tooling and cutting. Some other methods such as chemical and electrical polishing might also be used to achieve a smooth surface finish and tight tolerances. They can also be used to remove any surface defects introduced during sintering.

1.3 Challenges and the Need for Additive Manufacturing

The above-mentioned shaping, forming and manufacturing methods have undergone significant advancements in the ceramic industry in the past few decades. However, these methods face significant challenges. They often necessitate the use of molds, which can be demanding both in terms of process time and economic costs, and limit the production of highly complex or customized parts. Additionally, the inherent brittleness and hardness of ceramic materials present challenges in machining, also leading to difficulties in achieving intricate geometries and often resulting in tool interference during the processing of highly complex structures. The production of complex structures is particularly challenging due to the limitations in design imposed by the manufacturing process itself, often requiring simplification for manufacturability. Moreover, these methods can be time-consuming and may not be cost-effective for producing small batch sizes or one-off, customized items.

AM, also known as 3D printing, has emerged as a revolutionary approach to address these challenges. The discrete mold-free unit-by-unit (e.g. point-by-point, line-by-line, layer-by-layer, and volume-by-volume) [56] manufacturing process offers the ability to create complex geometries with a high degree of dimensional precision and shape complexity, without the need for expensive tooling [57]. This technology eliminates many traditional design constraints, enabling the fabrication of parts with unique shapes and functionalities that were previously unattainable with conventional methods. The discrete manufacturing nature of AM also allows to produce components with varying material properties, which is particularly beneficial in industries like aerospace and automotive where weight and performance are critical.

Furthermore, AM facilitates mass customization by reducing setup costs, making it economically viable to produce small lots or even single units for specific users or applications. This is especially advantageous for creating customized products such as prosthetics or implants, where each item needs to be tailored to individual specifications. AM supports both high and low-volume production, making it a versatile manufacturing solution. AM also allows engineers to create integrated and intricate structural or functional ceramic components with greater design flexibility, effectively shortening the lead times and saving costs. It also enables distributed manufacturing, which allows products to be manufactured closer to the point of need, further reducing transportation costs and lead times. The technology also promises to minimize waste during manufacturing due to its additive nature, aligning with sustainability goals.

In summary, while traditional ceramic forming methods have their merits, they are often hindered by the limitations in producing highly complex, customized parts and the challenges associated with machining hard and brittle ceramic materials. AM presents a viable solution to these issues, offering efficiency, design flexibility, customization, and the ability to create intricate structures with minimal material waste, thus paving the way for new possibilities in ceramic component manufacturing.

1.4 Brief Introduction to Ceramic Additive Manufacturing Techniques

AM has revolutionized the way ceramics are fabricated, offering a range of advantages over the conventional ceramic processing/shaping/manufacturing methods and other advanced manufacturing processes [48, 58–60].

The history of ceramic AM is a narrative of innovation and technological advancement. The earliest research in ceramic AM can be traced back to the late 20th century, with significant developments in the 1990s. Over the years, various techniques have been developed and refined for ceramic fabrication.

Here, different types of AM techniques used for ceramic fabrication are briefly summarized, according to the ISO/ASTM 52900:2021 standard [61] (Figure 1.4). Their working principles, controlling factors, and relevant research progress will be briefly covered and discussed in the following chapters.

- **Vat photopolymerization (VPP):** This method includes the linear scanning stereolithography (SL), planar photopolymerization such as digital light processing (DLP), two-photon polymerization (TPP), and volumetric additive manufacturing (VAM). These techniques use light to polymerize a ceramic slurry or resin in a vat. A dynamic mask or projection system is used to irradiate the light and cure the polymer in defined areas, building the part from the bottom up. They are suitable for creating parts with high precision and excellent surface quality.
- **Material extrusion (MEX):** MEX, such as direct ink writing (DIW) and fused deposition modeling (FDM), is an important part of ceramic AM. This method involves the extrusion of a ceramic material, often in a paste or filament form, through a nozzle to build up layers. The principle is based on the action of extrusion force during the printing process, which produces ceramic green bodies with relatively high solid content, eliminating the need for subsequent densification processes. Recently, a hybrid-process MEX technique such as DIW combined with UV photopolymerization as well as Embedded DIW has also been developed for high-quality printing of structures with geometries not achievable using common DIW approaches.

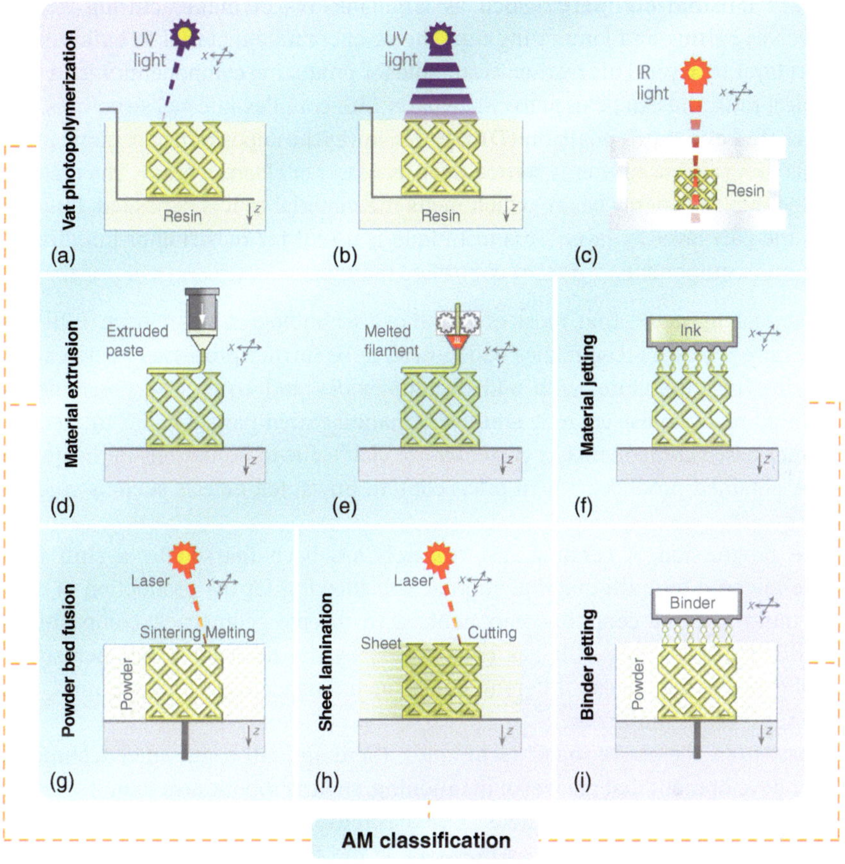

Figure 1.4 Major categories of additive manufacturing techniques suitable for ceramic fabrication. (a) SL. (b) DLP. (c) TPP. (d) DIW. (e) FDM. (f) IJP. (g) SLS/SLM. (h) LOM. (i) BJ3DP. Reference [62] / with permission of Elsevier.

- **Material jetting (MJT):** MJT involves the selective deposition of droplets of ceramic-binder mixed slurry onto a build platform. The droplets solidify to form a layer of the part, and the process is repeated until the part is complete. This method allows for the creation of complex geometries and can be used to produce parts with varying material properties or multi-material components.
- **Binder jetting (BJT):** BJT is a process in which a liquid bonding agent (i.e. binder) is selectively deposited to join the pre-deposited ceramic powder materials on the build platform. This process is different from the MJT method in that only the binder is released by the nozzles.
- **Powder bed fusion (PBF):** Techniques like selective laser sintering (SLS) and selective laser melting (SLM) fall under the category of PBF. These processes involve spreading a thin layer of ceramic powder (with secondary phase of binder powder in the case of SLS) in a bed and then using a laser to selectively fuse the powder based on the cross-sectional area of the desired part.

- **Sheet lamination (SHL)**, such as laminated object manufacturing (LOM), involves cutting and laminating thin sheets of ceramic material to build up the part layer by layer. This method is suitable for producing components with a high aspect ratio and can be used to create parts with complex internal structures.
- **Directed energy deposition (DED):** It involves the deposition of ceramic materials using a focused energy source, such as a laser or electron beam. The material is fed into the energy beam, which melts the material as it is deposited, building up the part layer by layer. This technique is useful for repairing or building up ceramic components in a more controlled manner.

It should be noted that most of the above technologies, apart from PBF and DED, only produce green bodies which need to be further processed by debinding and sintering, therefore, with added complexities and costs. The possibility of obtaining ready-to-use ceramic additively manufactured parts, similar to metal or polymer-based components, is currently not viable due to limitations in the quality of the obtained products, which often contain unwanted defects such as voids or cracks.

The progression of ceramic AM research has been marked by a shift from single-material to multi-material approaches, allowing for the production of complex multi-material ceramic components with diverse geometries, compositions, and distributions. This shift has been driven by the need for components with varying properties across different fields, including aerospace, electronics, and communications industries.

Apart from the above major techniques, there are still some other techniques under development that are worth mentioning, such as robotic arm manufacturing. Of course, hybrid printing methods as well as multi-material printing methods are currently under the spotlight, as they enable the fabrication of complex ceramic components with integrated functional materials in a single build process.

As the individual ceramic AM techniques are discussed, it should also be noted that the feedstock required for successful ceramic AM might require special treatment and preparation. Ceramic powder quality is an important basis and guarantee for the consistency and reliability of high-performance electronic ceramics, structural ceramics, and bioceramic materials and products, and especially nano and submicron ultrafine powders have become the key raw materials for today's advanced ceramic industry.

Table 1.2 shows the history and development of AM technologies suitable for the fabrication of ceramic products and their origin, the requirements of the feedstocks, as well as the pros and cons of the individual techniques.

The future of ceramic AM lies in the development of multi-material and multi-functional ceramics. Research is focusing on improving the precision, strength, and functionality of ceramic parts, as well as expanding the range of materials that can be used in AM processes. Associated with this, there is a need to develop suitable approaches for the efficient and precise detection of internal defects in the produced parts as well as spotting any quality issues during printing, to avoid further processing of unsuitable components. The integration of different ceramic materials with

Table 1.2 Brief chronicle of ceramic AM techniques and their pros/cons.

Category	Technique	Time first appeared	Applied to ceramics	Pros	Cons
Vat photopolymerization (VPP)	Stereolithography	1984 [63]	1993 [64]	Good resolution, large printing size	Scattering, large material volume is needed
	Digital light processing (for 3D printing)	1996~ [65]	2010~ [66]	Suitable speed	Medium size
	Two photopolymerization	1997~ [67]	2006 (for polymer-derived ceramics) [68]	High resolution	Very slow, small size
	Volumetric additive manufacturing	2017 [69]	2022 (for polymer-derived ceramics) [70]	No staircase effect	Small size
Material extrusion (MEX)	Direct ink writing	1998 [71]	2001~ [72]	Suitable for lattice structures	Low resolution, low surface quality, anisotropy
	Fused deposition modeling	1989 [73]	1995 [74]	High density	Low resolution
Material jetting (MJT)	Inkjet printing	1950~ [75]	1995~ [76]	Suitable for layered structures	Low solid content
Binder jetting (BJT)	Binder jetting	1989 [77]	1990 [78]	Versatility in material choices	Low resolution, low surface quality
Powder bed fusion (PBF)	Selective laser sintering	1986 [79]	1990 [80]	Versatility in material choices, large printing size	Low resolution, complicated post-treatments
	Selective laser melting (or direct laser sintering)	1996 [81]	1999~ [82, 83]	Versatility in material choices, large printing size	Low resolution, stress concentration
Sheet lamination (SHL)	Laminated object modeling	1984~ [84]	1994 [85]	High density	Layer bonding issues, anisotropy
Directed energy deposition (DED)	Laser-engineered net shaping	1996 [86]	2008 [87]	Large printing size	Low resolution, low surface quality

varying properties, as well as the combination of ceramics with other materials like metals or polymers, is also an area of expanding interest. This will enable the production of more complex and functional components for high-performance applications in aerospace, biomedical, and electronics industries.

Additionally, there is a growing focus on the development of equipment and processes that can handle multi-material ceramic AM. This includes improving feeding modes and coupling various process technologies (so-called hybridization) to create more advanced and versatile ceramic components. As the technology matures, we can expect to see more innovative applications of ceramic AM, pushing the boundaries of what is possible in terms of material properties, part complexity, and functionality.

In conclusion, AM of ceramics is a rapidly evolving field with significant potential for growth and innovation. As research and development continue to advance, we can anticipate a future where ceramic components are not only more precise and complex but also more functional and tailored to specific applications. Table 1.3 provides a detailed comparison between the features and characteristics of different types of ceramic AM techniques.

1.5 Summary of the Book

The continuous research and development in ceramics processing are driven by the need for materials and components with superior properties for increasingly demanding applications. As technology progresses, it is expected that new fabrication techniques will emerge, further enhancing the capabilities and performance of both traditional and advanced ceramics. The aim of this book is to comprehensively address the unique manufacturability challenges of ceramics through advanced AM technologies, bridging fundamental material science with process engineering. It provides an authoritative synthesis of AM principles and cutting-edge innovations to enable reliable fabrication of high-performance ceramic components across various applications and materials.

In this book, **Chapter 1** introduces the foundations of ceramic AM, which fostered ceramics' critical role across industries, highlighting their unique chemical stability, thermal resistance, and biocompatibility. It contrasts traditional manufacturing limitations with AM's capacity for geometric freedom, setting the stage for ISO/ASTM 52900 classification of AM processes. **Chapter 2** covers stereolithography (SLA) and DLP for ceramics using UV-cured ceramic suspensions. It compares top-down and bottom-up configurations, emphasizing high resolution and surface quality. Challenges include resin formulation, shrinkage during sintering, and support structures, while VAM revolutionizes production by curing entire 3D volumes in seconds, eliminating layer interfaces and surface features. **Chapter 3** discusses BJT and MJT technologies. BJT uses powder beds selectively bound by liquid agents, while MJT deposits ceramic suspensions droplet-by-droplet (e.g. via piezoelectric or thermal printheads). Applications include gypsum models, sand molds, and technical ceramics, with challenges in nozzle clogging, droplet fusion,

Table 1.3 Detailed features and characteristics of different types of ceramic AM techniques.

Category	Technique	Forming principle	Energy source	Feedstock form	Feedstock requirements	Printing envelope	Resolution	Support needed	Speed	Surface quality	Overall cost
Vat photopolymerization (VPP)	Stereolithography	Photopolymerization	Light	Slurry	• Low viscosity	50 μm–100 cm	μm	Yes	★★★	★★★	★★
	Digital light processing	Photopolymerization	Light	Slurry	• Low viscosity	100 μm–50 cm	μm	Yes	★★★	★★★	★★
	Two photopolymerization	Photopolymerization	Light	Slurry	• Low viscosity • Transparency	1 μm–1 mm	nm–μm	Yes	★	★★★	★★★
	Volumetric additive manufacturing	Photopolymerization	Light	Slurry	• Low viscosity • Transparency	100 μm–10 mm	μm–mm	No	★★★	★★★	★★★
Material extrusion (MEX)	Direct ink writing	Extrusion	Force	Slurry	• High solid content	100 μm–100 cm	μm–mm	Yes	★★★	★	★
	Fused deposition modeling	Extrusion	Heat/force	Filament	• Filament	100 mm–10 cm	mm	No	★★	★	★★
Material jetting (MJT)	Inkjet printing	Binding	Force	Slurry	• Low viscosity	100 μm–10 mm	mm	No	★★	★★★	★

(Continued)

Table 1.3 (Continued)

Category	Technique	Forming principle	Energy source	Feedstock form	Feedstock require-ments	Printing envelope	Resolution	Support needed	Speed	Surface quality	Overall cost
Binder jetting (BJT)	Binder jetting	Binding	Force	Slurry/ powder	• Binder required	10 mm–100 cm	μm–mm	No	★★★	★★	★★
Powder bed fusion (PBF)	Selective laser sintering	Fusion	Laser	Powder	• Binder required	10 mm–100 cm	μm–mm	No	★★★	★	★★★
	Selective laser melting	Fusion	Laser	Powder	• Good powder flowability	10 mm–100 cm	μm–mm	No	★★★	★	★★★
Sheet lamination (SHL)	Laminated object modeling	Binding	Laser	Sheet	• Sheet form	100 mm–100 cm	μm–mm	No	★★	★★	★
Directed energy deposition (DED)	Directed energy deposition	Fusion	Laser	Powder	• Good powder sphericity	10 mm–1000 cm	mm	Yes	★★★	★	★★★

Note: The number of ★ denotes the significance level.

and sintering shrinkage. **Chapter 4** details DIW and Fused Filament Fabrication (FFF) for ceramics. DIW relies on yield-stress ceramic inks for automated deposition, while FFF uses thermoplastic-ceramic filaments. Both methods enable complex geometries and multi-material structures but face challenges in shape fidelity, thermal gradients, and post-processing (debinding/sintering). Rheology and printability criteria are critically discussed. **Chapter 5** examines Selective Laser Sintering/Melting (SLS/SLM) and Electron Beam Melting (EBM) for ceramics. It distinguishes direct (full melting) and indirect (binder-assisted) approaches, addressing material constraints like electrical conductivity for EBM. Key challenges include dusting (powder bed instability), residual stresses, and achieving full density. Process optimization via parameter mapping is emphasized. **Chapter 6** explores laser-based DED for fabricating ceramic structures and composites. It highlights DED's ability to create monolithic ceramics (e.g. alumina and zirconia) and metal-ceramic functionally graded materials, enabling site-specific properties. Key challenges include thermal stress-induced cracking, limited material selection (excluding ultra-high-temperature melting ceramics), and scalability constraints due to inert atmosphere requirements. **Chapter 7** demonstrates hybrid AM (concurrent/sequential processes) and multi-material fabrication for ceramics, enabling freeform geometries, embedded features, and functional integration. It highlights strategies for co-sintering compatibility and stress management, with applications in heating elements, electronics, and SOFCs for next-generation multifunctional devices. **Chapter 8** comprehensively addresses computational design strategies, feedstock characterization protocols, and post-processing evaluation methods essential for ceramic AM. It details topology optimization for complex geometries, classifies feedstocks into powder-, liquid-, and bulk-solid-based systems with corresponding analytical techniques, and examines critical post-consolidation assessments, including anisotropic shrinkage control, thermomechanical stability, and defect minimization, to ensure structural integrity, especially in aerospace and biomedical applications. **Chapter 9** details industrial and consumer ceramic VPP differences, emphasizing industrial systems' material control, calibrated optics, and quality management for reproducibility. It explores multi-material LCM printing for enhanced mechanical properties and damage tolerance in complex ceramic components. **Chapter 10** explores diverse applications of ceramic AM, emphasizing its transformative impact across aerospace (lightweight satellite mirrors with integrated cooling channels), semiconductors (complex wafer-handling components), defense (high-performance radomes), energy (corrugated solid oxide cells for efficient hydrogen production), biomedical (customized implants and surgical tools), and environmental sectors (3D-printed coral substrates for reef restoration), leveraging technologies like stereolithography (SLA) to achieve unprecedented geometric complexity and functional integration. **Chapter 11** explores emerging frontiers in ceramic AM, focusing on green/smart design (recycled feedstocks, AI-driven process optimization), 4D printing (shape-morphing ceramics via differential sintering or hydrogel-ceramic bilayers), and 5D printing (multi-axis deposition for curved geometries), addressing sustainability challenges while enabling adaptive structures for advanced applications.

1.6 Conclusions

AM has fundamentally transformed ceramic fabrication, shattering geometric constraints and enabling unprecedented functional integration. From VPP's submicron precision to MJT's material gradients, each technology unlocks new possibilities in aerospace, biomedicine, energy, and other strategic sectors. While challenges like thermal cracking in PBF and debinding complexities in FFF persist, innovations in hybrid manufacturing and AI-driven process control demonstrate the field's adaptive resilience. The future lies in the convergence of sustainable material cycles, smart monitoring systems, and multifunctional designs – where ceramics evolve from passive components to active, adaptive elements, for instance, in quantum devices and personalized implants. As these technologies mature, they will not merely enable smart manufacturing of ceramics but will redefine their key role in an increasingly technologically sophisticated world.

AM has irrevocably transformed the fabrication landscape for advanced ceramics, overcoming intrinsic limitations of conventional methods such as geometric constraints, tooling costs, and machining brittleness. This book systematically discusses the major AM technologies: VPP, BJT, MJT, MEX, and PBF, demonstrating their unique capabilities in producing complex, high-performance ceramic components with tailored properties.

The future of ceramic AM research and development is likely to focus on improving the precision, strength, and functionality of ceramic parts, as well as expanding the range of materials that can be used in AM processes. The integration of different ceramic materials with varying properties, as well as the combination of ceramics with other materials like metals or polymers, is a growing area of interest, promising to open up extensive research opportunities in the realm of structural/functional ceramics.

- Key advancements highlighted include:
  - **High-resolution fabrication:** VPP techniques (TPP, VAM) achieve submicron features for micro-optics and biomedical devices, while MJT enables multi-material integration with droplet-level precision.
  - **Scalability:** BJT excels in large-format applications (e.g. architectural ceramics), and PBF supports high-temperature materials like ultra-high-temperature ceramics (UHTCs) for extreme environments.
  - **Material innovation:** Novel suspensions (DIW) and filaments (FFF) enhance sinterability, and hybrid processes (e.g. UV-assisted extrusion) combine speed with resolution and geometrical complexity.
- **Persistent challenges remain**
  - **Process-induced defects:** Thermal gradients in PBF cause cracking, requiring preheating or adaptive scanning strategies. Debinding/sintering complexities in BJT and FFF necessitate precise thermal profiles to avoid delamination. Furthermore, timely detection of these defects is of great importance to promote the upscaling of AM technologies into the industrial sector.
  - **Material limitations:** Non-oxide ceramics (SiC, Si_3N_4) demand inert atmospheres, while low-thermal-conductivity oxides (ZrO_2) struggle with thermal shock.

- **Standardization gaps:** Qualification protocols for critical applications (e.g. aerospace and implants) require further development.
- **Future directions emphasize**
  - **Smart manufacturing:** AI-driven real-time monitoring and closed-loop control to enhance reproducibility.
  - **Multifunctional systems:** Co-printing ceramics with metals or polymers for integrated sensors, energy harvesters, and bio-implants.
  - **Sustainability:** Recycling ceramic powders and bio-derived binders to reduce environmental impact.

In summary, ceramic AM bridges material science, process engineering, and digital design. It unlocks unprecedented opportunities for customized, high-value components across several key sectors, including aerospace, energy, and healthcare. As technology matures, interdisciplinary collaboration will be pivotal to harnessing its full potential, ultimately reshaping the future of advanced ceramics manufacturing.

References

1 Barsoum, M. (2019). *Fundamentals of Ceramics*. CRC Press.
2 Singh, S., Das, P.K., and Mondal, D. (2022). *Advanced Ceramics for Versatile Interdisciplinary Applications*. Elsevier.
3 Otitoju, T.A., Okoye, P.U., Chen, G. et al. (2020). Advanced ceramic components: materials, fabrication, and applications. *J. Ind. Eng. Chem.* 85: 34–65.
4 Dorey, R., Rocks, S., Dauchy, F. et al. (2008). Integrating functional ceramics into microsystems. *J. Eur. Ceram. Soc.* 28 (7): 1397–1403.
5 Drukteinis, S. and Camilleri, J. (2021). *Bioceramic Materials in Clinical Endodontics*. Springer.
6 Song, C., Ye, F., Cheng, L. et al. (2022). Long-term ceramic matrix composite for aeroengine. *J. Adv. Ceram.* 11 (9): 1343–1374.
7 Shim, J.H. (2018). Ceramics breakthrough. *Nat. Energy* 3 (3): 168–169.
8 Sengupta, P. (2020). *Refractories for the Chemical Industries*. Springer.
9 Justin, J. and Jankowiak, A. (2011). Ultra high temperature ceramics: densification, properties and thermal stability. *Aerosp. Lab* 3: 1–11.
10 Pastore, R., Delfini, A., Santoni, F. et al. (2021). Space environment exposure effects on ceramic coating for thermal protection systems. *J. Spacecr. Rocket.* 58 (5): 1387–1393.
11 Zhao, J., Pan, Z., Tang, L. et al. (2022). Greatly enhanced discharged energy density and efficiency of $BiFeO_3$-based ceramics by regulating insulation performance. *Mater. Today Phys.* 27: 100821.
12 Zhang, B., Ghassemi, M., and Zhang, Y. (2021). Insulation materials and systems for power electronics modules: a review identifying challenges and future research needs. *IEEE Trans. Dielectr. Electr. Insulat.* 28 (1): 290–302.
13 Wachtman, J.B., Cannon, W.R., and Matthewson, M.J. (2009). *Mechanical Properties of Ceramics*. Wiley.

14 Yu, Y., Wu, Y., Wang, B. et al. (2024). Review of the penetration resistance performance of lightweight ceramic composite armor under high-speed impact. *Mech. Adv. Mater. Struct.* 32: 1–23.

15 Shabani, M. and Gjoni, A. (2025). *Advancements in Hard Turning: Leveraging Ceramic Cutting Tools for Sustainable and High-Quality Metalworking, Advanced Ceramic Materials-Emerging Technologies.* IntechOpen.

16 McColm, I.J. (2013). *Ceramic Hardness.* Springer Science & Business Media.

17 Ikesue, A. and Aung, Y.L. (2008). Ceramic laser materials. *Nat. Photonics* 2 (12): 721–727.

18 Akinribide, O.J., Mekgwe, G.N., Akinwamide, S.O. et al. (2022). A review on optical properties and application of transparent ceramics. *J. Mater. Res. Technol.* 21: 712–738.

19 Hernandez, J.L. and Woodrow, K.A. (2022). Medical applications of porous biomaterials: features of porosity and tissue-specific implications for biocompatibility. *Adv. Healthcare Mater.* 11 (9): 2102087.

20 Kulkarni, S.V., Nemade, A.C., and Sonawwanay, P.D. (2022). An overview on metallic and ceramic biomaterials. *Recent Adv. Manuf. Process. Syst. Sel. Proc. RAM* 149–165.

21 Upadhyay, A., Pradhan, L., Yenurkar, D. et al. (2024). Advancement in ceramic biomaterials for dental implants. *Int. J. Appl. Ceram. Technol.* 21 (4): 2796–2817.

22 Evans, A.G. and Langdon, T.G. (1976). Structural ceramics. *Prog. Mater. Sci.* 21 (1–2): 171–425.

23 Padture, N.P. (2016). Advanced structural ceramics in aerospace propulsion. *Nat. Mater.* 15 (8): 804–809.

24 Noor, A.M., Abbas, M.R., Rajoo, S. et al. (2014). Review on ceramic application in automotive turbocharged engines. *Appl. Mech. Mater.* 660: 219–228.

25 Ikesue, A., Aung, Y.L., Taira, T. et al. (2006). Progress in ceramic lasers. *Annu. Rev. Mater. Res.* 36 (1): 397–429.

26 Khaliq, J. (2021). Ceramic matrix composites (CMCs). In Jibran Khaliq (Ed) *Advances in Machining of Composite Materials: Conventional and Non-conventional Processes*, 285–309. Springer.

27 Bansal, N.P. and Lamon, J. (2014). *Ceramic Matrix Composites: Materials, Modeling and Technology.* Wiley.

28 Kanyo, J.E., Schafföner, S., Uwanyuze, R.S., and Leary, K.S. (2020). An overview of ceramic molds for investment casting of nickel superalloys. *J. Eur. Ceram. Soc.* 40 (15): 4955–4973.

29 Bar-Cohen, Y. (2014). *High Temperature Materials and Mechanisms.* Boca Raton, FL: CRC Press.

30 Haubner, R., Herrmann, M., Lux, B. et al. (2003). *High Performance Non-Oxide Ceramics II.* Springer.

31 Katoh, Y. and Snead, L.L. (2019). Silicon carbide and its composites for nuclear applications–Historical overview. *J. Nucl. Mater.* 526: 151849.

32 Pushpakaran, B.N., Subburaj, A.S., Bayne, S.B., and Mookken, J. (2016). Impact of silicon carbide semiconductor technology in Photovoltaic Energy System. *Renew. Sust. Energ. Rev.* 55: 971–989.

33 Eray, E., Candelario, V.M., Boffa, V. et al. (2021). A roadmap for the development and applications of silicon carbide membranes for liquid filtration: recent advancements, challenges, and perspectives. *Chem. Eng. J.* 414: 128826.

34 Singh, S., Rathi, K., and Pal, K. (2018). Synthesis, characterization of graphene oxide wrapped silicon carbide for excellent mechanical and damping performance for aerospace application. *J. Alloys Compd.* 740: 436–445.

35 Behar-Lafenêtre, S., Louh, N., Grasset, P. et al. Silicon nitride for structural parts of space applications. *SPIE* 11101: 8–21.

36 Klemm, H. (2002). Corrosion of silicon nitride materials in gas turbine environment. *J. Eur. Ceram. Soc.* 22 (14–15): 2735–2740.

37 Blumenthal, D.J., Heideman, R., Geuzebroek, D. et al. (2018). Silicon nitride in silicon photonics. *Proc. IEEE* 106 (12): 2209–2231.

38 Ramachandran, K., Bear, J.C., and Jayaseelan, D.D. (2025). Oxide-based ceramic matrix composites for high-temperature environments: a review. *Adv. Eng. Mater.* 27: 2402000.

39 Lin, L., Liu, B., Zhang, T. et al. (2024). Analysis on carbonization evolution behavior of tungsten carbide powders and its application in cutting tools. *Int. J. Refract. Met. Hard Mater.* 118: 106447.

40 Sharath, B.N., Jeevan, T.P., Baig, M.A.A. et al. (2021). Machinability studies on boron carbide and graphite reinforced aluminium hybrid composites. *Mater. Today Proc.* 46: 8734–8741.

41 Wuchina, E., Opila, E., Opeka, M. et al. (2007). UHTCs: ultra-high temperature ceramic materials for extreme environment applications. *Electrochem. Soc. Interf.* 16 (4): 30.

42 Marques, A., Miranda, G., Silva, F. et al. (2021). Review on current limits and potentialities of technologies for biomedical ceramic scaffolds production. *J. Biomed. Mater. Res. B Appl. Biomater.* 109 (3): 377–393.

43 Chen, S., Ding, Y., Mu, H. et al. (2025). Research progress on multilayer ceramic capacitors for energy storage. *J. Mater. Sci. Mater. Electron.* 36 (1): 77.

44 Testino, A. (2013). Ceramics for electronics and energy: issues and opportunities. *Int. J. Appl. Ceram. Technol.* 10 (5): 723–730.

45 Nogawa, H., Hirao, A., Nishimura, Y. et al. *High Power IGBT Module with New AlN Substrate*, 1–8. VDE.

46 Guo, Q., Li, F., Xia, F. et al. (2021). Piezoelectric ceramics with high piezoelectricity and broad temperature usage range. *J. Mater.* 7 (4): 683–692.

47 Pohanka, R.C. and Smith, P.L. (2020). Recent advances in piezoelectric ceramics. *Electron. Ceram.* 45–145.

48 Hamza, M., Kanwal, Q., Hussain, M.I. et al. (2025). Recent progress in 3D printed piezoelectric materials for biomedical applications. *Mater. Sci. Eng. R. Rep.* 164: 100962.

49 Lim, Y., Park, J., Lee, H. et al. (2021). Rapid fabrication of lanthanum strontium cobalt ferrite (LSCF) with suppression of LSCF/YSZ chemical side reaction via flash light sintering for SOFCs. *Nano Energy* 90: 106524.

50 Chen, S.-P., Lv, D., Chen, J. et al. (2022). Review on defects and modification methods of $LiFePO_4$ cathode material for lithium-ion batteries. *Energy Fuel* 36 (3): 1232–1251.
51 Lin, X., Gong, H., and Chen, Z. (2023). Fabrication of SiCN (Fe)/Al_2O_3 wave-absorbing ceramics with enhanced electromagnetic performance. *Ceram. Int.* 49 (14): 23851–23863.
52 Lin, X., Long, Z., Jiang, L. et al. (2023). 3D printed crack-free SiOC (Fe) structures with pyrolysis-induced carbon nanowires for enhanced wave absorption performance. *Ceram. Int.* 49 (20): 33205–33213.
53 Kasana, R., Rathore, P., Shashikumar, U., and Chawla, S. (2024). Nanoceramics: fabrication, properties, and applications. *Ind. Appl. Nanoceram.* 19–36.
54 Thakur, A. and Kumar, A. (2023). *Self-Healing Ceramic Coatings, Advanced Ceramic Coatings for Emerging Applications*, 19–42. Elsevier.
55 Kuznetsova, Y.V. and Popov, I.D. (2022). Design and technological aspects of novel CdS quantum dots doped glass–ceramics. *Ceram. Int.* 48 (13): 18972–18982.
56 Khan, K., Hussain, M.I., Tareen, A.K. et al. (2026). Advances in vat photopolymerization 3D printing: multifunctional materials, process innovations, and emerging applications. *Mater. Sci. Eng. R. Rep.* 167: 101120.
57 Chen, Z., Li, Z., Li, J. et al. (2019). 3D printing of ceramics: a review. *J. Eur. Ceram. Soc.* 39 (4): 661–687.
58 Rasaki, S.A., Xiong, D., Xiong, S. et al. (2021). Photopolymerization-based additive manufacturing of ceramics: a systematic review. *J. Adv. Ceram.* 10 (3): 442–471.
59 Asghar, A., Khan, K., Rashid, M.S. et al. (2025). Extrusion-based additive manufacturing of carbonaceous and non-carbonaceous electrode materials for electrochemical energy storage devices. *Adv. Mater. Technol.* 10: 2400136.
60 Idrees, M., Batool, S., Din, M.A.U. et al. (2023). Material-structure-property integrated additive manufacturing of batteries. *Nano Energy* 109: 108247.
61 ISO/ASTM, ISO/ASTM 52900 (2021). Additive Manufacturing — General Principles — Terminology, 2021.
62 Chaudhary, R.P., Parameswaran, C., Idrees, M. et al. (2022). Additive manufacturing of polymer-derived ceramics: Materials, technologies, properties and potential applications. *Prog. Mater. Sci.* 128: 100969.
63 Hull, C.W. (1986). Methods and apparatus for production of three-dimensional objects by stereolithography. U.S. Patent (Ed.), USA.
64 Griffith, M.L. and Halloran, J.W. (1996). Freeform fabrication of ceramics via stereolithography. *J. Am. Ceram. Soc.* 79 (10): 2601–2608.
65 Nakamoto, T. and Yamaguchi, K. (1996). *Consideration on the Producing of High Aspect Ratio Micro Parts Using UV Sensitive Photopolymer, the Seventh International Symposium of Micro Machine and Human Science*, US, IEEE, pp. 53–58. Nagoya, Japan, October 2–4, 1996.
66 Choi, J.-W., Wicker, R., Lee, S.-H. et al. (2009). Fabrication of 3D biocompatible/biodegradable micro-scaffolds using dynamic mask projection microstereolithography. *J. Mater. Process. Technol.* 209 (15–16): 5494–5503.

67 Maruo, S., Nakamura, O., and Kawata, S. (1997). Three-dimensional microfabrication with two-photon-absorbed photopolymerization. *Opt. Lett.* 22 (2): 132–134.

68 Pham, T.A., Kim, D.P., Lim, T.W. et al. (2006). Three-dimensional SiCN ceramic microstructures via nano-stereolithography of inorganic polymer photoresists. *Adv. Funct. Mater.* 16 (9): 1235–1241.

69 Shusteff, M., Browar, A.E.M., Kelly, B.E. et al. (2017). One-step volumetric additive manufacturing of complex polymer structures. *Science* 3 (12): eaao5496.

70 Kollep, M., Konstantinou, G., Madrid-Wolff, J. et al. (2022). Tomographic volumetric additive manufacturing of silicon oxycarbide ceramics. *Adv. Eng. Mater.* 24: 2101345.

71 Cesarano, I. and Segalman, R. (1998). Robocasting provides moldless fabrication from slurry deposition. *Ceram. Ind.* 148 (4): 94–100.

72 Tuttle, B.A., Smay, J.E., Cesarano, J. et al. (2001). Robocast $Pb(Zr_{0.95}Ti_{0.05})O_3$ ceramic monoliths and composites. *J. Am. Ceram. Soc.* 84 (4): 872–874.

73 Crump, S.S. (1992). Apparatus and method for creating three-dimensional objects, Google Patents.

74 Danforth, S. (1995). Fused deposition of ceramics: a new technique for the rapid fabrication of ceramic components. *Mater. Technol.* 10 (7–8): 144–146.

75 Le, H.P. (1998). Progress and trends in ink-jet printing technology. *J. Imaging Sci. Technol.* 42 (1): 49–62.

76 Blazdell, P., Evans, J., Edirisinghe, M. et al. (1995). The computer aided manufacture of ceramics using multilayer jet printing. *J. Mater. Sci. Lett.* 14 (22): 1562–1565.

77 Sachs, E.M., Haggerty, J.S., Cima, M.J., and Williams, P.A. (1993). Three-dimensional printing techniques, USA. US5204055A.

78 Sachs, E., Cima, M., and Cornie, J. (1990). Three-dimensional printing: rapid tooling and prototypes directly from a CAD model. *CIRP Ann. Manuf. Technol.* 39 (1): 201–204.

79 Deckard, C.R. (1989). Method and apparatus for producing parts by selective sintering, Google Patents.

80 Lakshminarayan, U., Ogrydiziak, S., and Marcus, H. (1990). Selective laser sintering of ceramic materials. *1990 International Solid Freeform Fabrication Symposium*. Austin, TX, USA, August 6–8 1990.

81 Schleifenbaum, H., Meiners, W., Wissenbach, K., and Hinke, C. (2010). Individualized production by means of high power selective laser melting. *CIRP J. Manuf. Sci. Technol.* 2 (3): 161–169.

82 Harlan, N., Park, S.-M., Bourell, D.L., and Beaman, J.J. (1999). Selective laser sintering of zirconia with micro-scale features. *Proc. SFF Symp.*, Austin, pp. 297–302. Austin, TX, USA, August 9–11, 1999.

83 Klocke, F. and Wirtz, H. (1998). Selective laser sintering of zirconium silicate. *Proceedings of Solid Freeform Fabrication*, Austin, Texas. Austin, TX, USA, August 10–12, 1998.

84 Kunieda, M. and Nakagawa, T. (1984). Manufacturing of laminated deep drawing dies by laser beam cutting. *Adv. Technol. Plast.* 1: 520–525.

85 Griffin, C., Daufenbach, J., and McMillin, S. (1994). Desktop manufacturing: LOM vs. pressing. *Am. Ceram. Soc. Bull.* 73 (8): 109–113.

86 Griffith, M., Keicher, D., Atwood, C. et al. (1996). Free form fabrication of metallic components using laser engineered net shaping (LENS). *Solid Freeform Fabrication Proceedings, Proc. 1996 Solid Freeform Fabrication Symposium*, Austin, pp. 125–131.

87 Balla, V.K., Bose, S., and Bandyopadhyay, A. (2008). Processing of bulk alumina ceramics using laser engineered net shaping. *Int. J. Appl. Ceram. Technol.* 5 (3): 234–242.

2

Vat Photopolymerization (VPP)

Vincent Pateloup and Gweni Alonso Aruffo

University of Limoges, CNRS, Institute of Research for Ceramic (IRCER), UMR 7315, European Ceramic Center, 12 rue Atlantis, F-87068, Limoges Cedex, France

2.1 Introduction

Photopolymerization is a process in which a photopolymeric resinous raw material is exposed to a localized UV light source in order to selectively cure it. This process finds extensive application in additive manufacturing, enabling the production of complex structures characterized by high levels of detail, high resolution and smooth surface finishes. The first introduction of light-curing additive manufacturing processes was in the 1980s by Chuck Hull, who pioneered the development of stereolithography (SLA) and filed the first patent [1] for polymer-based materials. This breakthrough can be regarded as the inception of vat light curing as a viable technique for three-dimensional manufacturing. Vat photopolymerization (see Figure 2.1) is a process that utilizes a layer-by-layer polymerization mechanism, whereby three-dimensional objects are generated through the sequential deposition and curing of thin layers of a ceramic suspension paste that exhibits a photosensitive response. The initiation of the polymerization reaction is triggered by exposure to a light source (UV laser in the case of SLA or UV light digital projector in the case of digital light processing (DLP)). This results in a change of the physical and chemical state of the material. The process is highly automated and ensures that each newly formed layer adheres to the previous one, leading to the gradual formation of a fully dense three-dimensional structure. A salient benefit inherent to the vat photopolymerization process pertains to its capacity to produce objects with elevated feature resolutions and intricate geometric patterns, thereby rendering it particularly well-suited for applications necessitating meticulous dimensional control and a concomitant minimization of post-processing requirements. In the specific case of ceramic additive manufacturing, the object that is printed – and which is known as the "green part" – must undergo a number of post-processing steps such as solvent cleaning, debinding, and sintering in order to achieve full densification and functional mechanical properties. Notwithstanding the advantages of the process, there are also challenges to be addressed, such as shrinkage

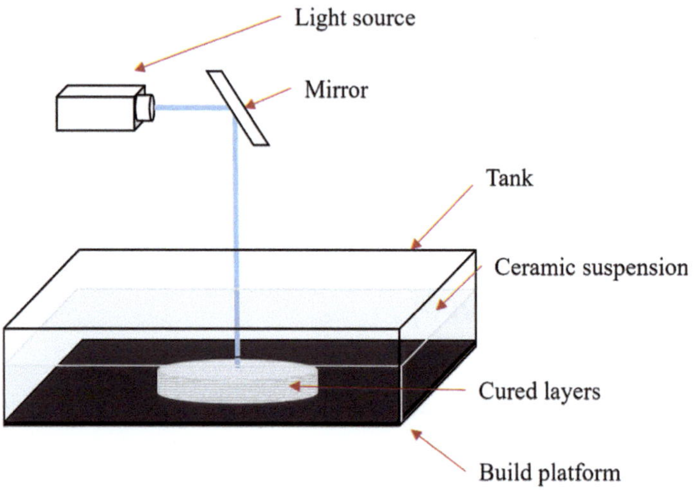

Figure 2.1 SLA process.

during printing and sintering, stress accumulation, and material formulation constraints. These issues must be carefully addressed through the optimization of processing parameters and numerical modeling techniques. Advancements in research are facilitating the refinement of vat photopolymerization of ceramics, paving the way for the production of functional components for high-performance applications. Vat photopolymerization was initially restricted to polymeric materials, primarily due to the challenges associated with handling highly viscous, particle-rich suspensions. Nevertheless, during the 1990s, researchers initiated the exploration of the potential for the extension of this technology to the realm of ceramic materials [2]. The reliability and scalability of this technique have been enhanced by recent advancements in resin formulation, process control, and numerical simulations. Consequently, it is regarded as a vital instrument in modern ceramic additive manufacturing. Ceramic materials are characterized by superior mechanical, thermal, and chemical resistance when compared with polymers. The feasibility and effectiveness of photopolymerization additive manufacturing for ceramic suspensions were demonstrated by research teams led by Griffith et al., Brady et al., and Chartier et al., respectively [2–4, 32, 33]. This development paved the way for the development of advanced applications in diverse domains, including biomedicine, aerospace, and electronics.

2.2 Photopolymerization 3D Printing Processes for Ceramic Parts

The authors would like to specify that in the continuation of this chapter, only studies concerning SLA processes applied to ceramics will be detailed.

It is acknowledged that a series of common steps and/or materials to the SLA and DLP processes exist, from the digital object to the final product. These include the heat-treatment steps (where the printed substrate undergoes debinding and sintering) and the properties of the ceramic suspension. In light of these observations, the authors propose a presentation of the steps in question that is unified, while simultaneously highlighting the nuances that differentiate them.

2.2.1 Stereolithography Process

Stereolithography, otherwise referred to as SLA, is a vat polymerization 3D printing process. Introduced in the 1980s, the process is based on the photopolymerization of suspensions [1]. In the context of ceramic three-dimensional printing, this process involves the application of a photosensitive paste, loaded with ceramic powder, to the desired substrate. Further details pertaining to the composition of ceramic suspensions can be found in Section 2.2.4.

The photopolymerization procedure consists in the reaction of the suspension toward UV light. A UV laser source (see Figure 2.1) selectively scans the surface of a suspension bed. This interaction between the laser source and the ceramic suspension leads to a chemical reaction, causing the hardening of the material. SLA might be configured in either a bottom-up or a top-down manner (see Section 2.2.5).

In the initial phase of the printing procedure, the building platform is immersed in a suspension of UV-sensitive ceramic particles. This suspension, which is photosensitive, is subsequently deposited on the platform in layers of variable thickness, ranging from 10 to 200 μm, using a surface scraper. A scanning head composed of mobile and motorized mirrors is then used to direct the laser beam along a specific trajectory to polymerize the suspension at specific locations according to the numerical model of the 3D object to be printed [6], forming the first hardened layer. Subsequent to the completion of this layer, the building platform undergoes a slight descent, enabling the deposition of a new suspension layer. This sequence is repeated until the formation of the entire object is complete. Once the printing process is concluded, the final object is immersed in a tank and covered by a non-cured ceramic suspension. A subsequent cleaning stage is then required to obtain the final cured or green object. The object can then undergo thermal treatment for further consolidation (a more detailed discussion of this aspect can be found in Section 2.2.5).

2.2.2 Digital Light Processing Process

DLP, the so-called mask-based SL, was initially proposed by Nakamoto et al. [7]. This process is analogous to the SLA process, yet it is distinguished by its significantly enhanced printing speed, a feature that has contributed to its widespread adoption. A distinctive feature of DLP technology is the suspension solidification stage. In contrast to the SLA, where the local solidification of each point of the slice is achieved by means of the interaction between the laser beam and the suspension, the DLP adopts a more global approach, employing a UV digital light projector that illuminates an entire layer through a patterned mask (numerically controlled by a digital micromirror device). This process is repeated until the entire desired surface is cured, thus enabling the simultaneous realization of all the required slices [5].

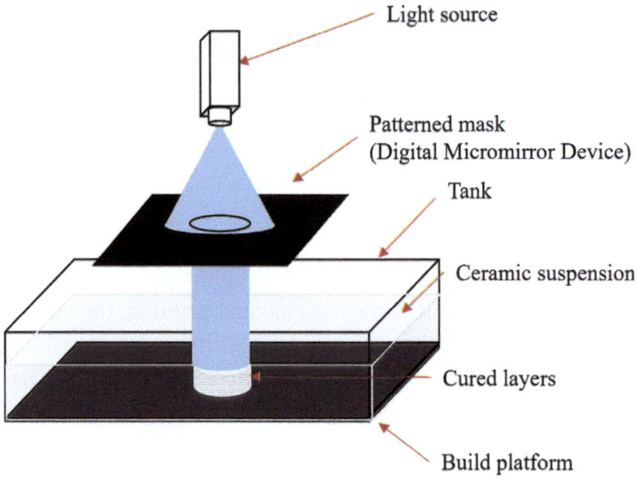

Figure 2.2 DLP process.

As illustrated in Figure 2.2, the printing process is initiated with the submersion of a building platform in a ceramic suspension. Depending on the configuration of the device, the resin is solidified, either from the bottom of the tray (bottom-up approach) or from the top (top-down approach). This specific difference will be addressed in more detail in Section 2.2.3. The light projector, as illustrated in Figure 2.2, projects a numerical image corresponding to a horizontal section of the object directly onto the suspension layer. The totality of this illuminated region subsequently hardens under the effect of the UV light. Following this process, a novel layer of suspension is deposited on the surface, and this sequence is repeated until every constituent layer of the object has been adequately cured.

2.2.3 Top-Down and Bottom-Up Approaches

The top-down configuration represents the most conventional method for both processes, SLA and DLP. The latter strategy is illustrated in Figure 2.1 and Figure 2.2; in this case, the light source originates from above. Consequently, the building platform descends within the tank, concurrently with the progressive construction of the 3D object.

Contrary to the initial approach, the bottom-up strategy entails the gradual ascent of the building platform during the formation of each layer of the 3D object. The process entails the utilization of light (laser and projection) to illuminate the surface of the suspension from beneath, facilitated by the transparency of the tank's base. A schematic representation of the difference is depicted in Figure 2.3. A notable characteristic of this approach is the option to utilize a relatively compact suspension storage tank.

The nature of the structure has also demonstrated its effect depending on whether the bottom-up or top-down approach is employed. Specifically, DLP methodology has been examined [8] for its impact on lattices and structures with high aspect ratios (length/diameter >2), as shown in Figure 2.4. The conclusions drawn from

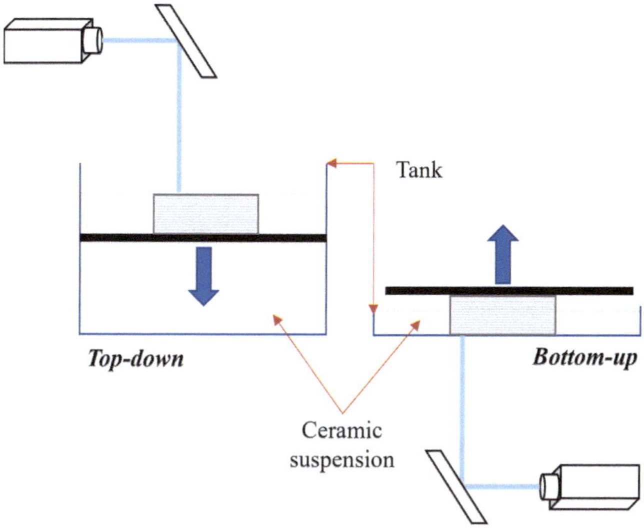

Figure 2.3 Comparison between top-down and bottom-up configurations.

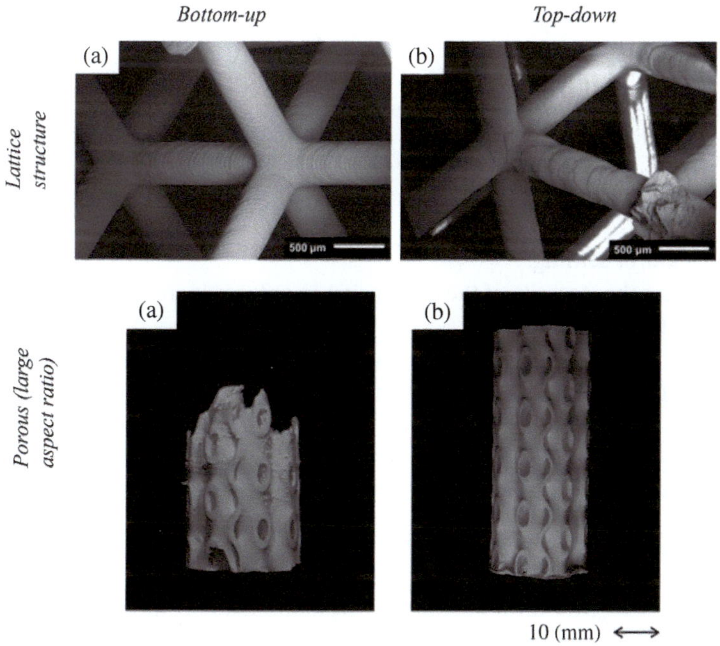

Figure 2.4 Effect of bottom-up (a) and top-down (b) procedures for different structures – AL_2O_3 ceramic. [8] / with permission of Elsevier.

Table 2.1 Properties comparison between top-down and bottom-up configurations.

Comparison criteria	Top-down	Bottom-up
Resin consumption	High	Reduced
Adhesion to building platform	Less strength required	Strong strength required
Printability	Adapted for large objects with a specific height/width ratio [8]	Ideal for smaller objects

this study indicate that the top-down method yields superior results for structures with high aspect ratios (length/diameter >2) and that it produces a smoother surface in the case of lattice structures. In the context of lattice structures, the study noted the impact of the bottom-up or top-down procedure on the final relative density of the sintered objects, highlighting a 2.5% variation.

A balanced consideration of both bottom-up and top-down settings is essential, taking into account factors such as resin usage, adhesion strength of the 3D object to the building platform, and printability. A comprehensive overview of the distinguishing characteristics of these approaches is provided in Table 2.1.

2.2.4 Feedstock Formulations for VAT Photopolymerization Processes

In order to be employed in an SLA process, a photosensitive ceramic suspension must be composed of the following elements:

- A ceramic loading agent in the form of a powder with a specific particle size (depending on the process parameters).
- A dispersant to stabilize the solid particles in suspension and prevent them from agglomerating and/or settling.
- A diluent to reduce the viscosity of the suspension in order to increase the level of ceramic powder filler.
- A functional photoinitiator within a UV wavelength (varying between 350 and 410 nm) that triggers polymerization of the monomers or oligomers by converting light energy into chemical energy.
- A photosensitive resin composed of monomers/oligomers with high reactivity, high mechanical strength, low viscosity, and low polymerization shrinkage.

An illustration of the final ceramic-loaded suspension composition is provided in Figure 2.5.

Despite the existence of a multitude of monomer/photoinitiator combinations that can be solidified by exposure to a source of energy, such as electromagnetic radiation of varying wavelengths (examples of such include gamma rays, X-rays, ultraviolet, and visible light), SLA machines essentially utilize an ultraviolet wavelength for two primary reasons: first, to minimize cost, and second, to simplify the management of the radiation source. In such instances, the monomer is typically derived from the acrylate or methacrylate family.

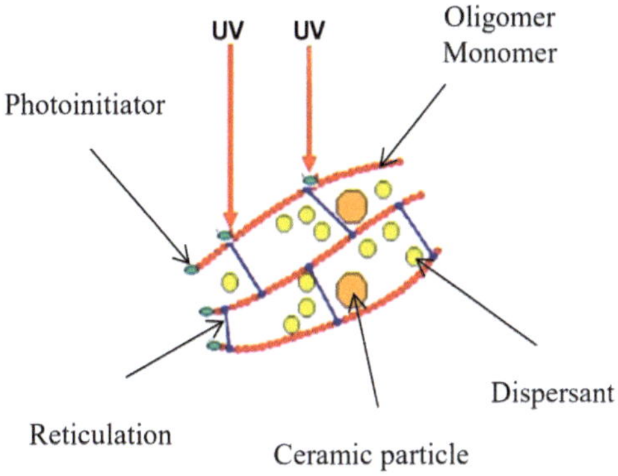

Figure 2.5 Composition of a reactive suspension.

The operational process of photopolymerization can be succinctly summarized as comprising three fundamental stages: first, the absorption of light by the substrate, followed by the generation of reactive species, and finally, the initiation of polymerization.

During the initial phase of light absorption, the photoinitiator absorbs the light energy emitted by the 3D printer's UV source. In response to this energy, the photoinitiator undergoes decomposition, thereby generating free radicals and reactive species. Finally, the radicals attack the double bonds of the monomers, triggering a polymerization chain reaction.

These reactive suspensions must meet the following conditions:

- The employment of a high amount of ceramic filler, typically in excess of 50 vol%, is imperative to ensure the maintenance of the necessary cohesion of the object during the process of debinding and sufficient reactivity during sintering.
- The material must possess a rheology (i.e. viscous behavior) that is suitable for the deposition of thin layers (i.e. 10 μm).
- It has been established that a high polymerization rate of the reactive suspension (named reactivity) is observed, despite the high concentration of ceramic particles. It is evident that any reactive suspension employed in SLA adheres to the principles of Beer-Lambert law, thereby reflecting the rate of absorption of energy supplied by the UV source within the photosensitive material [35, 36]. The Beer-Lambert law is given by the following equation:

$$E_p = D_p \cdot \ln\left(\frac{E_0}{E_c}\right)$$

In this equation, E_c is defined as the critical polymerization energy, below which the photopolymerization process does not initiate, and D_p corresponds to the penetration coefficient of the laser beam into the suspension.

This final parameter is dependent on the characteristics of the system, which include the size of the ceramic particles, the volume fraction, and the optical index. Additionally, it considers the diffusion phenomena that control both the polymerized depth and the dimensional resolution. Consequently, it is advantageous to adapt the values of E_c and D_p according to the desired final characteristics of the printed part. The printing time is directly linked to these two parameters.

The following imperative characteristics have been identified:

- The restriction of diffusion phenomena is imperative in order to achieve high-dimensional resolution.
- The mechanical properties of the green object must be appropriately calibrated (i.e. Young's modulus and yield strength) to ensure that it can be handled without risk of destruction or deformation. Furthermore, this calibration is necessary to facilitate the cleaning and heat-treatment of the object without causing damage.
- The final object must be suitable for debinding in order to eliminate the appearance of cracks and minimize deformation caused by the evacuation of gases resulting from the combustion of its various organic components.

2.2.5 Thermal Treatment

As mentioned before, the green specimen is a ceramic and polymeric composite. The rate of organic volumetric concentration is typically in the range of 40–50% [3]. Subsequent to the cleaning process, it is imperative to subject the specimen to thermal treatment, with the objective of eradicating the residual organic material. The debinding procedure and sintering are two main phases that must be considered. The initial heating phase (debinding) facilitates the extraction of the organic composition from the green part. The debinding stage can typically be followed as depicted in Figure 2.5. The heating slope must be selected considering the material under study and should be moderate in order to minimize thermal stresses on the specimen and, consequently, the likelihood of cracking. The subsequent holding stages (highlighted in orange in Figure 2.6) are determined by preliminary TG-DSC analyses, which identify the thermal transitions of each polymer contained within the ceramic resin [9].

The maximal permissible debinding temperature is determined by several factors, including the volume rate and the type of ceramic AL_2O_2 (alumina 400/600 °C),

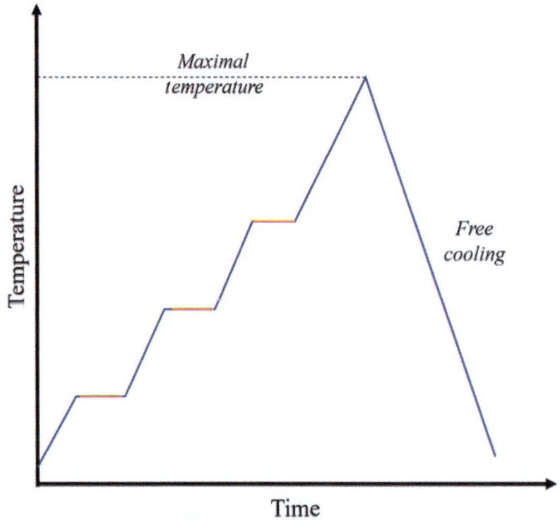

Figure 2.6 Standard debinding profile.

ZrO_2 (zirconia 400/600°C), SiO_2 (silica 300/600°C), the controlled atmosphere (air, pure oxygen, gas), and the heating rate [10].

Subsequently, the cohesive, yet remarkably fragile debinded, object is sintered [27]. The purpose of this thermal treatment is to close the porosities created by the extraction of the organic phase. This process results in an enhancement of the object's density, which approaches the solid density of the ceramic material used. The specific parameters of temperature and duration of this process are linked to the nature of the ceramic material in question, with the sintering temperature ranging between 900 °C and 2500 °C and the duration spanning a range between one hour and several hours [28–31] and in a specific atmosphere [26]. This stage of sintering is known to enhance the mechanical properties of the specimen. A geometrical shrinkage is thus observed, typically ranging from 10% to 30% of the volume in each direction [11].

2.2.6 Summary of Both Processes

It is evident that both processes bear a certain degree of similarity, yet they differ in several notable aspects. Indeed, these two processes are 3D-printing technologies that rely on layer hardening through UV photosensitivity, a process that involves light-induced polymerization. The salient differences between the two processes are outlined in Table 2.2.

The primary difference in resolution between the two processes is evident in the dependence of the external geometry of the slice on the diameter of the laser beam in the SLA process and on the pixilation of each light point by a mask (DLP), resulting in a "pixelated" perimeter, as demonstrated in Figure 2.7.

The selection of a method is linked to the desired final outcome, the dimensional parameters of the intended 3D object, the temporal constraints imposed during printing, and the economic considerations (i.e. whether the approach is bottom-up or top-down). In the case of the DLP process, the concurrent exposure of each layer leads to a substantial reduction in fabrication time. Nevertheless, it should be noted that this method results in a decrease in surface smoothness and, consequently, a

Table 2.2 Brief comparison between SLA and DLP processes.

	SLA	**DLP**
Light source	Local UV laser source and galvanometer head for orientation	UV light projection and DMD – digital micromirror device
Time	More time-consuming	Faster
Resolution and quality	Better precision, homogeneous layers, dependent on the laser beam thickness	Less precise, more significant staircase effect (up to several micrometers [5]) depends on the projector resolution
Layer thickness	10–150 μm [12]	10–100 μm [34]

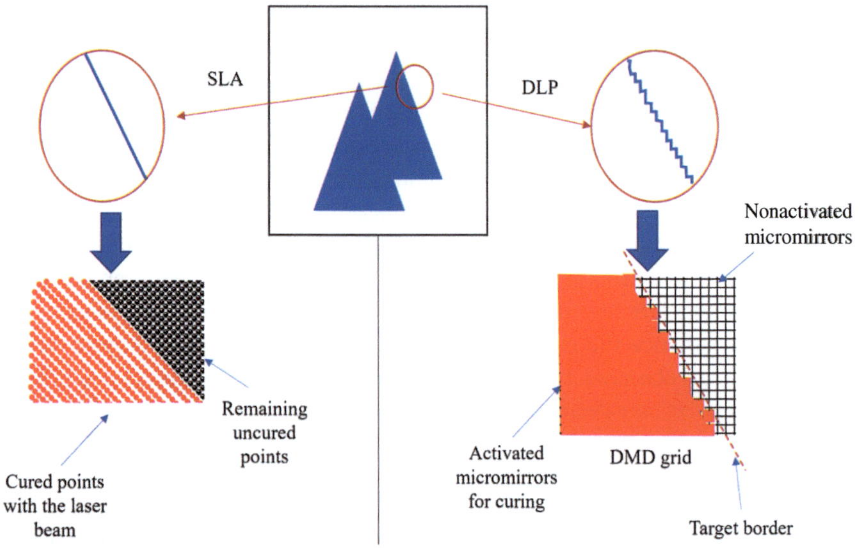

Figure 2.7 Resolution difference between SLA and DLP processes.

reduction in the precision of the final product. This is a consequence of the pixel size resolution of the projection screen, and consequently, in most cases, the precision of the details is comparable between the two processes.

2.3 Process Modeling and Simulation

The modeling and simulation of such printing processes are a current research priority, as they offer the prospect of enhanced prediction capabilities and address existing challenges. In fact, incorporating a numerical approach into physical processes can offer valuable insights into material behavior. For instance, it has been demonstrated that numerical modeling can provide significant understanding of the behavior of ceramic suspensions during the solidification process and layer-by-layer construction.

2.3.1 Context

Several numerical approaches can be enumerated herewith. Through an optical and/or chemical approach, with the modeling of absorption and diffusion of UV light, the amount of energy absorbed by the resin during polymerization can be determined. Furthermore, polymerization kinematics or even thermo-mechanical analyses for debinding and sintering can be conducted to predict shrinkage. The chemical and optical approaches will not be elaborated upon in this chapter; however, further information can be found in [13, 14]. The modeling of shrinkage presents significant challenges, particularly those associated with phase changes (solid to liquid to gas), necessitating a multiphysical approach [15]. To the best of our knowledge, no such approach has yet been applied to ceramics, and this remains a significant challenge.

An additional approach, the details of which are delineated in Section 2.3.2, is based on a purely mechanical modeling. Its objective is the prediction of the strain and residual stresses induced by the process itself on the constructed object. This approach facilitates the estimation of the viability of the printing process. Furthermore, the analysis of the suspension during displacements of the building plate, as well as during scraping operations, could be facilitated by incorporating a fluid dynamic method.

The numerical approach in the context of 3D printing is advantageous for process optimization. However, it should be noted that the process parameters, such as printability and scalability, necessitate a considerable number of trial-and-error tests to ensure good reproducibility. The integration of numerical methods into investigations could significantly enhance the analysis of new or more technically complex ceramics.

The following modeling approaches can be applied to both DLP and SLA procedures, depending on the hypotheses utilized in the model. To date, only a few studies have focused on the prediction of induced strains and residual stresses on green-cured ceramic objects [16, 18, 19].

2.3.2 Mechanical Modeling

The mechanical modeling approach by the finite element method (FEM) is predicted using the same principle for SLA or DLP impression types. The purpose of this approach is to calculate residual stresses and strains caused by the curling phenomenon that occurs during the curing of each layer. The occurrence of curling is directly correlated with the shrinkage of the material during the polymerization phase. This phenomenon, in turn, contributes to the development of residual stresses within the manufactured component.

In practice, current modeling proposals have a tendency to emulate the DLP method. Nevertheless, with regard to the estimation of residual stress and strain, this model may also be utilized for the purpose of modeling the SLA process. The first studies on this subject appeared in the 1990s and were conducted by Budega et al. [20] and Chambers et al. [21], who described SLA as a process of selective hardening in which only the exposed region is solidified. In their analyses, these studies sought to capture and explain the solidification process of a photosensitive polymer material, accounting for various influencing factors, such as scanning rate, the extent of curing depth, structural linkage, and material behavior over time. Consequently, these pioneering studies constituted the basis for the development of predictive models concerning the phenomenon of curling occurring in successive layers as a consequence of the process.

Specifically, the study by Huang et al. [22] utilized a dynamic finite element simulation to examine the curling process in photopolymer resin. As Figure 2.8 illustrates, the slice activation and remeshing process represents a significant challenge in this numerical approach, which is based on the control of meshing during layer activation. It has been established that the activation phase is associated with numerical

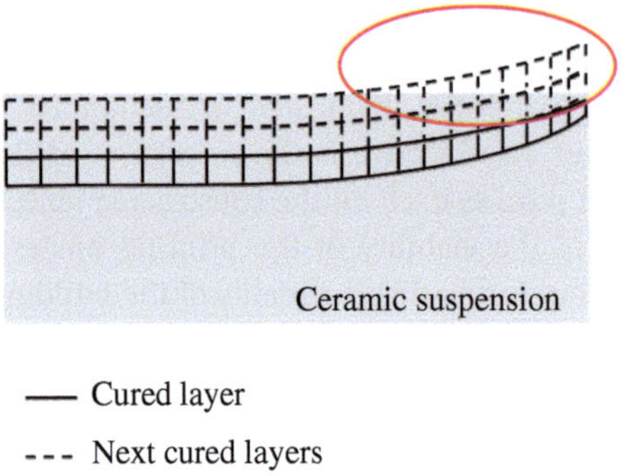

Figure 2.8 A detailed overview of the numerical finite element difficulties caused by meshing issues. Figure adapted and modified from Figure 9 in [22] to highlight the zone (in red) where the remeshing process must be applied to modify the element thickness. Modified from [22].

difficulties arising from the variable thickness of the finite elements that constitute the layer. This study identifies the initial process parameters that exert an effect, including the temperature rise, inhomogeneous curing properties in individual layers, and the optimization of the scanning path.

More recently, on an unfilled polymer suspension, Westbeek et al. [17] confronted predicted cured objects with DLP printed ones. Their results show a great prediction potential from the numerical approach, as their final morphology matches very well the final shape of printed objects obtained by X-ray tomography on complex shapes (Figure 2.9). These findings indicate a great potential for numerical premanufacturing.

In the field of ceramic suspensions, the study conducted by Wu et al. [18] provides a comprehensive investigation into the mechanical behavior of the material under consideration of the residual stresses induced by the layer-by-layer printing process known as large area maskless photopolymerization (LAMP). The outcomes of this

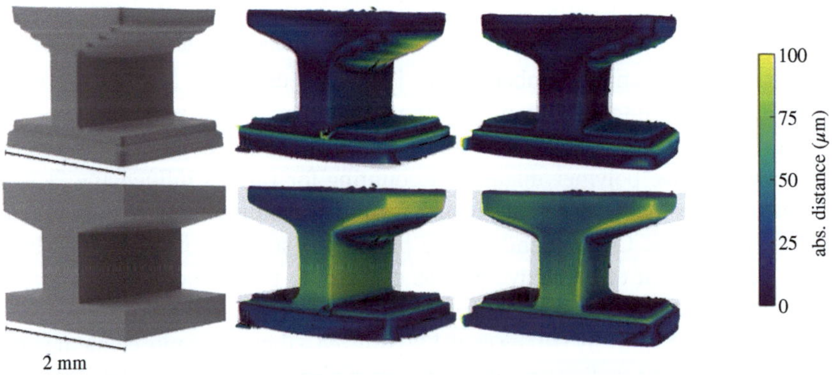

Figure 2.9 Comparison between simulated and printed geometries for a complex shape. Top row: comparison of simulated geometry and printed samples. Bottom row: target geometry with the printed samples. [17] / with permission of Elsevier.

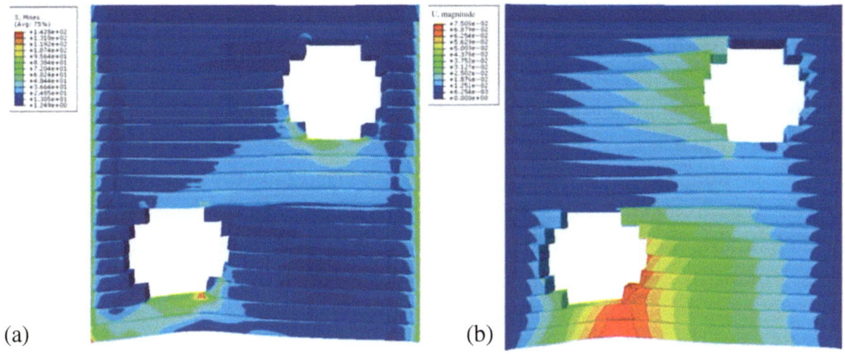

Figure 2.10 Visualization of residual Von Mises stress and calculated displacement after printing process. (a) Von Mises stress. (b) Displacement magnitude. [18] / The University of Texas at Austin.

study demonstrate a substantial enhancement in Von Mises stress with an increase in the number of layers, peaking at up to 120 MPa. Furthermore, the investigation underscores the impact of boundary conditions on the occurrence of curling effects. The insights derived from these findings are presented in Figure 2.10.

2.3.3 Limits of Actual Modeling Approaches

In the field of material science and engineering, the numerical approaches employed to simulate layer deposition and solidification represent a significant area of research endeavor. The estimation of numerous sophisticated model parameters, including material properties (such as curving and thermo-elastic-viscous-plastic properties), poses significant difficulties. As previously referenced, the numerical methodology required to activate layers remains challenging due to a multitude of factors, including the type of element, the layer thickness, the extent of curling in preceding layers, and numerous additional elements.

The utilization of numerical simulation employing finite elements constitutes a robust methodology for the anticipation of structural imperfections and the refinement of the precision of ceramic components fabricated by SLA and DLP processes. The application of finite element analysis confers three principal benefits. First, it enables the optimization of manufacturing conditions; second, it is devoted to the optimization of the scanning path in order to limit distortions prior to the construction of the component; and third, it reduces the costs associated with experimental testing and increases the success rate of complex printing, particularly in the demanding sectors of aerospace, biomedicine and high-performance electronics.

2.4 Advantages and Limits of VAT Processes

SLA and DLP have become prevalent in the field of 3D-printing ceramic components due to their ability to achieve high levels of precision, facilitate the fabrication of intricate geometries, and deliver surface quality that is superior to many alternative methods. A salient benefit of these technologies is their capacity to produce ceramic structures with fine detail and very tight dimensional tolerances, rendering them optimal for applications in the medical, aerospace, and electronics industries. A notable advantage of SLA and DLP over conventional ceramic forming methods, which frequently necessitate the use of molds and complex machining operations, is that they can be produced directly from a digital model, reducing the time and cost of prototyping. The utilization of photopolymer resins, comprising ceramic particles, facilitates the fabrication of components with sophisticated internal structures that cannot be achieved by conventional methods, opening up promising prospects for optimized design, lighter structures, and improved mechanical properties. Specifically, DLP offers a distinct advantage in terms of printing speed, whereby an entire layer is polymerized in a single light exposure, in contrast to the SLA method, which utilizes point-by-point scanning with a laser. This characteristic renders DLP particularly well-suited for large-scale manufacturing or the simultaneous production of multiple components without a loss in print quality.

Nevertheless, it should be noted that these processes are also subject to significant limitations, particularly with material and post-processing aspects. Indeed, some strongly UV-absorbent materials (case of SiC) are not recommended for SLA printing. Subsequent to the printing process, the components are designated as "green parts," signifying composite structures comprising ceramic and organic binder. Thermal debinding, a critical stage in the manufacturing process, involves the gradual elimination of the binder through pyrolysis. However, if the process is not adequately controlled, it can result in undesirable outcomes such as deformation, cracking, or the collapse of the structures. Following debinding, the parts are required to undergo sintering at elevated temperatures (typically between 1200 and 1800 °C) to densify the ceramic and establish their final mechanical properties. This stage is accompanied by a substantial dimensional shrinkage (typically 15–30%), a factor that necessitates compensation in the initial design to ensure precise final dimensions.

This technical compromise can impose limits on the range of printable ceramics available, thus influencing the eventual properties of the components manufactured. Notwithstanding these constraints, the combination of SLA and DLP techniques with innovative ceramic materials continues to advance, thereby paving the way for increasingly complex and high-performance applications, most notably in the manufacture of miniaturized electronic components, customized medical implants, and high-temperature-resistant parts for industry. Illustrative examples are presented in Figure 2.11 (see [23–25]).

A significant benefit of ceramic 3D printing is the production of customized parts. A particularly noteworthy instance within this domain pertains to the fabrication of biocompatible cranial implants that are customized to the patient's anatomical structure, directly from the scanner of the intended reconstruction area. This development signifies a substantial increase in the scope of possibilities.

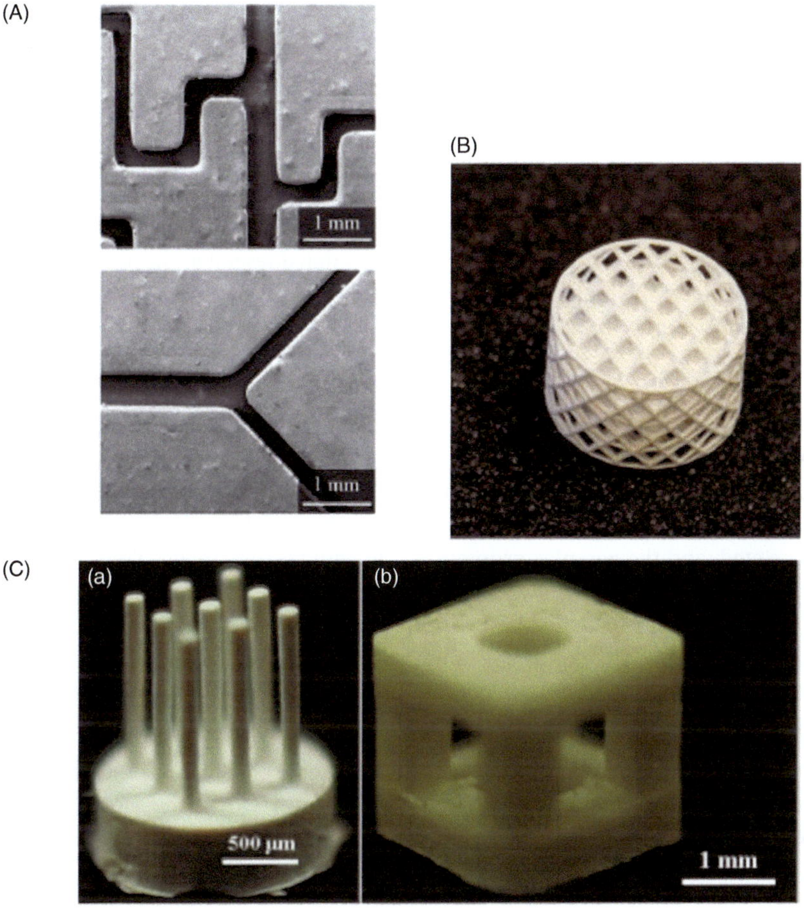

Figure 2.11 Examples of applications of the photopolymerization process for high-added-value uses. (A) Hydroxyapatite microchannel geometry for bone cell culture. [23] / Elsevier / CC BY-NC-ND 4.0. (B) Al_2O_3 10-mm long structure for heat transfer maximization. [24] / Elsevier / CC BY-NC-ND 4.0. (C) $BaTiO_3$ piezoelectrical applications. [25] / with permission of Elsevier.

2.5 AM Hybridization Opportunities

In the context of the increasing availability of 3D printing, the development of hybrid processes is facilitating the utilization of multiple materials in a manner that optimizes their performance for specific applications [37]. Lauwers et al. [38] defined two main types of hybrid processes. First, intrinsic hybridization (narrow hybrid manufacturing), which involves the concurrent utilization of two distinct processes in a single location. Second, open hybrid manufacturing, which involves the combination of two processes for the purpose of innovation.

Among the three main additive processes (Selective Laser Sintering [SLS], SLA, and Fused Feedstock Deposition [FFD]), Baranowski et al. [39] studied the hybridization of SLA and fused wire deposition. Hybridization enables the final added value to be increased by combining the advantages of each process. Here, metal inserts are integrated into polymer structures manufactured using FFD in order to stiffen the overall structure. In 2021, Rajamani et al. [40] highlighted the potential of additive processes when combined with other traditional processes. The case studied is the hybridization of molten wire deposition and plastic injection. Despite its large manufacturing scale, plastic injection does not allow products to be customized, due to the high cost of manufacturing molds. In this case, hybridization with an additive process makes it possible to offer easy customization of products on a large scale.

In the research carried out by M. Bernard [41] and J. Raynaud [42], a combination of two different additive processes was considered for the production of a ceramic substrate containing metal tracks. The fabrication process involved the utilization of SLA for the substrate and micro-extrusion for the metal components. The principal criteria considered were the achievable resolution, the quantities of material to be printed, and the complex geometries that could be envisaged. This process is known as open hybridization, whereby the two processes work successively to produce the final object. As a result, it is now possible to produce High-Temperature Cofired Ceramic (HTCC) and Low-Temperature Cofired Ceramic (LTCC) components on demand. Some examples of HTCC-manufactured objects are presented in Figure 2.12. The following process is applied to each printed layer:

1) Printing the ceramic shape of the substrate using the classic SLA process.
2) Cleaning of the non-polymerized areas of the layer where the metal is to be deposited by blowing/machining.
3) Metal deposition in the desired areas of the layer.

As an example of process hybridization, Figure 2.13 presents an innovative SLA printing device. The SLA machine has undergone a process of hybridization,

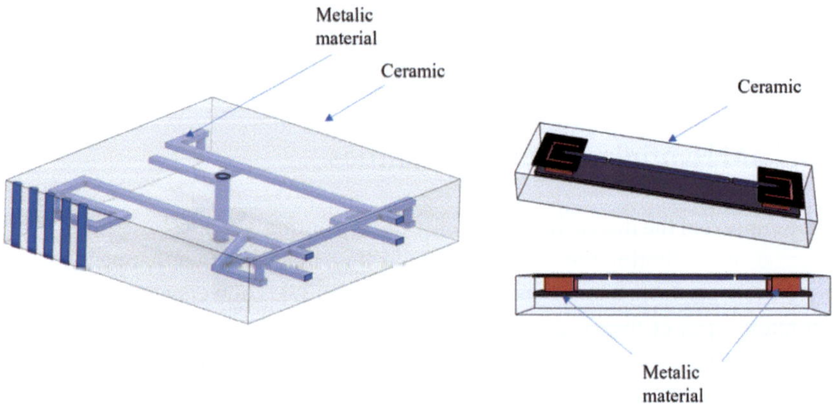

Figure 2.12 Example of HTCC parts, manufactured at the IRCER laboratory by means of process hybridization.

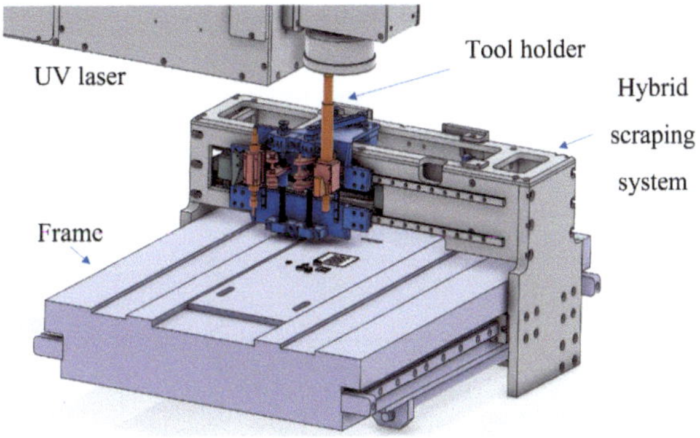

Figure 2.13 Design of a hybrid printing device developed at the IRCER laboratory.

facilitated by a hybrid scraper, in conjunction with two tool holders that incorporate a metal suspension deposition system for the micro-extrusion process and a machining tool spindle. This integration enables the fabrication of the desired geometries with tools ranging from 50 to 500 μm in diameter.

2.6 Next Generation Ceramic Printing: Two-Photon Photopolymerization and Volumetric Additive Manufacturing Methods

As mentioned, until now, the development of photopolymer-based additive manufacturing has evolved significantly, particularly in the field of ceramics. Conventional methodologies such as SLA and DLP have now become prevalent in the domain of printing ceramic-loaded resins, owing to their satisfactory resolution and layer-by-layer fabrication approach.

However, both SLA and DLP have inherent limitations. First, they require support structures, and second, they can be time-consuming for large or detailed parts (namely resolution). Finally, they exhibit anisotropy due to their layer-by-layer nature. In order to address these challenges, advanced techniques such as two-photon polymerization (TPP) and volumetric additive manufacturing (VAM) have emerged [43].

The TPP technique facilitates submicron resolution through the utilization of nonlinear optical absorption, rendering it particularly well-suited for microscale ceramic structures [44]. This method allows the fabrication of intricate features that conventional photopolymerization cannot realize, although its throughput remains limited to small volumes. In contrast, VAM technology has been developed to an extent that it is now capable of curing an entire 3D shape within seconds by projecting light patterns from multiple angles into a photosensitive ceramic slurry. The process is characterized by its support-free nature, isotropy, and capacity to produce complex parts with smooth surfaces.

Collectively, these methodologies signify the advent of a novel generation of ceramic additive manufacturing techniques. TPP places emphasis on achieving levels of precision that are unparalleled to date, while VAM facilitates the expeditious fabrication of large, intricate structures.

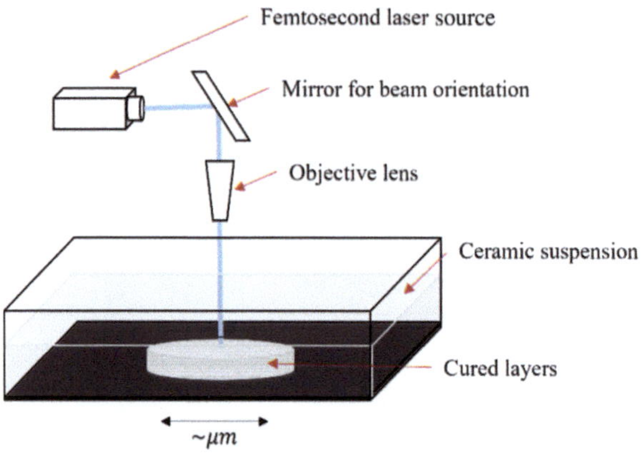

Figure 2.14 TPP process.

2.6.1 Two-Photon Polymerization

TPP is a high-resolution 3D-printing technique that uses a focused femtosecond laser to initiate polymerization at a very small focal point through a nonlinear optical process. This process stands out in the fields of micro-optics, MEMS, nanoelectronics, and biomedical applications and presents a really high resolution of under 1 μm [48]. In this context, the photopolymerization is activated by the simultaneous absorption of two photons of near infrared (780 nm) or green (515 nm) [43]. This solution is therefore analogous to the SLA method, including thermal treatment (detail in Section 2.2.5), with the exception of the light source, which is different (Figure 2.14).

Although the application of this process has historically been focused on the domain of polymeric materials, there have been notable advancements in the field that have enabled the exploration of its potential for use with ceramic materials. It is imperative to note that the final product is intended for specific applications in which the environment is characterized by complexity, encompassing chemical and thermal elements.

In the domain of ceramics, TPP is conventionally utilized in conjunction with a ceramic-loaded photosensitive resin, wherein the laser selectively cures the polymer matrix, thereby embedding ceramic nanoparticles. An example of 3D woodpile structures [45] printed with the TPP process is presented in Figure 2.15. The present study sets out the methodology for the fabrication of ceramic structures with a high spatial resolution. It is important to note that in these types of micro and nano SLA processes, thermal post-treatment of pyrolysis must be carried out with the greatest care.

It has been demonstrated by further authors, such as Combo et al. [46] and Koroleva et al. [47], that the process is capable of building high-precision 3D objects on technical ceramics, such as SiOC and ZrSi, for the purpose of bone tissue engineering, both before and after thermal treatment.

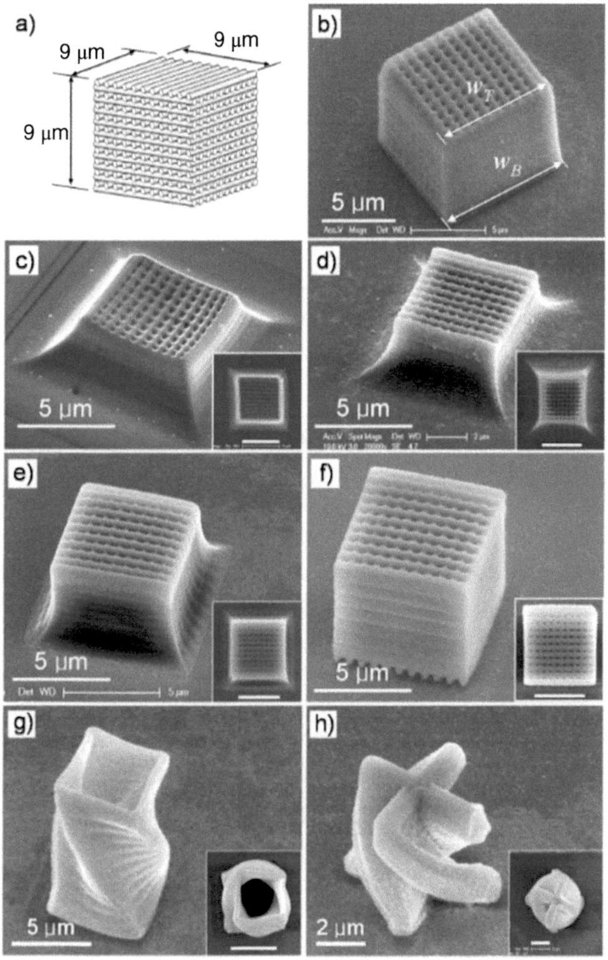

Figure 2.15 Example of micro-scaled woodpile structures (b-h) of SiCN. (a) Giving the information on the geometry. [45] / with permission of John Wiley & Sons.

It is important to note that the reduced dimensions of the objects manufactured using the TPP method present particular difficulties with the detachment of the printed object from the building platform [46]. Nevertheless, it remains an essential tool where precision and miniaturization are paramount, and continues to drive innovation in the field of ceramic microfabrication.

2.6.2 Volumetric Additive Manufacturing

The VAM process was first introduced by Hideo Kodama in 1981 [49]. A distinctive feature of the process is its ability to produce three-dimensional objects in a single step, as opposed to the more conventional layering technique. The process is

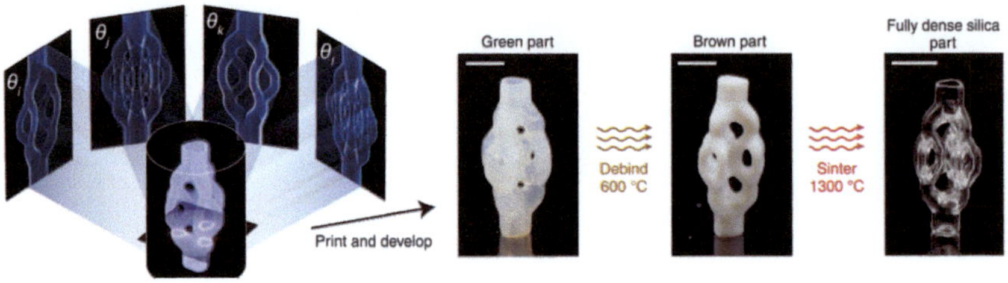

Figure 2.16 Superposition of tomographic data for the generation of a 3D dataset to selectively polymerize a resin into a specific geometry. In order to obtain the final printed object, the cured object undergoes thermal treatment (debinding and sintering) – the case of silica glass. Reference [51] / American Association for the Advancement of Science – AAAS.

predicated on the direct formation of three-dimensional components through the use of multi-angle exposure of photosensitive materials. It can be found in various configurations, including adaptive light superposition, subtractive light superposition and tomographic VAM [50]. In comparison to the previously mentioned VAT photopolymerization processes, the VAM is a more complex approach, as it is based on the employment of volumetric exposure algorithms. These algorithms facilitate the conversion of the 3D model into the input data required for the aforementioned process, specifically the exposure distribution functions [50]. The process is illustrated in Figure 2.16.

The system developed by Toombs et al. [51] enabled the fabrication of 3D objects with resolutions of 20 μm for polymeric materials and 50 μm for silica materials. The system demonstrated good geometric reproducibility, good surface roughness and the desired mechanical and optical properties.

Kollep et al. [52] worked on the production of isotropic objects that are fully dense and entirely free of defects. The resin utilized in the study was a transparent, preceramic resin. It is imperative to emphasize the indispensability of transparent media and transparent tanks for the success of the printing process. The final printed objects exhibited a resolution of 10 μm and demonstrated relatively good isotropy during the thermal post-treatment process. The final object exhibited an enhanced level of resistance to thermal stresses and chemical environments.

While VAM introduces a breakthrough in the field of additive manufacturing – offering rapid fabrication, support-free printing, and isotropic properties [52] – it also faces several significant challenges, particularly when applied to ceramic materials.

One of the most critical issues lies in the formulation of the resin. To produce high performance ceramic parts, the resin must contain a high-volume fraction of ceramic particles. However, these particles tend to scatter and absorb light, making it difficult to achieve the uniform and precise light exposure required for volumetric curing. Additionally, ensuring the optical clarity and homogeneity of the ceramic slurry over time is a major concern, as particle sedimentation or agglomeration can severely impact print quality. Precise formulation protocols were described by [51, 52].

Another important limitation is the resolution. VAM typically achieves feature sizes in the range of 10–100 μm [52, 53], which, although sufficient for many functional applications, is lower than what can be achieved using DLP or TPP. This resolution is further affected by light diffusion, especially in thicker resins or larger print volumes.

Moreover, the VAM process requires highly accurate projection systems and calibration. The reconstruction of the 3D shape relies on synchronizing multiple light projections from different angles. Any misalignment, inconsistency in light dose, or optical aberration can lead to deformed geometries or incomplete polymerization. Therefore, precise control of projection parameters and optimization of tomographic algorithms are essential.

Lastly, the technology is still in a relatively early stage of development. Few commercial systems currently support VAM (Xolo3D); and standardized materials, equipment, and processing protocols are still being developed. This limits the widespread adoption of the technique in industrial settings, although research and innovation are progressing rapidly, namely for its application on ceramic materials.

Together, TPP and VAM represent cutting-edge techniques in ceramic additive manufacturing, enabling high-resolution and high-speed production of intricate ceramic components with superior design freedom.

2.7 Conclusion

In this chapter, vat photopolymerization techniques – SLA, DLP, TPP, and VAM – were examined in the field of ceramic 3D printing. The operational principles of these systems were examined, with a comprehensive analysis highlighting both the similarities and distinctions in their respective mechanisms. A comprehensive overview of the feedstock suspensions employed in these processes was provided, placing particular focus on the critical role of the formulation in achieving optimal performance. Furthermore, the importance of thermal post-treatment steps was discussed, shedding light on how they contribute to enhancing the final mechanical properties and dimensional stability of printed parts.

It is evident that both SLA and DLP are distinguished among additive manufacturing techniques for their remarkable precision and surface quality. These processes are especially well-suited for applications where fine feature resolution and surface smoothness are paramount, such as in biomedical devices, dental prosthetics, microfluidics, and precision engineering components. The potential to produce complex geometries with a high degree of precision is a key competitive advantage of this technology, in contrast to other 3D-printing methods that often prioritize either accuracy or operational efficiency over precision.

Despite their shared foundation in vat photopolymerization, SLA and DLP have different strengths and weaknesses. SLA, with its laser-based scanning method, tends to offer slightly higher resolution and is more adaptable to larger build volumes, albeit at the cost of longer build times. In contrast, DLP uses digital projection to expose entire image slices, enabling faster fabrication.

The progression of these technologies is still being pushed forward by improvements in photopolymer chemistry, projector and laser technology, process engineering, and thermal processing techniques. As research progresses, improvements in printing speed, resolution, material diversity, and mechanical performance can be anticipated. Furthermore, integrating process monitoring, real-time control systems, and numerical modeling is likely to improve reliability and reproducibility, both of which are essential for industrial-scale use.

Beyond SLA and DLP, emerging vat photopolymerization techniques such as TPP and VAM are pushing the boundaries of what is currently achievable. To illustrate, the TPP facilitates the fabrication of 3D micro- and nanostructures with submicron resolution, rendering it a potent instrument for advanced photonics, biomedicine, and microscale engineering. In contrast, VAM facilitates the volumetric polymerization of entire 3D objects within seconds, as opposed to the conventional layer-by-layer approach, utilizing computed light fields that are projected into a rotating or stationary resin volume. Despite its early stage of technological implementation, VAM shows potential for rapid fabrication of complex geometries with minimal post-processing and reduced mechanical anisotropy. The application of the aforementioned concept to ceramic suspensions is characterized by a multitude of challenges.

In conclusion, SLA and DLP are considered to be two of the most mature and refined approaches in the field of vat photopolymerization. In conjunction with innovative methodologies such as TPP and VAM, these approaches serve to illustrate the considerable potential inherent in photopolymer-based additive manufacturing processes. The capacity of these technologies to produce parts with both high precision and functionality is such that they are considered essential tools in the additive manufacturing landscape. In order to select the most appropriate technique for a given application, it is first necessary to understand the respective strengths, material requirements, and post-processing needs of the materials under consideration. This understanding will then facilitate further innovation in this dynamic and rapidly evolving domain.

Acknowledgments

The authors would like to express their gratitude to Prof. Zhangwei Chen and Prof. Paolo Colombo for the kind invitation to contribute to this work.

References

1 Hull, C.W. (1986). Apparatus for production of three-dimensional objects by stereolithography, US4575330A, 11 mars 1986.

2 Griffith, M.L. and Halloran, J.W. (1996). Freeform fabrication of ceramics via stereolithography. *J. Am. Ceram. Soc.* 79 (10): 2601–2608.

3 Brady, G.A. and Halloran, J.W. (1997). Stereolithography of ceramic suspensions. *Rapid Prototyp. J.* 3 (2): 61–65.

4 Brady, G.A., Chu, T.-M., and Halloran, J.W. (2024). Curing behavior of ceramic resin for stereolithography, 1996, Consulté le, 25 novembre 2024.

5 Chen, Z., Li, Z., Li, J. et al. (2019). 3D printing of ceramics: a review. *J. Eur. Ceram. Soc.* 39 (4): 661–687.

6 Jacobs, P.F. (1992). *Rapid Prototyping & Manufacturing: Fundamentals of Stereolithography*. Society of Manufacturing Engineers.

7 Nakamoto, T., Yamaguchi, K., and Abraha, P. (1996). Consideration on the producing of high aspect ratio micro parts using UV sensitive photopolymer. *MHS'96 Proceedings of the Seventh International Symposium on Micro Machine and Human Science*, October 1996, pp. 53–58.

8 Santoliquido, O., Colombo, P., and Ortona, A. (2019). Additive manufacturing of ceramic components by digital light processing: a comparison between the "bottom-up" and the "top-down" approaches. *J. Eur. Ceram. Soc.* 39 (6): 2140–2148.

9 J.-H. Sim, B.-K. Koo, M. Jung, et D.-S. Kim, Study on debinding and sintering processes for ceramics fabricated using digital light processing (DLP) 3D printing, *Processes*, vol. 10, no 11, Art. no 11, 2022.

10 Zhou, S., Liu, G., Wang, C. et al. (2024). Thermal debinding for stereolithography additive manufacturing of advanced ceramic parts: a comprehensive review. *Mater. Des.* 238: 112632.

11 Deckers, J., Vleugels, J., and Kruth, J.P. (2014). Additive manufacturing of ceramics – a review. *J. Ceram. Sci. Technol.* 5: 245–260.

12 Andjela, L., Abdurahmanovich, V.M., Vladimirovna, S.N. et al. (2022). A review on vat photopolymerization 3D-printing processes for dental application. *Dent. Mater.* 38 (11): e284–e296.

13 Zakeri, S., Vippola, M., and Levänen, E. (2020). A comprehensive review of the photopolymerization of ceramic resins used in stereolithography. *Addit. Manuf.* 35: 101177.

14 A. Bove, F. Calignano, M. Galati, et L. Iuliano, Photopolymerization of ceramic resins by stereolithography process: a review, *Appl. Sci.*, vol. 12, no 7, Art. no 7. 2022.

15 Kakanuru, P. and Pochiraju, K. (2022). Simulation of shrinkage during sintering of additively manufactured silica green bodies. *Addit. Manuf.* 56: 102908.

16 Classens, K., Hafkamp, T., Westbeek, S. et al. (2021). Multiphysical modeling and optimal control of material properties for photopolymerization processes. *Addit. Manuf.* 38: 101520.

17 Westbeek, S., Remmers, J.J.C., Van Dommelen, J.A.W. et al. (2021). Prediction of the deformed geometry of vat photo-polymerized components using a multi-physical modeling framework. *Addit. Manuf.* 40: 101922.

18 Wu, T. and Das, S. (2012). Theoretical modeling and experimental characterization of stress development in parts manufactured through large area maskless photopolymerization.

19 Westbeek, S., Remmers, J.J.C., van Dommelen, J.A.W., and Geers, M.G.D. (2020). Multi-scale process simulation for additive manufacturing through particle filled vat photopolymerization. *Comput. Mater. Sci.* 180: 109647.

20 Bugeda, G., Cervera, M., Lombera, G., and Onate, E. (1995). Numerical analysis of stereolithography processes using the finite element method. *Rapid Prototyp. J.* 1 (2): 13–23.

21 Chambers, R.S., Guess, T.R., and Hinnerichs, T.D. (1995). A phenomenological finite element model of part building in the stereolithography process, Sandia National Lab. (SNL-NM), Albuquerque, NM (United States), SAND-94-2565C; CONF-9506149-2, mars 1995.

22 Huang, Y.-M. and Jiang, C.-P. (2003). Curl distortion analysis during photopolymerisation of stereolithography using dynamic finite element method. *Int. J. Adv. Manuf. Technol.* 21 (8): 586–595.

23 Tang, Q., Li, X., Lai, C. et al. (2021). Fabrication of a hydroxyapatite-PDMS microfluidic chip for bone-related cell culture and drug screening. *Bioact. Mater.* 6 (1): 169–178.

24 Pelanconi, M., Barbato, M., Zavattoni, S. et al. (2019). Thermal design, optimization and additive manufacturing of ceramic regular structures to maximize the radiative heat transfer. *Mater. Des.* 163: 107539.

25 Wang, W., Sun, J., Guo, B. et al. (2020). Fabrication of piezoelectric nano-ceramics via stereolithography of low viscous and non-aqueous suspensions. *J. Eur. Ceram. Soc.* 40 (3): 682–688.

26 Coble, R.L. (1962). Sintering alumina: effect of atmospheres. *J. Am. Ceram. Soc.* 45 (3): 123–127.

27 Rahaman, M.N. (2007). *Sintering of Ceramics*. CRC Press.

28 Liu, X., Zou, B., Xing, H., and Huang, C. (2020). The preparation of ZrO_2-Al_2O_3 composite ceramic by SLA-3D printing and sintering processing. *Ceram. Int.* 46 (1): 937–944.

29 Ji, S.H., Kim, D.S., Park, M.S., and Yun, J.S. (2021). Sintering process optimization for 3YSZ ceramic 3D-printed objects manufactured by stereolithography. *Nanomaterials* 11 (1): 192.

30 Li, W.-K., Wu, J.-M., Zhou, F.-L. et al. (2025). Influence of sintering temperature and holding time on the properties of DLP-fabricated Si_3N_4/SiAlON ceramics. *Ceram. Int.* 51 (3): 3198–3208.

31 Redl, M. (2019). Lithographic 3D printing of SiC based ceramics. Diploma Thesis, Technische Universität Wien. repositUm.

32 Chaput, C. and Chartier, T. (2007). Fabrication of ceramics by stereolithography. *RTe J.* 79: 2601–2608.

33 Chartier, T., Dupas, C., Lasgorceix, M. et al. (2014). Additive manufacturing to produce complex 3D ceramic parts. *J. Ceram. Sci. Technol.* 6 (2): 95–104.

34 Young, J.C., Brinckmann, S.A., Fertig, R.S. III, et al. (2024). Influence of layer thickness and exposure on mechanical properties of additively manufactured polymer-derived SiOC ceramics. *Open Ceram.* 19: 100652.

35 Tomeckova, V. and Halloran, J.W. (2010). Predictive models for the photopolymerization of ceramic suspensions. *J. Eur. Ceram. Soc.* 30: 2833–2840.

36 Chartier, T., Badev, A., Abouliatim, Y. et al. (2012). Stereolithography process: Influence of the rheology of silica suspensions and of the medium on polymerization kinetics - Cured depth and width. *J. Eur. Ceram. Soc.* 32: 1625–1634.

37 A. Torres Marques, S. Esteves, J. P. T. Pereira, and L. M. Oliveira, Éd., *Additive Manufacturing Hybrid Processes for Composites Systems. Advanced Structured Materials*, vol. 129. Cham: Springer International Publishing, 2020.

38 Lauwers, B., Klocke, F., Klink, A. et al. (2014). Hybrid processes in manufacturing. *CIRP Ann.* 63 (2): 561–583.

39 M. Baranowski, Tristan Schlotthauer, Markus Netzer, Philipp Gönnheimer, Hybridization of fused filament fabrication components by stereolithographic manufactured thermoset inserts, in *Recent Advances in Manufacturing Engineering and Processes* R. K. Agarwal, Éd., in Lecture Notes in Mechanical Engineering. Singapore: Springer Singapore, 2022, p. 3-14.

40 Rajamani, P.K., Ageyeva, T., and Kovács, J.G. (2021). Personalized mass production by hybridization of additive manufacturing and injection molding. *Polymers* 13 (2): 309.

41 Bernard, M., Pateloup, V., Passerieux, D. et al. (2022). Feasibility of manufacturing of Al_2O_3–Mo HTCC by hybrid additive process. *Ceram. Int.* 48 (11): 14993–15005.

42 Raynaud, J. et al. (2021). Hybridization of additive manufacturing processes to build ceramic/metal parts: Example of HTCC. *J. Eur. Ceram. Soc.* 41 (3): 2023–2033.

43 Chen, Z., Li, Z., Li, J. et al. (2019). 3D printing of ceramics: a review. *J. Eur. Ceram. Soc.* 39 (4): 661–687.

44 Maruo, S., Nakamura, O., and Kawata, S. (1997). Three-dimensional microfabrication with two-photon-absorbed photopolymerization. *Opt. Lett.* 22 (2): 132–134.

45 Pham, T.A., Kim, D.P., Lim, T.W. et al. (2006). Three-dimensional SiCN ceramic microstructures via nano-stereolithography of inorganic polymer photoresists. *Adv. Funct. Mater.* 16 (9): 1235–1241.

46 Colombo, P., Schmidt, J., Franchin, G. et al. (2017). Additive manufacturing techniques for fabricating complex ceramic components from preceramic polymers. *Am. Ceram. Soc. Bull.* 96 (3): 16–23.

47 Koroleva, A., Deiwick, A., Nguyen, A. et al. (2015). Osteogenic differentiation of human mesenchymal stem cells in 3-D Zr-Si organic-inorganic scaffolds produced by two-photon polymerization technique. *PLoS One* 10 (2): e0118164.

48 Zhang, F., Li, Z., Xu, M. et al. (2022). A review of 3D printed porous ceramics. *J. Eur. Ceram. Soc.* 42 (8): 3351–3373.

49 Kodama, H. (1981). Automatic method for fabricating a three-dimensional plastic model with photo-hardening polymer. *Rev. Sci. Instrum.* 52 (11): 1770–1773.

50 Mathur, V., Dsouza, V., Srinivasan, V., and Vasanthan, K.S. (2025). *Volumetric Additive Manufacturing for Cell Printing: Bridging Industry Adaptation and Regulatory Frontiers.* ACS Biomaterials Science & Engineering.

51 Toombs, J.T., Luitz, M., Cook, C.C. et al. (2022). Volumetric additive manufacturing of silica glass with microscale computed axial lithography. *Science* 376 (6590): 308–312.

52 Kollep, M., Konstantinou, G., Madrid-Wolff, J. et al. (2022). Tomographic volumetric additive manufacturing of silicon oxycarbide ceramics. *Adv. Eng. Mater.* 24 (7): 2101345.

53 Huang, K., Franchin, G., and Colombo, P. (2024). Volumetric additive manufacturing of SiOC by xolography. *Small* 20 (37): 2402356.

3

Binder Jetting (BJT) and Material Jetting (MJT)

Andrea Zocca[1], *Jens Günster*[1], *Anna-Katharina Hofer*[2], *and Stefan Waldner*[2]

[1]*Advanced Multi-materials Processing, Bundesanstalt für Materialforschung und -prüfung (BAM), Unter den Eichen 87, 12205, Berlin, Germany*
[2]*D3-AM, Durst Group, Julius Durst-Strasse 2c, 39042, Brixen, Italy*

Chapter 3 is dedicated to two separate classes of additive manufacturing (AM) technologies, binder jetting (BJT) and material jetting (MJT). Since inkjet printing (IJP) is a key enabling technology at the core of both BJT and MJT, a general introduction to IJP will be given first in Section 3.1, before delving into BJT in Section 3.2, and into MJT in Section 3.3. Finally, Section 3.4 will shortly present the peculiarities of BJT and MJT regarding the post-processing of AM ceramic parts.

3.1 Inkjet Printing

IJP is a term that many individuals are familiar with and is commonly associated with home and small office printers. However, the scope of IJP extends far beyond the domain of conventional paper printing. It has found significant applications in various fields, including the printing of electronically functional patterns for solar cells, micro lenses, solid oxide fuel cells, batteries, and radio frequency identification (RFID) tags [1–4]. In addition, the display market has seen significant contributions from IJP, particularly in the manufacturing of flat panel displays (FDPs), liquid crystal displays (LCDs), polymer light-emitting diodes (PLEDs), and flexible discs [5, 6]. Furthermore, IJP has found applications in the chemical, medicine, and life science markets, where it facilitates the precise dispensing of liquids, polymers, DNA, and proteins [7–17].

It should be noted that these applications primarily involve the deposition of a single layer of droplets on a substrate. However, recent advancements by multiple research groups and companies have demonstrated the potential of IJP to extend beyond two dimensions, paving the way for its application in the AM of polymeric, metallic, and ceramic components in three dimensions.

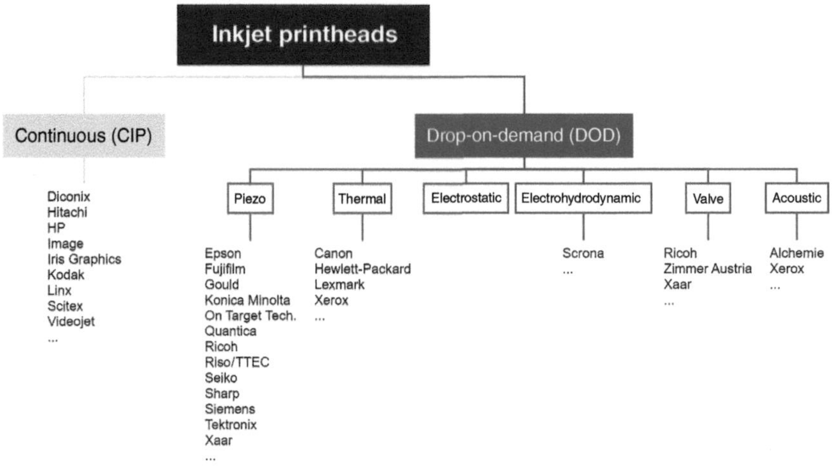

Figure 3.1 Organigram of the different inkjet print concepts (continuous and drop-on-demand) to generate droplets with their corresponding approaches.

But how did IJP originate, and what ways do inkjet printhead concepts differ?

The advent of IJP can be traced back to the 1850s, when William Thomson devised a siphon recorder, thereby facilitating the automated documentation of telegraph communications through the selective application of ink droplets from a continuous ink stream, guided by a moving siphon [18].

This development laid the foundation for the initial concept of IJP, namely **continuous inkjet printing** (*CIP*). For an extended period, CIP served as the underlying principle of inkjet technologies, subsequently giving way to the **drop-on-demand** (*DOD*) concept. For both printing concepts, several approaches have been developed that differ in their working principle. Figure 3.1 provides an overview of established IJP approaches, printhead developers/manufacturers as well as applicants of certain approaches.

3.1.1 Continuous Inkjet Printing

CIP relies on the principle of the Plateau-Rayleigh instability. This phenomenon pertains to the liquid dynamics of a liquid stream. Perturbations in the stream and the driving force of reducing the surface area, and consequently the surface energy, first lead to a wave-like behavior of the stream and finally to the breakup of individual droplets, as illustrated in Figure 3.2 [19–21].

In order to reach uniformity in size and shape of the separated droplets, R.G. Sweet used a magnetostrictive transducer to apply a pressure wave pattern to an orifice where the ink was ejected. This approach resulted in a controlled droplet breakup and the ability to selectively charge and deflect droplets onto a substrate by passing through an electric field [22]. Uncharged droplets, on the other hand, followed the straight path of motion and were subsequently collected and returned to the original ink system. To date, this working principle underlies the contemporary

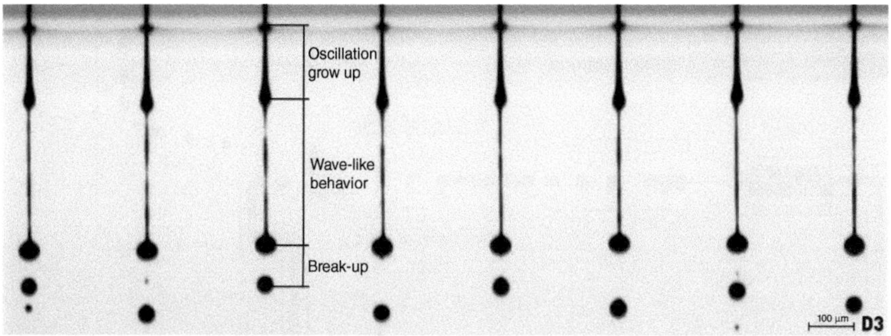

Figure 3.2 Representative image of a nozzle array illustrating the Plateau-Rayleigh instability at its different states. At the exit of the droplet from the orifice, oscillation grows, leading to a wave-like behavior and finally to a droplet breakup.

methodology of CIP. However, various approaches have emerged as binary, multiple, Hertz, and μDot.

In contemporary applications, piezoelectric crystals are utilized to induce oscillation in a continuous stream, resulting in droplet breakup. An applied electric field onto the generated droplet induces charging. As a consequence, these charged droplets can be deflected by passing through a second electric field. Uniform droplet sizes are generated, with a typical droplet size ranging from 20 to 150 μm [12]. Figure 3.3 presents the binary (a) and the multiple printing approach (b) of CIP. In the case of the binary approach (Figure 3.3a), as in Sweets' initial apparatus, charged droplets are deflected onto the substrate, while uncharged droplets follow the straight trajectory and are collected and recycled. This configuration, however, limits printing to a single dot position. The multiple printing approach (see Figure 3.3b) enables the printing of a more complex image. The application of different voltages on charged droplets, causes a variation in the angle of deflection, resulting in different printing positions. The Hertz approach operates on the principle of binary printing, yet it facilitates gray-scaling through the selective variation in droplet number at a given position [23, 24]. The μDot approach stands out for its capacity to generate small droplets. The μDot approach utilizes the same nozzle to generate droplets of varying sizes, with only the smaller droplets entering the electric field and being ejected onto the substrate [25].

CIP facilitates high-speed printing for industrial applications, such as bar code labeling. The continuous stream eliminates concerns regarding nozzle clogging. However, CIP exhibits limitations regarding material selection, as it can only utilize liquids that are electrically chargeable. Additionally, CIP necessitates a costly setup, including drop selection, a recycling system, and intricate maintenance procedures.

3.1.2 Drop-on-Demand Inkjet Printing (DOD)

In addition to CIP, the DOD concept has emerged and is currently predominant in the domain of IJP. In the case of DOD, independent droplets are formed through distinct approaches and are ejected one by one from the nozzle(s) onto a substrate.

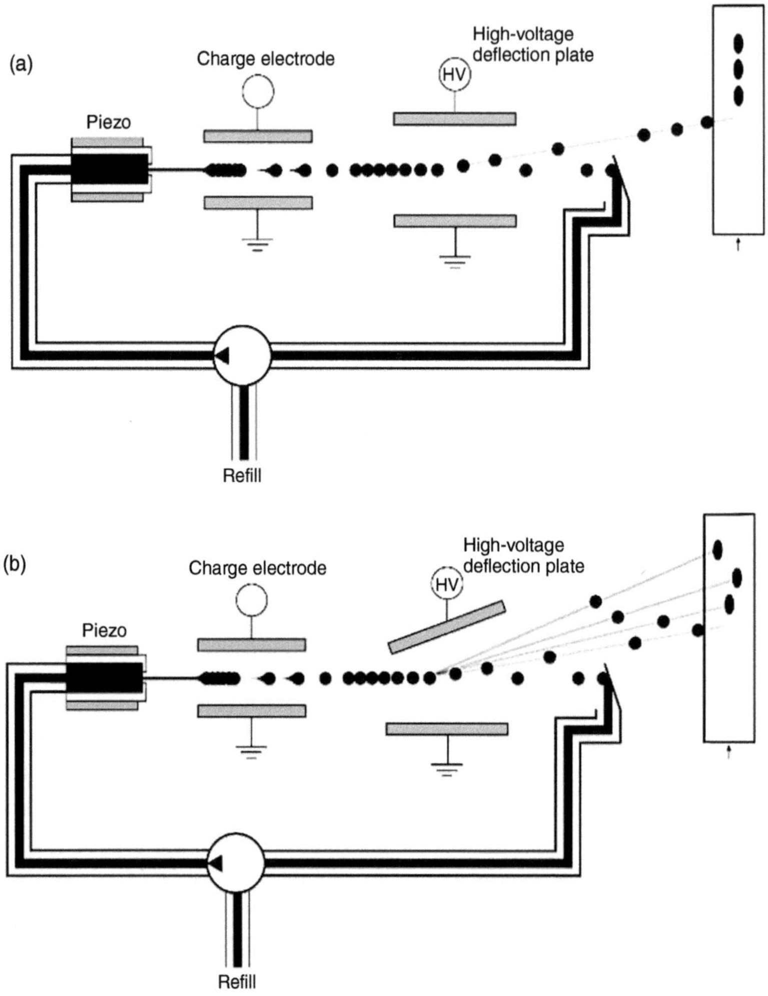

Figure 3.3 Schematics of two continuous inkjet printing modes: (a) binary and (b) multiple mode.

Thus, the necessity for complex droplet charging, deflection, and recycling systems, as required in a CIP process, is eliminated. In addition, smaller droplet sizes, less than 20 μm [12], can be realized with a higher placement accuracy. A multitude of approaches have been developed and implemented to generate these independent droplets, including piezoelectric, thermal, electrostatic, electrohydrodynamic, acoustic, and valve. Of these, the piezoelectric and thermal approaches have garnered particular attention and commercialization.

3.1.2.1 Piezoelectric DOD Inkjet Printing (PIJ)

The earliest experiments with DOD as a writing mechanism took place in the 1950s, employing a piezoelectric disc to project individual droplets from a nozzle [26, 27]. This configuration, however, did not ultimately result in commercialization.

A significant milestone in the evolution of DOD IJP was reached two decades later when four inventors, Zoltan (Clevite company) [28], Stemme (Chalmers University) [29], Kyser and Sears (Silonics company) [30], submitted three patents that would become the foundation for the subsequent advancements in the field. These patents similarly utilize the piezoelectric effect to facilitate the ejection of a droplet from a nozzle. A piezo crystal is attached to the ink chamber, with a diaphragm acting as a barrier to prevent direct contact with the ink. An applied electrical driving voltage on the piezo crystal is converted into a mechanical deformation, thereby generating a pressure wave within the ink chamber that results in the ejection of a droplet through the nozzle. The deactivation of the electrical driving voltage, in conjunction with the absence of pressure, facilitates an ink refill [31, 32]. Four different modes can be utilized, according to the orientation of the pressure wave in relation to the ink channel. Figure 3.4 illustrates these four different modes of piezoelectric IJP: (a) squeeze, (b) bend, (c) push, and (d) shear. Zoltan applied the squeeze mode (Figure 3.4a) to generate pressure in the ink chamber for droplet ejection, while both Stemme and Kyser and Sears followed the approach of the bend mode (Figure 3.4b). Subsequent developments by Howkins (Exxon company) and Fischbeck led to two additional piezoelectric mode configurations: push (Figure 3.4c.) and shear (Figure 3.4d). The volume of ink that is ejected with piezoelectric IJP currently ranges from 1.5 to 240 pl [31]. Table 3.1 presents a comprehensive list of commercially available piezoelectric DOD printheads, accompanied by detailed specifications, including ink type, viscosity, drop volume, print width, color, and resolution. The high degree of freedom in ink selection and the long lifetime of piezoelectric printheads make them indispensable for certain applications despite their high cost.

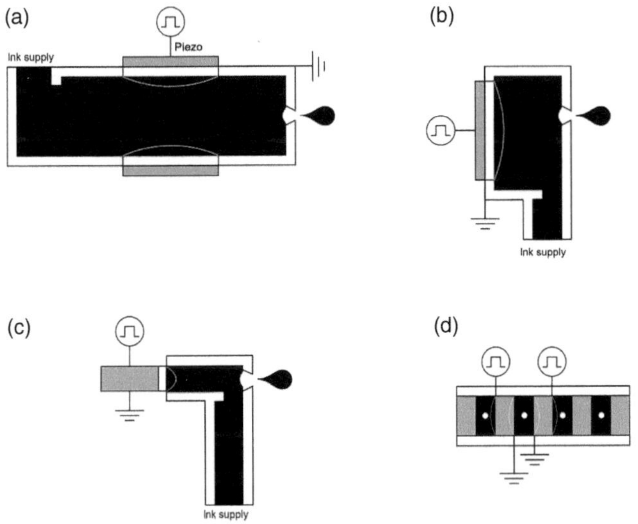

Figure 3.4 Schematics of the different modes for piezoelectric DOD inkjet printing: (a) squeeze, (b) bend, (c) push, and (d) shear mode.

Table 3.1 List of exemplary printheads from different suppliers and some selected specifications (gray).

Manufacturer	Model	Ink Type	Viscosity min	Viscosity max	Drop volume min	Drop volume[a)] max	Print width	Colors	Resolution
			(cP/mPas)	(cP/mPas)	(pL)	(pL)	(mm)	Maximum	max (dpi)
Konica Minolta	KM1024aLHG-RC	Aqueous, solvent, UV, oil	10	14	27	240	72	1	360
Seiko	RC1536-L	Aqueous, solvent, UV, oil	n.a.	n.a.	25	215	108.3	1	360
Fujifilm Dimatix	Starfire SG1024/LC	Aqueous, solvent, UV, oil	8	20	72	200	64.96	1	400
Xaar	Xaar 1003 Amp	Solvent, UV, oil	n.a.	n.a.	1	3	70.5	1	360
Kyocera	KJ4B-1200	Aqueous	5.5	6	1.5	5	112.42	1	1200
Memjet	DuraBolt	Aqueous	2	4	2.1	n.a.	324.4	4	1600
Fujifilm Dimatix	Samba G3L	Aqueous solvent, UV	2	10	2.4	13,2	43	1	1200
Quantica	NovoJet 1	Aqueous solvent, UV, oil	1	380	30	180	121.92	1	11.5
Xaar	Nitrox Pro	Oil, solvent	6	100	6	40	70.5	1	360
Xaar	Aquinox 12	Aqueous	6	100	12	n.a.	70.5	2	720
Seiko	RCE2560 – S	Aqueous, solvent, UV	9	30	25	225	108.3	2	600

Adapted from https://imagexpert.com/the-ultimate-industrial-inkjet-printhead-comparison-chart//last accessed on November 04, 2025.
a) Gray scale mode included.

3.1.2.2 Thermal DOD Inkjet Printing (TIJ)

Thermal DOD IJP is arguably the most commercially successful IJP approach, being employed in the majority of home and office printers at present. The development of thermal IJP occurred concurrently at two prominent companies, Canon and Hewlett-Packard (HP), during the 1970s and 1980s [26, 32]. The technology was pioneered by Endo and Hara at Canon, who invented the so-called "bubble jet" [33]. The printhead developed by Hewlett-Packard functioned on the same principle to form individual droplets, and the company was the first to bring a commercial, low-cost printhead to market [34]. The TIJ printhead's cost-effectiveness and miniaturization capabilities, coupled with its continual enhancement in nozzle packing density, led to its predominance over the PIJ printhead for a substantial period. Consequently, the TIJ printhead was regarded as the superior approach for printhead technology at that time.

Thermal IJP operates by leveraging the natural process of boiling water, where water bubbles are formed. Two modes of thermal IJP exist, depending on the position of the heating element: (a) side-shooter and (b) top-shooter. In Figure 3.5, the working principle of a thermal inkjet printhead in the side-shooter mode is illustrated. A heating element, attached to the wall of the ink chamber (**I**), rapidly increases the temperature of the ink (350–400 °C [31]). This increase in temperature causes the ink to vaporize, resulting in the formation of a bubble at the position of the heating element (**II**). The generation of this bubble produces a pressure pulse (**III**), which, in turn, initiates the ejection of an ink droplet through the nozzle. As the vapor bubble collapses as the droplet is ejected (**IV**), a reverse force results in ink refill. This entire process, from bubble generation to droplet ejection, occurs in less than 10 μs. The volume of ink contained within a single droplet is typically between 150 and 200 pl. In contrast to piezoelectric IJP, thermal IJP is constrained to vaporizable and thermally stable inks. Consequently, TIJ is best suited for water-based inks but is not compatible with certain ink formulations, such as nonvolatile, highly viscous,

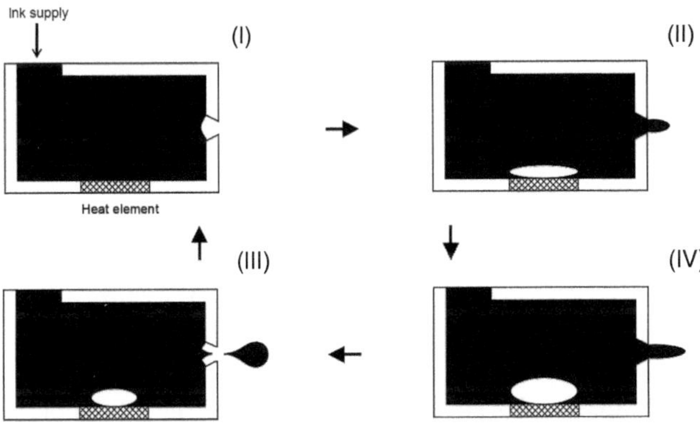

Figure 3.5 Schematic of the working principle of thermal DOD inkjet printing in the side-shooter mode. (**I**) Initial/Refill state, (**II**) temperature increase causes evaporation and bubble formation, (**III**) a pressure pulse is generated, resulting in a droplet formation, and (**IV**) the bubble collapses as the droplet is ejected from the orifice.

or thermally unstable inks. Additionally, solid inks (e.g. hot-melt or phase-change inks) require specialized modifications to be used in TIJ systems. The organigram in Figure 3.1 catalogues established printhead suppliers of the thermal IJP concept.

3.1.2.3 Other Approaches of DOD Inkjet Printing

As previously mentioned, in addition to piezoelectric and thermal DOD IJP, there are several other approaches (electrostatic, electrohydrodynamic, acoustic, and valve) that have been developed and are still in the research and development stage or applied for very specific applications.

Electrostatic DOD inkjet printing (EIJ) enables smaller droplet sizes compared to PIJ and TIJ. Similar to PIJ, EIJ utilizes an electric pulse to generate and eject individual droplets. The EIJ printhead is composed of a pressure plate and an electrode, which are positioned on one side of the ink chamber. The application of a direct current (DC) voltage between these components results in the expansion of the pressure plate and the increase of the ink chamber's volume, thereby causing a refill of ink. Upon the release of the voltage, the pressure plate reverts to its flat state, leading to a reduction in the chamber's volume and the subsequent ejection of a droplet. The magnitude of the applied DC voltage dictates the pressure plate's expansion, thereby regulating the ink chamber's volume and the dimensions of the ejected droplets [35, 36].

Another IJP approach is the **electrohydrodynamic DOD inkjet printing** (EHD). While EHD may have been among the earliest DOD IJP approaches, it later attracted minimal attention due to suboptimal printing quality and reliability during its early stages. However, subsequent research has revived interest in EHD, particularly within the domain of bioprinting [26, 37, 38], as, for example, by Scrona AG [39] in Switzerland. Among the various IJP approaches, EHD has the potential to obtain the smallest droplet sizes, being just a fraction of the nozzle diameter. In the electrohydrodynamic IJP approach, an ink meniscus is formed at the nozzle tip with a certain back pressure. The application of an external electric field between the nozzle tip and the substrate causes the ink meniscus to deform into a conical shape due to the reduction of surface tension caused by the electric field. Subsequent to the attainment of a critical electrostatic stress level that supersedes the surface tension at the conical orifice, a droplet is ejected. The shape and the volume of the ejected ink are determined by the magnitude of the applied electric field. High voltage results in a continuous stream, intermediate voltage causes the ejection of individual droplets, and low voltage determines soft dripping. As aforementioned, a notable advantage of EHD is its ability to produce small droplets, and in contrast to other DOD approaches, EHD is less susceptible to variations in nozzle quality or contamination, ensuring consistent printing quality. However, it should be noted that EHD is limited to conductive inks and does necessitate high voltages, which can incur higher costs compared to piezo or thermal IJP. Further, the formation of the meniscus imposes limitations on the packing density of nozzles [26, 36, 40].

Contrary to EHD, the **valve DOD inkjet printing** (VIJ) method generates comparatively large droplets. Figure 3.6 displays the underlying working principle of VIJ. It involves the back-and-forth/up-and-down movement of a piston/needle, which

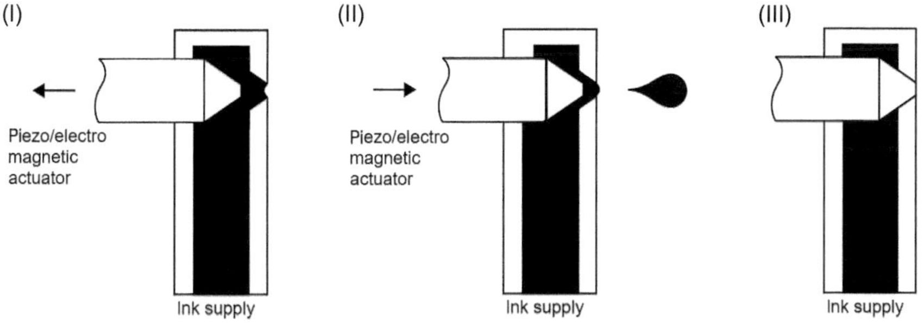

Figure 3.6 Schematic of the working principle of valve DOD inkjet printing. (**I**) The nozzle is opened, causing the ink to be refilled (**II–III**). The nozzle is closed, resulting in droplet ejection.

opens and closes the nozzle (**I–III**). During closure, a droplet is ejected. This motion is governed by an electromagnetic or piezoelectric actuator (**II**). The VIJ has found application in the textile, carpet, and packaging industries, with the electromagnetic actuator being particularly useful in these contexts [41]. This configuration yields droplet sizes ranging from 10 nl to 3 μl, according to the textile industry applications [42]. Additionally, valve printing has been adopted in the semiconductor industry, where a configuration of needle-valve IJP utilizes piezoelectric forces to accelerate the needle [43, 44].

Acoustic DOD inkjet printing (AIJ) is the only DOD printing approach that functions without a nozzle. This eliminates the prevalent concern of nozzle clogging that is often associated with piezoelectric and thermal IJP. In this regard, Xerox has pioneered the development of an acoustic printhead, marking a significant advancement in the field. The acoustic wave generation process entails the utilization of a high-frequency transducer, which generates acoustic waves that are subsequently transferred and focused through an affixed acoustic lens, positioned between the transducer and the ink chamber. Focusing the acoustic energy of the waves results in the induction of a pressure wave and the subsequent expulsion of a droplet from the ink chamber [31, 45]. This noncontact transfer of ink enables droplet volumes ranging from 2 nl to 1 μl. The AIJ printing method is reputed to be a highly gentle approach, making it of significant interest for life science applications, such as the printing of DNA or live cells [40].

3.1.3 Theory of Ink Printability

The printability of inks is contingent on numerous parameters, including density, viscosity, surface tension, nozzle diameter, and nozzle shape. To ascertain the printability of an ink, dimensionless groupings, such as the Reynolds number (Re), the Weber number (We), the Ohnesorge number (Oh), and the Z number (Z), have been established. The Reynolds number is defined by Eq. (3.1):

$$Re = \frac{\rho v L}{\eta} \tag{3.1}$$

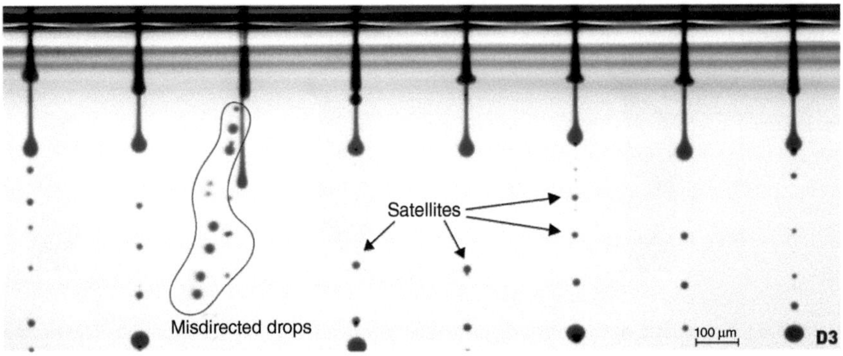

Figure 3.7 Representative image of a nozzle array illustrating the separation of small particles, known as "satellites." These satellites form due to the occurrence of Rayleigh-Plateau instability and a high Re number.

where ρ is the density of the ink in (g/ml), v is the velocity in (m/s), L is the characteristic linear dimension, typically diameter of the nozzle, drop or jet in (m), and η is the dynamic viscosity of the ink in (Pa s). When the *Re* number is too low, it can be challenging to actuate and separate a droplet from the nozzle, as is the case with viscous inks. Conversely, when *Re* exceeds a certain threshold, the ink may form satellites and splash over the substrate, leading to suboptimal printing precision. Satellites describe small droplets that separate from the leading droplet and form due to Rayleigh-Plateau instability, as shown in Figure 3.7. As the Reynolds number exhibits a considerable correlation with viscosity, it may be modified by regulating the temperature of the ink.

The *Weber* number, defined by Eq. (3.2), includes the important parameter of surface tension:

$$We = \frac{v^2 \rho L}{\gamma} \tag{3.2}$$

where γ is the surface tension in (N/m). The magnitude of surface tension dictates the energy required to eject a droplet from the nozzle. It is recommended to attain a *We* number greater than four to ensure optimal printing conditions. However, as the *We* number rises, surface tension proportionally decreases. This can lead to nozzle plate wetting, potentially impeding or complicating the printing process due to clogging, or it may cause splashing [31, 46].

The ratio of the square root of the Weber number to the Reynolds number is defined as the Ohnesorge number (Oh), which is expressed in Eq. (3.3) [31, 47]:

$$Oh = \frac{\sqrt{We}}{Re} = \frac{\eta}{\sqrt{\gamma \rho L}} \tag{3.3}$$

The Ohnesorge number integrates two predominant ink parameters: viscosity and surface tension. These parameters regulate the droplet ejection from the nozzle, its shape during transferring, and the impact on the substrate. The inverse of the

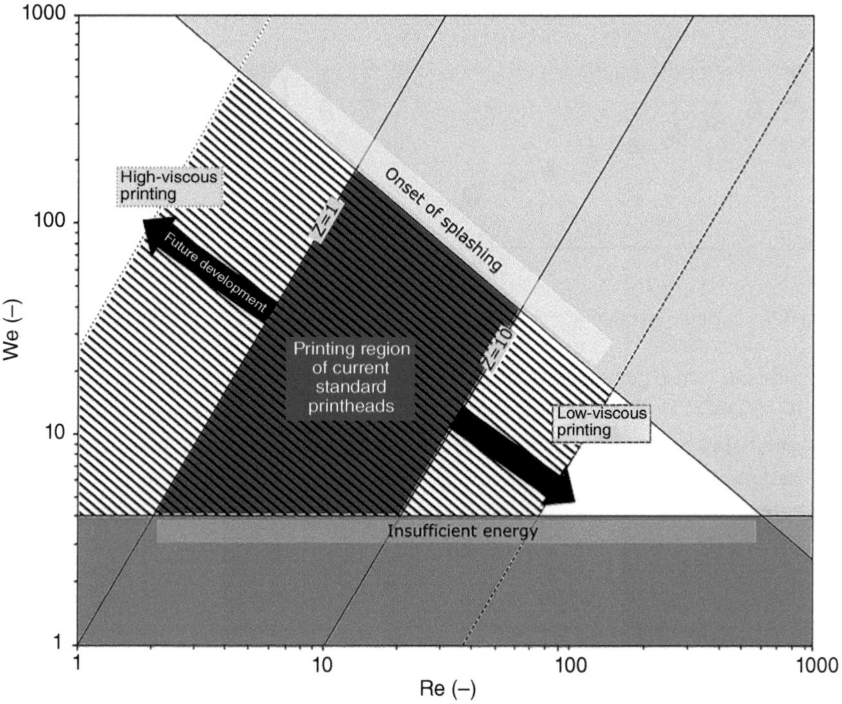

Figure 3.8 Diagram putting the *We* and *Re* numbers in correlation. The different areas of the diagram provide an indication of the range of inks that are printable with the standard printheads that are currently available.

Ohnesorge number, the *Z* number (*Z*) (Eq. (3.4)), is a widely used metric for determining the printability of an ink:

$$Z = \frac{1}{Oh} = \frac{\sqrt{\gamma \rho L}}{\eta} \tag{3.4}$$

The range of *Z* extending from 1 to 10 has been established as the interval within which an ink should yield a stable drop formation when used with standard printheads [48]. Figure 3.8 presents a graphical representation of the Weber number plotted against the Reynolds number, delineating the regions of printability and those that require further investigation.

The dimensionless numbers discussed are primarily suited for Newtonian fluids. However, most industrial inks contain polymers or pigments that introduce visco-elastic properties, altering their droplet formation behavior. To better assess printability for non-Newtonian fluids, parameters such as the Deborah number, the elasto-capillary number, and the elasticity number are being explored. While these offer more precise characterization, further research is needed to establish a definitive *Z* range for visco-elastic inks [47, 49–51]. Figure 3.9 compares the ejection, formation, and transfer for (a) a Newtonian and (b) a non-Newtonian (visco-elastic) fluid on the example of piezoelectric DOD IJP. In the Ohnesorge number and the Reynolds number, the dynamic viscosity can be included.

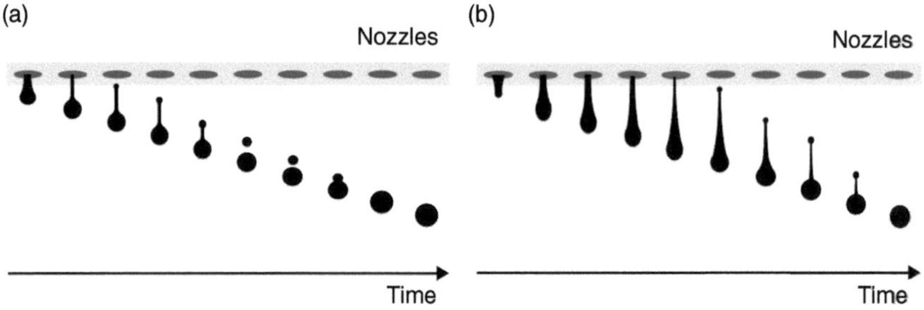

Figure 3.9 Droplet ejection and formation of (a) a Newtonian and (b) a visco-elastic fluid.

In addition to the parameters regarding the ink and printing parameters (e.g. ink density, viscosity and surface tension; printing nozzle diameter and velocity), the stand-off distance between the nozzle and the substrate, and humidity (considering evaporation – drying) and jetting voltage or temperature in the case of piezoelectric or thermal DOD inkjet printing, respectively, have a significant impact on the printing quality.

3.2 Binder Jetting (BJT)

3.2.1 Definition of the Technology (ISO/ASTM) and Working Principle

ISO/ASTM 52900 defines BJT as an AM "process in which a liquid bonding agent is selectively deposited to join powder materials."

From the definition, it follows that BJT uses powders as feedstock and a liquid bonding agent to selectively bind the particles in each layer, based on a chemical and/or thermal reaction bonding mechanism. The liquid bonding agent is thus often referred to as "binder,"[1] from which the name BJT originates.

Figure 3.10 shows a schematic of two possible configurations of a BJT 3D-printer: (a) roller recoater with a powder reservoir lifted by a feed piston and using a roller to spread and level the layers, and (b) blade recoater with a funnel reservoir.

The printing process is summarized as a repetition of the following steps:

1) Feeding of fresh powder from the reservoir and spreading of a powder layer.
2) Printing of the liquid bonding agent to define the cross section of the part/s being built.
3) Drying and/or curing of the liquid bonding agent.
4) Lowering of the build platform (Z-axis) by the layer thickness.

1 In the context of BJT of technical ceramics, alternative terms may be used such as "ink," "cross-linker," "printing liquid," etc., to avoid confusion with additives mixed within the powder feedstock, since such additives in traditional ceramic process are also considered as "binders." In the context of BTJ of construction materials, "binder" is often used to indicate a binding cementitious phase mixed within the powder feedstock, and alternative terms can be used for the printing liquid.

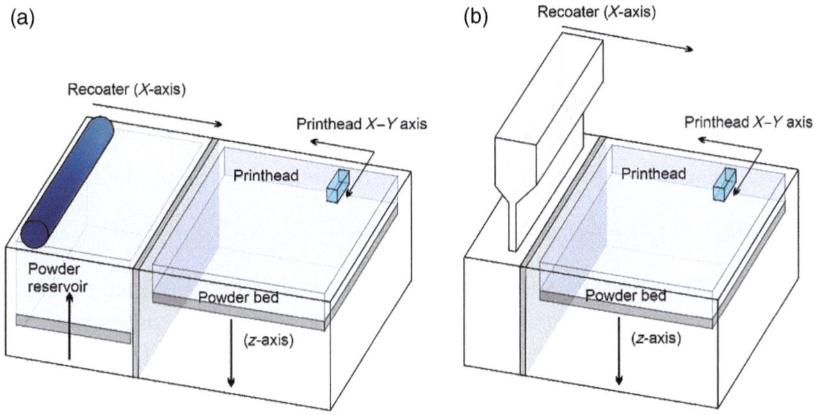

Figure 3.10 Schematic of a binder jetting 3D-printer with (a) powder reservoir and roller recoater or (b) blade recoater and funnel reservoir.

Several substeps may be involved in the process, depending on the configuration of the equipment and on the process settings, for example, flushing, cleaning, and capping of the printhead.

3.2.2 Historical Development and Applications

BJT is one of the first invented AM technologies, developed at the beginning of the 1990s the Massachusetts Institute of Technology (MIT) and patented in 1993 by Sachs et al. [52, 53]. The original name of BJT was three-dimensional printing, from which the widespread term "3D-printing" was coined.

Interestingly, BJT is one of the AM technologies that was invented with specific consideration for ceramic materials. In a way, this marks a difference from most of the first AM techniques such as vat photopolymerization and selective laser sintering (powderbed fusion), which originate from the processing of polymeric powders or of organic resins. Nevertheless, nowadays BJT has become a flexible AM method and finds application for polymeric, metallic, and ceramic materials. This flexibility derives from the fact that most materials can be used in BJT, as long as they can be processed from a flowable powder feedstock. While this book chapter will focus on ceramic materials, it should be noted that metal BJT is a growing field of research and of industrial innovation [54], with printing solutions offered by some of the larger companies on the AM market.

For ceramic (inorganic, non-metallic) materials, BJT finds application in four major fields:

- Gypsum plaster models.
- Sand molds and cores for metal casting.
- Large-scale BJT (concrete and other building materials).
- Ceramics for technical applications.

3.2.2.1 BJT of Gypsum Plaster Models

BJT of plaster models is chronologically one of the first applications and commercial milestones of BJT, also due to the simplicity of the approach and low cost of the materials used. Plaster powder, mixed with starch or other additives, is consolidated by IJP using a water-based binder to form a gypsum model.

The binding mechanism is based on a setting reaction, which for plaster of Paris is:

$$\mathrm{CaSO_4} \cdot \frac{1}{2}\mathrm{H_2O} + \frac{3}{2}\mathrm{H_2O} \rightarrow \mathrm{CaSO_4} \cdot 2\mathrm{H_2O}$$

The use of other additives, such as starch, further stabilizes the printed part.

Printers for this application date back to Z Corp., a company commercializing equipment under license of the MIT patents. After the acquisition of Z Corp by 3D Systems in 2012, these systems are still on the market under the trade name "Colorjet Printing." BJT of gypsum is well-suited to produce architectural models, small sculptures, and colorful visual prototypes. Indeed, colored ink can be printed, reproducing the full CMYK color space in the parts. Due to the high porosity and low strength of the gypsum models, the material is typically post-processed by liquid infiltration, for example, with an epoxy resin. This, however, restricts the use of these parts to non-technical applications, which motivated the further specialization of the BJT technology in different industrial fields, as introduced in the following paragraphs.

3.2.2.2 BJT of Sand Casting Molds and Cores

The production of sand molds and cores for metal casting is an established application of BJT. Foundry molds and cores can traditionally cost tens of thousands of USD and take several weeks to manufacture. For small and flexible production series, it can thus be a significant advantage to produce such parts by BJT.

The printing systems can be some of the largest overall on the AM market, with, for example, a build volume of 1260 l ($1.8 \times 1 \times 0.7\ \mathrm{m^3}$) for the S-Max printer by company ExOne and up to 8000 l ($4 \times 2 \times 1\ \mathrm{m^3}$) for the VX 4000 by company Voxeljet. These systems are a testimony to some of the main advantages of BJT, which are its scalability and productivity. The VX 4000, for example, can print a volume up to 139 l of material per hour, at a resolution of 200 dpi (dots per inch), which allows printing relatively fine details.

For these applications, typically three main types of binders are used:

- Furane binders, which are chemically hardened with an acidic activator solution. The activator can be premixed with the powder feedstock.
- Phenolic binders, with ester or acidic activators. These binders are based on a phenol resol resin and are thermally hardened. Hardening is achieved partially during the printing process (e.g. with an IR lamp as heat source), but typically an additional thermal post-process of the part is needed in an oven at moderate temperatures (100–200 °C). Alternatively, cold curing phenolic resin-based systems have also been developed.
- Inorganic binders, which provide environmentally friendlier alternatives to furane and phenolic binders. For foundry applications, a water solution of sodium silicate (also known as "water glass") is often used as binder, which,

in the post-processing, can be hardened thermally and/or by flow of CO_2 gas through the part.

Printed sand parts can be used in a foundry similarly to traditional molds and cores, including the application of a wash coat before casting. As in the traditional method, the cores are removed after casting.

Other interesting applications have been proposed for printed sand components produced by BJT. Some of these offered by company ExOne are molds for thermoforming of plastics, layup molds for fiber-reinforced composites, and washout sacrificial cores for winding of fiber-reinforced composites [55].

Other companies, including Voxeljet, are exploring the use of sand BJT in architecture and arts. A recent example from 2021 is the "Meristem Wall," a project led by Swedish architects David Andreen and Ana Goidea at the bioDigital matter lab of Lund University, in collaboration with German company Sandhelden. This object, displayed at the Biennale in Venice in 2021, demonstrates the use of BJT to showcase a complex wall structure, which includes features such as the integration of utilities (electric wiring and water pipes) and the control of the air flow in the ventilation system. The printed sand parts were infiltrated with a resin (epoxy) to structurally reinforce the material.

BJT is receiving increasing attention also for the AM of other construction materials, including the printing of concrete structures. These applications will be discussed in more detail in the next section on large-scale BJT.

3.2.2.3 Large-Scale BJT (Concrete and Construction Materials)

According to a 2024 report of The World Economic Forum [56], concrete production is one of the largest contributors to climate change, responsible for about 6% of global greenhouse gas emissions, with a global annual production of concrete forecast to grow from 14 billion m^3 (cubic meters) today to 20 billion m^3 by mid-century.

Motivated by this challenge, research groups and industrial players are looking at AM as a possibility to decrease the environmental impact of the construction industry. This can be approached by lightweight design enabled by AM, by processing materials with lower energy and CO_2 impact compared to ordinary Portland cement (OPC), or by a combination thereof.

The flexibility of BJT allows combining a wide selection of feedstocks with the geometrical freedom offered by AM. Compared to extrusion-based AM of concrete, BJT does not need to rely on the complex rheological behavior of the paste to achieve stable geometries. Rather, the support that the powder bed provides to the printed parts allows for the manufacturing of complex-shaped pieces, including undercuts and unsupported structures.

In a 2018 review paper, Lowke et al. classified concrete BJT according to the technique used [57]. Figure 3.11 shows a classification adapted from Lowke et al.

Sand BJT, as described in the previous section, can be used to produce temporary or permanent formwork, including formwork for concrete casting.

An advantage of BJT for arts and architectural pieces is the ability to print at high resolution, even large-sized parts. This can be appreciated in large and geometrically

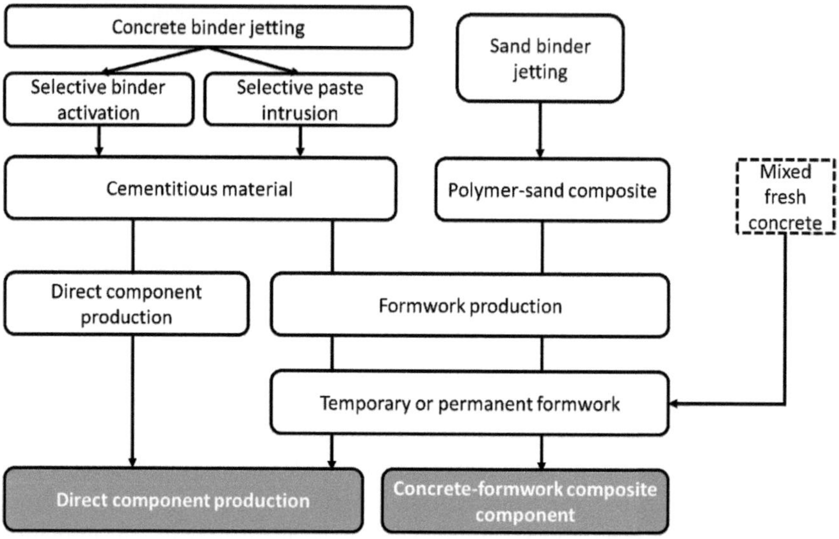

Figure 3.11 Classification of concrete BJT techniques. [57] / with permission of Elsevier.

complex structures such as "Digital Grotesque" by architects Michael Hansmeyer and Benjamin Dillenburger, which is printed from sand at 200 dpi resolution with a VX4000 printer (Voxeljet) [58].

At a smaller scale, Dutch company CONCR3DE commercializes BJT printing systems for stone-like materials with a print box size up to $1 \times 2 \times 1\ m^3$ and printhead resolution of 400 or 720 dpi.

BJT can also be used to directly print concrete components at a large scale. Concrete is a composite material made of an aggregate bound by a binder phase (cement). The cementitious phase thus takes the role of "binder" in BJT and is responsible for selectively binding the aggregates in the powder bed.

In principle, there are three possible configurations for concrete BJT according to how the binder phase and aggregates are delivered and mixed:

1) The cementitious reactive phase is fully mixed with the powder feedstock and is activated by selectively printing water.
2) Alternatively, one (or more) components of the binding phase are mixed within the powder feedstock, and one (or more) components are delivered through the printhead with the printing liquid (e.g. in the form of salt solution in water).
3) Alternatively, the reactive components of the binding phase (e.g. cement paste) are all delivered by the printhead, and the powder bed is only composed of aggregates.

Configuration 1) is technologically the easiest to implement, although formulating a flowable powder mix of the aggregates and of fine cement powder can be challenging. Technologies based on configuration 1) are often named "selective binder activation" (SBA).

Configurations 2) and 3) are more challenging in regard to issues related to clogging of the printhead nozzles and corrosion of the printhead materials by the chemically aggressive binder, but offer more possibilities for increasing the final density of the printed part beyond the packing density of the powder bed. Configuration 2) is usually considered as a variation of the SBA technology, while technologies based on configuration 3) are often denominated "selective paste infiltration" (SPI).

Based on the SBA technique, there has been some interest, both at the research and commercial levels, in developing low-resolution, easy-to-scale, BJT equipment for cementitious materials. Enrico Dini founded the company D-Shape on the premises of a 3D printer with two robust robotic gantries spreading powder layers directly on the ground and selectively depositing the printing liquid (activator) layer-by-layer. For flat structures, the powder bed is unconfined and relies on the angle of repose of the powder. For taller structures, a lateral support is built alongside the pile-up of the layers. The printhead is composed of single nozzles arranged at a distance of several millimeters, which are used to deposit jets or drops of liquid by controlling the opening of valves, according to the image to print and the movement of the gantry. This technology is easily scalable and has been applied to print structural elements such as a pedestrian bridge in Madrid [59] (assembled from up to 2 m long printed modules) and a concept for a small house (4 m side × 2 m height) produced in a single print and displayed at the Triennale of Milano in 2010 [60]. Investigations have also been carried out in a project founded by the European Space Agency (ESA) to print a habitat module with lunar regolith simulants, with the vision of building a lunar base with this technology [61].

Italian company Desamanera, active until 2024, followed a similar approach by commercializing large-scale BJT printers with tailored size, from 0.5 m side length up to several meters. These printers work with a layer thickness typically between 2 and 5 mm (also depending on the granulometry of the aggregates) and with a nozzle-to-nozzle distance of 3 or 5 mm. The range of applications varies from art and design pieces, artificial coral reefs, molds for molten glass, to fireproof panels and more.

Figure 3.12 shows a 3D-printed mold for artistic glass and the resulting glass pieces, developed in the frame of project D3Vero – "La stampa 3D nel settore del VEtro artistico per Rilanciare ed innOvare la filiera produttiva." The project, funded by the Veneto region in Italy, aims to bring innovation and technology to the Murano (Venice) artistic glass traditional field. The project was promoted by the R.I.R. Euteknos with the participation of the companies: Desamanera, Barovier & Toso, Foscarini, Lavai and Data Veneta Computers, and the research centers: UNIPD DII, UNIVE DAIS, UNIVR, and ABAVR.

Different types of cement have been printed by BJT at the research scale on smaller platforms (<1 m side length), including magnesium phosphate cement (MPC) and OPC [62]. OPC concrete is challenging to print, mostly due to its long setting time, which is unfavorable for the layer-by-layer consolidation of BJT [62]. MgO-based cements such as magnesium oxychloride cement (MOC) and MPC are commonly used instead. The D-Shape technology originally uses MOC cement as the binding phase. Selective hardening is activated by printing a water salt solution

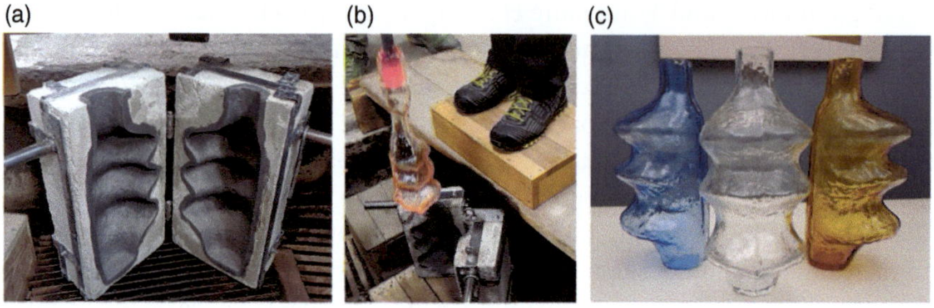

Figure 3.12 3D-printed glass mold and the resulting artistic glass pieces. Project D3Vero was supported by Regione del Veneto, POR FERS 20214-2020, Asse 1: Ricerca, Sviluppo Technologia e Innovazione; Azione 1.1.4 – COD. Bando 802 – ID domanda: 10317102. (a) 3D-printed mold. (b) 3D mold being used. (c) Artistic glass pieces produced with the mold. Images reproduced with permission from the D3Vero consortium (https://d3vero.it/), design by Benedetta Corporente (ABAVR, Italy).

of $MgCl_2$ into the powder bed, which contains coarse aggregates mixed with fine MgO powder. The reaction between $MgCl_2$ and MgO leads to the formation of magnesium oxychloride phases, called "5-phase" ($5Mg(OH)_2{\cdot}MgCl_2{\cdot}8H_2O$) and "3-phase" ($3Mg(OH)_2{\cdot}MgCl_2{\cdot}8H_2O$) [63].

For MPC cement, MgO and potassium phosphate (KH_2PO_4) are mixed with the aggregate (e.g. silica sand <2 mm in [64]), and only water is printed. The stoichiometric reaction can be schematized as:

$$MgO + KH_2PO_4 + 5\,H_2O \rightarrow MgKPO_4 \cdot 6H_2O$$

The strength of the printed concrete can be optimized by studying the interaction of the process and feedstock parameters, including layer thickness, water-to-cement ratio, spreading speed, granulometry of the powder, and amount of binder. Salari et al., for example, concluded that the modulus of rupture (MOR, 3-point bending) "was determined by the layer thickness, the amount of binder, the particle size, and the interaction between these factors [65]. When the aim was to print better mechanical strength blocks, the feed rate and the velocity of powder spread were not significant factors."

Nevertheless, a major challenge is achieving values of strength suitable for structural applications and approaching those of traditional OPC cement.

Weger et al. provided an overview of BJT concrete until 2021, showing that most research works obtained a compressive strength $f_{ck} < 20$ MPa, often around 15 MPa. It should be noted that these values are for testing with the load perpendicular to the layers' plane, and due to anisotropy in the printed material, even lower values are achieved with the load parallel to the layers' plane [66].

An alternative to traditional cements that has been explored in recent years, also for BJT, is the use of geopolymer binders. Geopolymers have been introduced by Davidovids at the end of the 1970s as amorphous or semi-crystalline (at room temperature) three-dimensional silicon-oxo-aluminate networks, formed by polycondensation of alumino-silicate units that are obtained from alumino-silicate precursors usually in alkaline conditions [67]. One of the main motivations for the use of geopolymer is the reduced carbon dioxide equivalent emissions, which are 9–80% lower than for OPC concrete depending on the raw materials and estimation methods used [68, 69]. Nevertheless, geopolymers offer further useful properties for BJT, such as a rapid initial reaction and a good development of strength, which can approach and even surpass that of OPC.

The use of geopolymer binders has been investigated by several authors, including Xia and Sanjayan [70], Xia et al. [71], Nematollahi et al. [72], Voney et al. [73], Rehman and Sglavo [74], and Elsayed et al. [75]. Xia and Sanjayan obtained specimens with a compressive strength up to ~29 MPa; however, this could be achieved only after immersion in a solution containing sodium silicate and sodium hydroxide for seven days at 60 °C.

In 2022, Elsayed et al. [75] achieved a compressive strength close to 20 MPa after storing the specimens in air for 7 or 28 days. The mixture used was based on fine metakaolin (MK) powder mixed with river sand as inert aggregate and printed with an alkaline solution of sodium silicate and NaOH in water as geopolymer alkaline activator. The addition of up to 30 wt% metakaolin powder, without compromising its flowability, was made possible by granulating in a mixer to coat the coarse sand particles with a layer of fine MK particles. The authors also observed no strength degradation after leaving the parts submerged in water for 1 week, concluding that the developed formulation could be compatible with outdoor uses or with applications where contact with water occurs.

With respect to extrusion-based AM of concrete, BJT offers distinct advantages, but also some limitations. One of the most relevant limitations of BJT is the reduced strength of the printed concrete, especially for SBA techniques. The reasons can be traced back to the difficulty in obtaining a homogeneous hydration of the cement in the powder bed, which, alongside the large porosity in the powder bed, results in a significant final porosity (typically >30%) in the final material. Depending on the technology and combination of materials used, compressive strengths <10 MPa are not unusual. In comparison, Portland cement has a compressive strength at 28 days in a range from ≥32.5 to ≥52.5 MPa according to European Standard EN-197-1. Despite these recent progresses, concrete BJT is an active field of research, and further improvements aimed at reducing residual porosity and increasing the strength of the printed material will be necessary to extend the field of applications of this technology in the construction sector.

An interesting alternative to BJT, based on the SBA, is the selective paste intrusion (SPI) technique. SPI is an AM process specifically developed to produce concrete components. Essentially, it involves bonding thin layers of aggregate with a cement paste. This cement paste is selectively injected into the aggregate layers. This method has the potential to deliver products with higher density and strength, even though it poses some additional challenges in the development of the printhead, as well as in tailoring the rheology of the paste and its penetration in the powder bed [66]. Nevertheless, promising results have been achieved both at research level and commercially [76].

Figure 3.13 Photos of façade elements produced by selective paste intrusion (SPI). Courtesy of Progress Group (Brixen, Italy).

Figure 3.13 shows a printed and colored SPI façade produced by Progress Group (Brixen, Italy) 3D-printing company. The material achieves strengths and exposure classes that are required for such projects.

3.2.2.4 BJT of Ceramics for Technical Applications

The manufacture of advanced ceramics by BJT is particularly challenging because it involves a complex process chain composed of multiple steps, in which the part's shaping by BJT needs to be seamlessly integrated. Consistent with the traditional ceramic processing terminology, the printed part is denominated "green part" and is subject to one or multiple (thermal) post-processing steps. In most cases, a debinding process is needed to eliminate all organic residues from the binders, generating a "brown part," which is then subject to densification by sintering or infiltration in a high-temperature furnace. Often, the challenge in BJT is not to print a part with stable geometry, but rather to achieve a green part that can be successfully post-processed to achieve its ultimate desired properties in terms of density, microstructure, mechanical and thermal properties, etc.

The following section will introduce some of these key areas and challenges that need to be considered in the BJT of advanced ceramics. For a detailed background introduction of specific aspects, the reader is referred to Sections 3.2.3 (Feedstock properties and process parameters) and Section 3.4 (Post-processes).

One of the first key aspects to consider in BJT is the selection of a suitable ceramic powder. Reproducible deposition of defect-free layers, in particular, depends on the flowability of the powder used. The flowability of a powder describes its ability to flow under the action of forces (gravity and other external forces applied in a process). The physical background of flowability is discussed in Section 3.2.3.1. In short, good flowability is generally ensured with coarse-grained powders with a particle size greater than 50 μm for a ceramic powder. From approximately this particle size, the attraction between particles, which is generally independent of particle size, is less than the gravitational force acting on each particle, which is a function of particle volume. This basic model is valid for powders with a nearly spherical particle shape. Elongated particles or even fibers may have a different particle size threshold for good flowability or may never flow well. Powders used in the manufacture of advanced ceramics, such as alumina, zirconia, and silicon carbide, to name the most common, are generally roughly spherical in shape or equiaxed due to their synthesis route.

The given particle size threshold of about 50 μm for a powder with sufficient flowability to deposit defect-free layers is generally not suitable for the production of advanced ceramics. Advanced ceramics are typically manufactured from submicron to nanopowders. Sophisticated recoating technologies (see Section 3.2.3.1) can process powders finer than 50 μm, but definitely not submicron powders. Another problem is the low packing density of free-flowing fine powders. Neither attractive nor repulsive interaction of the particles will help in this respect. While an attractive interaction results in the formation of aggregates whose random shape is not favorable for dense packing, a repulsive interaction prevents dense packing by arranging the particles in the powder bed to minimize particle-to-particle contact. This fact poses an inherent problem for the use of BJT in the processing of advanced ceramics, and it is fair to say that some of the process chains for manufacturing advanced ceramics may not be compatible with the basic concept of BJT. On the other hand, in the following section, an example is given that provides a workaround to this problem by using suspensions of ceramic particles instead of dry powders. Moreover, there are unique cases where the process chain of certain ceramic materials, still considered as advanced ceramics, differs from the traditional one and makes them compatible with the BJ process. This is the case for silicon-infiltrated silicon carbide. The process chain for this material starts with the formation of a rather coarse-grained silicon carbide (SiC) powder compact, which is infiltrated with liquid silicon in a second step to obtain silicon-infiltrated silicon carbide (SiSiC), a material that finds applications as a structural ceramic and in high-temperature applications due to its good thermal conductivity and low coefficient of thermal expansion. However, its application is limited to temperatures well below the melting point of metallic silicon of about 1450 °C. An example of a successful application of 3D-printed silicon carbide is the Schunk Group's IntrinSiC® product. Figure 3.14 shows a BJT 3D-printed silicon carbide satellite mirror structure with a diameter of 300 mm. Such mirrors are manufactured at

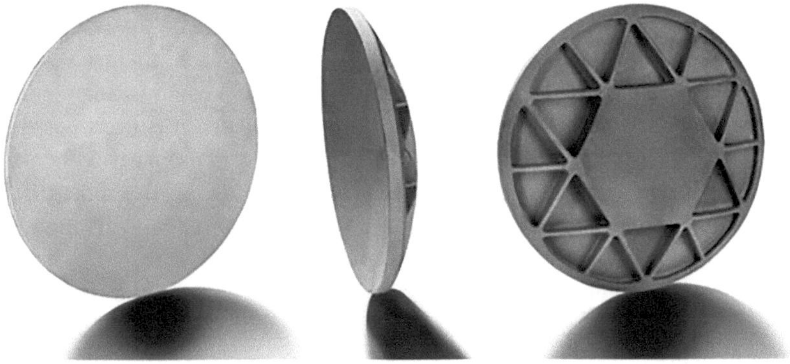

Figure 3.14 BJT 3D-printed silicon-infiltrated silicon carbide satellite mirror structure with a diameter of 300 mm. Courtesy of Schunk Ingenieurkeramik (Germany).

Schunk Ingenieurkeramik using the patented IntrinSiC process. In this process, reaction-bonded SiC functional parts are produced based on a BJT method. The process enables the production of highly complex and weight-efficient satellite mirrors with a diameter of up to 2 m. The high optical requirements for the mirror surface are achieved by applying a subsequent CVD coating process and polishing of the surface.

Another area of intensive research and development of BJT of advanced ceramics is the field of bioceramics. Bioceramics can be divided into bioinert and bioactive ceramics for endoprosthetic applications and bone grafting. Bioinert ceramics are used as implants to replace the function of degenerated or traumatized hard tissues of the human body due to their excellent chemical stability, biocompatibility, mechanical strength, corrosion resistance, and wear resistance. The processing of these ceramics follows the classical powder process chain, that is, raw material processing, shaping, sintering, and finishing, and their shaping by BJT is difficult due to the aforementioned particle size problems. On the other hand, bioactive ceramics that bond directly to bone are in interaction with the human body environment, and designed or random porosity is beneficial for their function. Designed porosity is taken into account when creating the virtual model and is translated into the real part by the BJT process. With a typical resolution limit of about 100 μm for BJT, this porosity helps promote vascularization and bone ingrowth into the implant. Random porosity, which results from the aforementioned need to use coarse-grained powders combined with their limited sintering activity, is finer and can be used to enhance the resorption of resorbable ceramics such as calcium phosphates and stimulate bone growth, thereby promoting osteoinduction, osteoconduction, and osseointegration. From this perspective, the process characteristics of BJT are well-suited for the production and application of bioactive ceramics, which explains the fact that much research is devoted to this topic. Figure 3.15 shows parts manufactured by BJT from an alkali calcium phosphate material for the regeneration of critical-sized mandibular defects (mini-pig model).

Silicate ceramics generally exhibit high sintering activity due to the formation of a liquid phase during sintering, which is beneficial for rapid material transport and densification. This, in turn, can allow the use of coarse-grained powders with good flowability. There is a considerable amount of literature on the BJT of silicate ceramics, and various companies and start-ups have dedicated themselves to this topic. It can be summarized that parts with good external contour finish and reasonable density can be produced, but not fully dense parts. It is even possible to decorate and glaze the printed parts. However, only fully dense parts will have the mechanical strength and aesthetics expected for tableware. Furthermore, the low mechanical strength associated with reduced density has frustrated all attempts to use such ceramics in technical applications.

Granulation has been considered as a workaround to increase the flowability of fine powders in the BJT process. In the granulation process, fine particles are prepared as a generally water-based suspension and spray-dried to form granules. Alternatively, fluid bed granulation can be used. Granules have good flowability, and fine powders can be processed in this way. However, the problem is that a powder bed

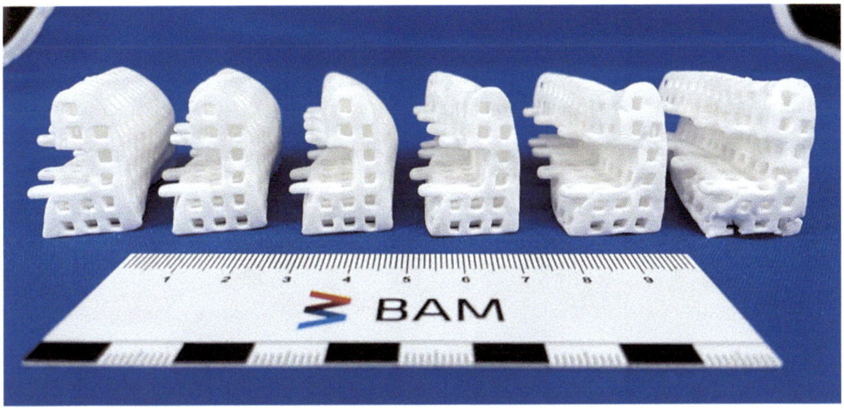

Figure 3.15 Bioceramic implants fabricated by BJ from a bioactive alkali calcium phosphate ceramic material for *in vivo* testing for regeneration of critical-size mandibular defects (mini-pig model). A visible feature is the designed porosity. Not visible is the fine random porosity of the material itself. It can be seen that the parts show an individual distortion due to a nonuniform support during sintering. [77] / Clausthal University of Technology.

formed by granules has inhomogeneous particle packing. While the particle packing density in the granules is relatively high and the pores are small, that is, in the range of the primary particle size, there are large pores between the granules as their size is typically larger than 50 μm. When a binder is sprayed onto the surface of a layer formed by granules, the binder is first drawn into the granules because the capillary force of the smaller pores in the granules is greater than that of the larger pores between individual granules. After the granules are saturated with binder, the binder remains between the granules to form an intergranular bond. The latter is required to form a part. Therefore, the amount of binder required when using granulated powders is very high, which causes problems during debinding. In addition, during sintering, the granules will sinter faster due to the higher density within the granules, and eventually, a network of almost dense granules connected by sinter necks will be formed. At this stage, sintering stops, and it is impossible to sinter this network into a dense ceramic body.

In order to make the BJT process compatible with the production of advanced ceramics using submicron and nanopowders, a suspension of ceramic particles can be used. The use of ceramic-based suspensions for ceramic slip casting is a simple and cost-effective molding technology that has long been established and used for various types of advanced ceramics such as alumina, silicon nitride, and silicon carbide. The technology is based on the formulation of a stable suspension of ceramic particles (slip) and a porous mold wetted by the suspension's liquid phase, ideally water. The water is driven out of the suspension by capillary forces, and a so-called cast is formed on the surface of the mold in contact with the suspension. The thickness of the cast follows a square root time law, that is, it grows more slowly as the cast thickness increases. The use of suspensions makes it easy to handle very fine powders, even for layerwise deposition in the BJT process, and makes it possible to

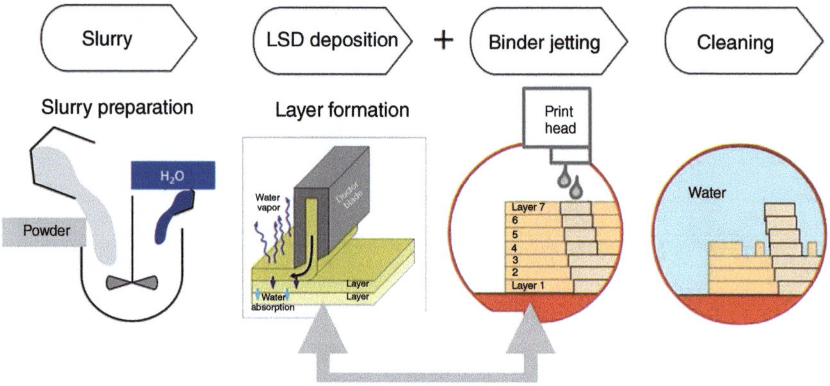

Figure 3.16 Process chain of the LSD-print process. The deposited layers form a powder bed that is strong enough to support itself. A washing step is required to remove the printed parts from the powder bed.

produce layers with high particle packing densities of up to 65% of the theoretical density of the material.

Figure 3.16 illustrates the basic concept of the LSD-print (layerwise slurry deposition) process. The uniqueness of the LSD-print process lies in the build-up of the powder layers, which combines elements of the tape casting and slip casting processes. Similar to tape casting, a doctor blade is used to spread a ceramic slurry or slip into a thin layer; while similar to slip casting, this layer is bonded to a porous support formed by previously deposited and dried layers. Powder beds produced by LSD-print have a final green density comparable to a conventional powder compact produced by slip casting. A critical issue is the complete drying of the deposited layers. It has been demonstrated that drying of a freshly deposited layer can be achieved in less than 10 seconds. There are two main mechanisms involved in the dehydration of a deposited layer: first, the solvent is drawn by capillary forces into the supporting powder compact formed by the previously deposited and dry layers, and second, by rapid evaporation of the solvent facilitated by a powder-bed temperature higher than the boiling point of the solvent system. In fact, for rapid drying of the deposited layers, the powder bed can be preheated to temperatures well above the boiling point of water. This is possible because the uptake of solvent by the powder bed is fast enough to effectively prevent boiling and thus bubble formation in a freshly deposited layer.

To build 3D structures, the process steps of deposition, drying, and BJT are repeated. Typical layer thicknesses range from 30 to 200 μm. The width of the layer stack (powder bed) is defined by the width of the doctor blade used, while the length is controlled by the movement of the blade parallel to the support surface. Thus, within the confines of the support platforms, the width of the doctor blade, and the trajectory of its movement determine the area of material deposition. The height of the powder bed is simply the sum of all deposited layers. This feature is sometimes helpful when only small amounts of material are available, and the total volume of the powder bed needs to be reduced to a reasonable ratio of width, length, and height of the processing volume. In addition, a free-standing powder bed allows the concept

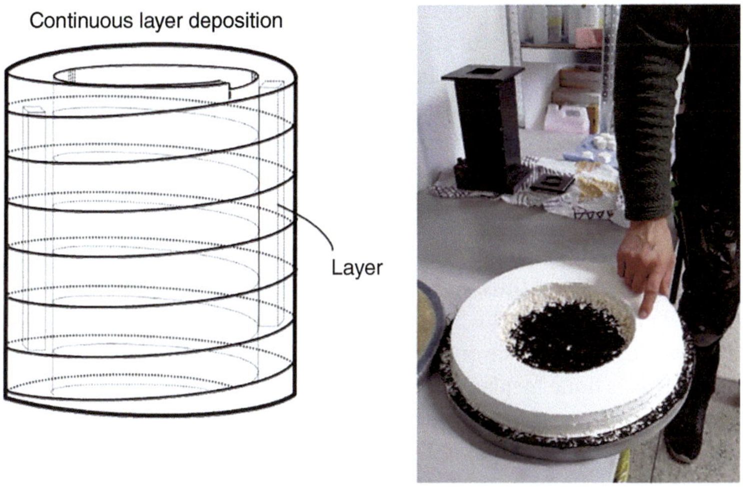

Figure 3.17 Left: Illustration of the continuous layer deposition concept; right: circular powder bed with embedded parts.

of continuous powder deposition on a rotating circular base plate to be developed. Figure 3.17 illustrates this concept. Instead of alternating layer deposition, drying and BJT over time, these three steps are continuous and separated along the circular powder bed. The distortion of the parts due to the circular movement of the powder bed can be compensated for by a correction of the virtual model, which results in a slightly lower printing resolution toward the outer circumference of the circular base plate. On the other hand, the continuous and parallel layer deposition and BJT increase the productivity of the process by a factor of almost 10. It is noteworthy that continuous deposition processes with dry powders are also planned for BJT and that the corresponding technologies have been developed and tested [78, 79].

The production of advanced ceramics by slurry-based deposition combined with BJT combines the advantages of high powder particle packing density in the powder bed, the ability to use very fine down to nanopowders, with the convenience of BJT. The ability to use printheads with hundreds to thousands of nozzles working in parallel makes BJT relatively fast in writing the layer information. The dense and mechanically strong powder bed provides good support for the printed parts, and overhangs and other complex geometries can be achieved without the need for additional support structures. On the other hand, depowdering becomes more complex as the density of the powder bed increases. In the case of the slurry-based deposition process, the powder bed has an intrinsic strength comparable to a conventional powder compact, and depowdering requires immersion of the powder bed in water. The water helps to dissolve the powder bed. The stability of the powder bed can be tuned by additives, such as binders, in the slurry, and ideally, these binders are readily soluble in water. The binder that defines the part, which comes from the printhead, is not water-soluble and ensures the integrity of the part during washing.

The example of using a water-based suspension to deposit layers in the BJT process shows that once the problem of handling fine powders has been solved and high powder packing densities are achieved in the powder bed, the BJT process is very well-suited to the production of advanced ceramic parts. There may be other technological solutions to this problem in the future.

An overview of strategies to handle fine powders in the BJT process and to increase the density of the powder bed can be found in Ref. [80].

With regard to the binder, the following properties are of particular importance for the BJT of advanced ceramics:

- It must bind the particles together and form a strong interparticular bond. Adhesion is the main function of the binder, allowing the printed part (green body) to be depowdered and handled. The bonding reaction is usually triggered by a heat treatment of each individual layer immediately after the layer information has been printed. This is done by placing a heat lamp over the layer to cross-link the binder. In addition, most printers have the powder bed housed in a job box, which can be removed from the printer after the print job has been completed and placed in an oven for a period of time for further curing. This additional curing is then followed by depowdering.
- It must wet the powder and penetrate the surface of the powder bed. It is necessary for the binder to penetrate the powder bed to just over a layer thickness so that the individual layers will bond together. Moistening the powder bed is not an easy task for the binder, as it is often heated to temperatures of around 50–70 °C and is therefore very dry. The rough surface formed by loose particles is not easy to wet, and the droplets generated and ejected by the printhead with their kinetic energy can form small impact craters on first contact with the powder bed. It can be observed that it sometimes takes a few seconds for the droplets to start wetting the powder bed. Further details on the interaction between binder and powder bed can be found in Ref. [81, 82].
- It must completely decompose and pyrolyze without residue during thermal treatment of the printed bodies. This property is essential in order to obtain advanced ceramic components with their unique physical and chemical properties. Particularly for bioceramics, it is important that no contamination from residual binder affects biocompatibility. On the other hand, typical binders based on phenolic resins are easily pyrolyzed and burned in ambient air. For non-oxide ceramics such as silicon nitride or silicon carbide, the atmosphere must be kept oxidizing until all the binder has been burned (debinding step) and then switched to an inert atmosphere for sintering the ceramic material.
- It should be printable by the printhead in use, that is, perfect drop formation. There is a great deal of literature discussing the properties of binders that make them printable by the printhead technology used. The most important properties of the binder are its viscosity and surface tension. Both are critical for forming droplets of equal volume without satellite droplets, which are generally smaller than the primary droplet and follow it at random. For a detailed background on the topic, see Section 3.1.
- It should have a reasonable shelf life: This is not only important for the supply of binder, but also for the binder in the printhead to avoid clogging of the nozzles if the printer is not used for a long time. If the printer is not used for a really long

time, the binder must be removed from the printhead and flushed with a cleaner or solvent, but for short breaks, it is helpful to let the binder in the printhead rest, as cleaning and recommissioning a cleaned printhead is time-consuming, and it takes time to condition the printhead so that it reaches its full performance. Sometimes bubbles and solvent residues lead to faulty printouts after a restart.

Together with the injection of the binder, the powder bed can be impregnated with additional ceramic material dispersed in the binder [83]. As the dispersed particles must be very fine, down to nanoscale powders, to prevent clogging of the printhead nozzles, they help with sintering and can act as an inorganic binder. Such particles must be about 10 times smaller than the nozzle opening of the printhead to avoid clogging. According to this strategy, alumina powder beds can be impregnated with nanosized alumina powders to form monolithic and dense alumina parts during sintering, even though the primary alumina particles are not extremely fine.

3.2.3 Feedstock Properties and Process Parameters

Even though the BJT can be schematically represented as merely the repetition of two steps (layer recoating and IJP), each machine has tens of single parameters, which, along with the properties of the feedstock, determine the success of a process run. In the following paragraphs, the influence of the main machine and feedstock parameters will be discussed in detail, including typical post-processing of the printed parts.

3.2.3.1 Powder Feedstock Properties

The recoating of densely packed and homogeneous powder layers is key for obtaining not only a stable printed part but also to achieve the desired properties of the final component.

In principle, a suitable feedstock must have all of the following properties:

- The material should have the necessary chemo-physical properties (both in the bulk and on the surface) in terms of purity, structure (crystallinity and crystal structure), amount of dopants or additives, etc.
- In case of additives mixed with the powder, these should not lead to unwanted contaminations of the material or cause unwanted reactions.
- The powder must be sufficiently fine to be "reactive," for example, leading to a fast setting reaction in a cement, or to a fine microstructure after sintering a technical ceramic (see Section 3.4 "Post-processes").
- The powder should be sufficiently flowable, which means, able to flow from the reservoir and be spread homogeneously to form a layer. The microstructure of the powder bed is closely related to the microstructure of the final component: total porosity, particle and pore size, and presence of clustering or agglomeration play an important role, reducing the homogeneity of the parts.

The last two points are especially critical since they are often in contradiction. Using fine powders is often necessary to achieve reactivity in most processes, due to the large surface area of the particles. On the other hand, it is well accepted that the flowability of a powder decreases with decreasing particle size. In a powder, the flow behavior is governed by an interplay of gravitational force (F_g) and the sum of interparticle forces (F_{pp}).

F_g is straightforward to calculate, assuming spherical particles in Eq. (3.5):

$$F_g = \rho V_g = \rho \frac{4}{3} \pi R^3 \tag{3.5}$$

where ρ is the density of the particles, V is the volume, g is the gravitational acceleration, and R is the radius.

F_{pp} is more difficult to estimate, since interparticle forces can be of different natures, including capillary forces and electrostatic forces, which greatly depend on the type of material and on the ambient conditions. However, even when taking all necessary precautions to minimize these forces, Van der Waals forces still have a contribution that can be approximated (again for identical spherical particles, see Figure 3.19a) as in Eq. (3.6):

$$F_{pp} \approx \frac{AR^*}{12z^2} \tag{3.6}$$

A is the Hamaker constant, R^* is the radius of particle or of asperities, and z is the distance between the spheres.

Plotting the ratio F_g/F_{pp} results in a curve such as shown in Figure 3.18, here calculated for α-Al_2O_3 particles ($\rho \approx 3.9\,g/cm^3$, $A \approx 1.5 \cdot 10^{-19}$ J). There is a small issue in the calculation of the Van der Waals force in that we do not know the value of the distance z; however, it is often estimated that a maximum of the adhesive force is

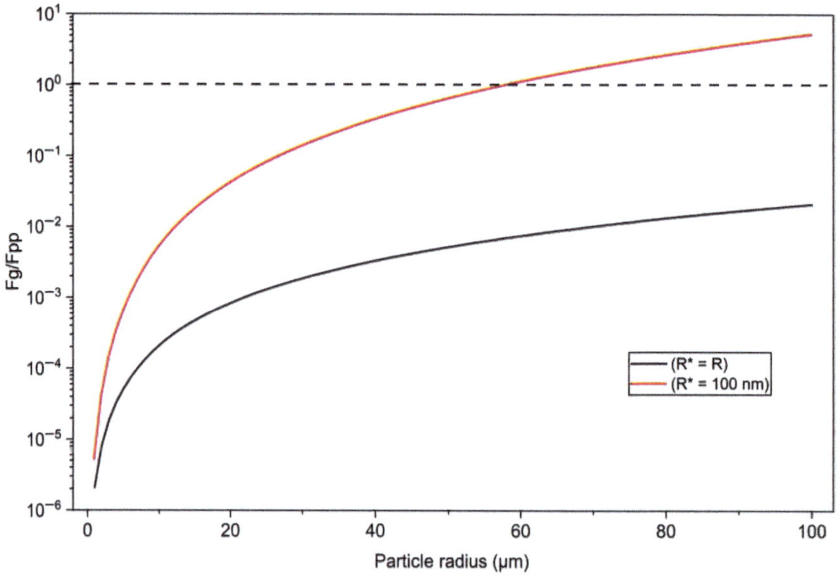

Figure 3.18 Ratio F_g/F_{pp} between gravitational force and Van der Waals interparticle force for two particles as a function of the particle radius.

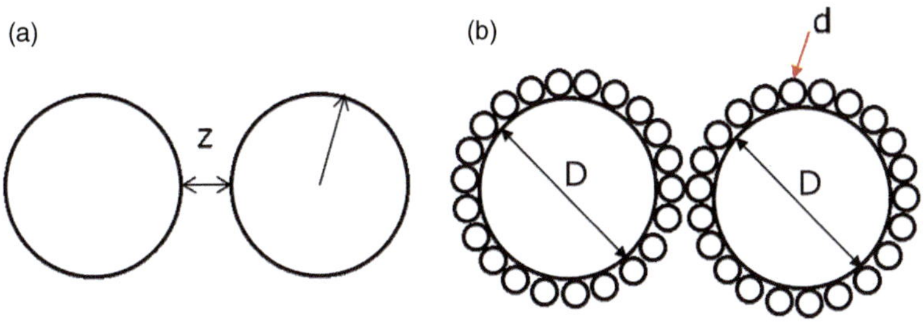

Figure 3.19 (a) Schematic of two particles of diameter $D = 2R$ separated by distance Z. (b) Schematic of two particles of diameter D coated by smaller particles of diameter d.

obtained at 0.4 nm [84]. In reality, the system will also be more complex, since there will be particles with different diameters and with irregular shapes, or with asperities at the surface. For micron-sized particles, it is known (see, e.g. O. Molerus [85]) that using the particle radius (R) as R^* (black curve) overestimates the interparticle forces and considering R^* as the radius of the asperities provides a more reasonable result. Krupp [86] estimated $R^* \approx 100$ nm, so this is the approximation used in the graph (red curve).

Even with these approximations, following the shape of the curves is sufficient to visualize a trend, which is, that the contribution of the gravitational forces decreases as the particle radius decreases, following a function proportional to the power of 2 (black curve) or power of 3 (red curve) of the radius R.

Below a certain diameter, which often is in the range ≈20–40 μm for ceramic powders, the interparticle forces become predominant, and the powder becomes cohesive and not flowable. There exist different strategies to improve the flowability of a powder, which can also be applied to BJT. A possibility is to coat the particles with a "flow agent," which can be, for example, silica nano-particles. This situation is visualized in Figure 3.19b, with larger particles of diameter $D = 2R$ coated with smaller particles of diameter d. It can be calculated that the Van der Waals force in this case is estimated as in Eq. (3.7) [87]:

$$F_{\text{pp,coated}} \approx \frac{A}{12} \frac{dD^*}{d + D^*} \frac{1}{z^2} + \frac{A}{12} \left[\frac{D^*}{2(2z + d)^2} \right] \tag{3.7}$$

The second term can be neglected as it is much smaller than the first term, from which it can be approximated in Eq. (3.8) that:

$$\frac{F_{pp,coated}}{F_{pp,uncoated}} \approx 2 \frac{d}{D^*} \tag{3.8}$$

Again, here, D^* is usually the radius of asperities, $D^* \approx 200$ nm, rather than $D^* = D$. For nano-particles with $d = 10$–20 nm coated on the surface, the attractive force F_{pp} can be reduced by up to one order of magnitude.

Another possible strategy is to mix (in this case, homogeneously mix, not coat as above) particles of different size fractions. The advantage of this approach can be twofold: (i) by adding a fraction of coarser particles, flowability is improved; and (ii) the size fractions can be selected in a way that optimizes the maximum packing density of the powder.

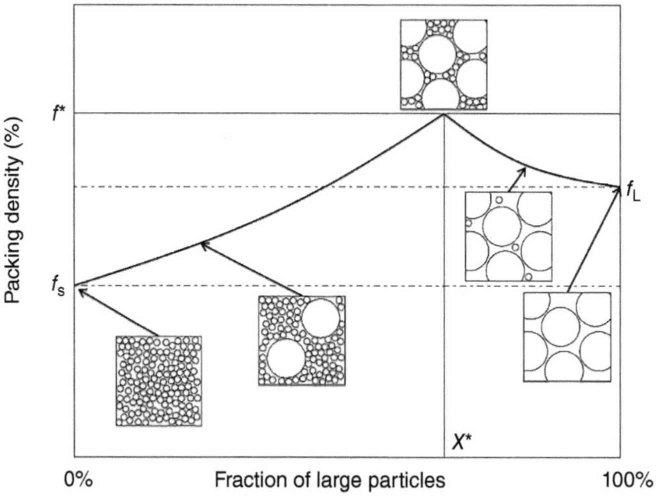

Figure 3.20 Packing density f of a bimodal powder mix of small and large particles, as a function of the weight fraction *X* of large particles.

The latter advantage is based on the simple concept that small particles can fit in the empty volume between larger particles, which usually requires the small particles to be ≈15% of the diameter of the large ones. The theoretical packing density of the mixture increases up to a maximum f^* at fraction X^* of large particles, as described by R. M. German [88] and schematized in Figure 3.20.

With reference to Figure 3.20, $f^* = f_L + f_s(1 - f_L)$ and $X_L^* = \frac{f_L}{f^*}$.

Mixing two spherical powers, each with a packing density of 64% (which corresponds to the random packing of spheres), results in a maximum packing density of 87% at 26.6 wt% of small particles. In theory, even higher packing can be achieved with three or more size fractions. This can be a valid and often used approach for BJT of cementitious materials, for which it is common to mix larger particles (aggregates) with a finer binding phase (cement). Possible issues regarding flowability and separation of the fractions should, however, be considered.

For BJT of technical ceramics, there are some additional limitations, because the selection of the particle size is constrained by the ability to sinter the product, for which having a fine and monomodal powder is often advantageous. Nevertheless, for specific ceramic systems, it has been shown that the printed parts from bimodal powder achieved higher green densities than those from the unimodal powder, for example, by Du et al. for silicon carbide powders [89].

3.2.3.2 Layer Recoating

Different recoater designs are implemented depending on the overall machine design and size, technology provider, and processable materials.

Based on the elements shown in Figure 3.10, that is, hopper, blade, and roller, many different combinations have been tested for the deposition of powder layers in BJT. The most common systems have been schematized by Diener et al., from which Figure 3.21 is reproduced [90].

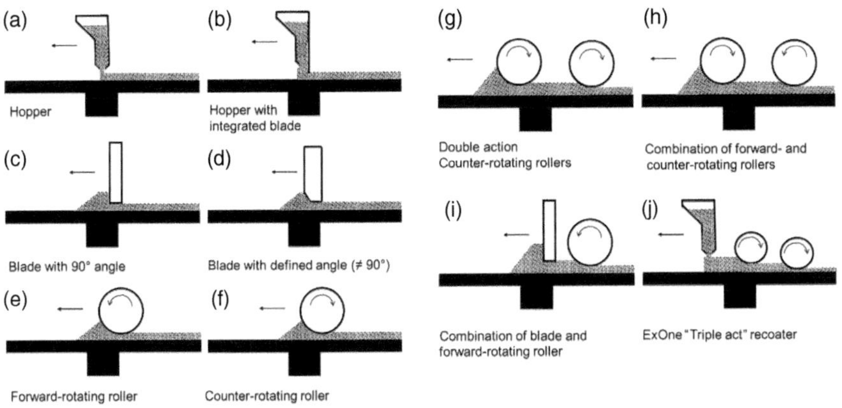

Figure 3.21 (a–f) Schematic of basic recoater designs and (g–i) advanced recoater designs. [90] / with permission of Elsevier.

Basic designs include hoppers (a), often integrated with a blade (b).

Of the basic designs, blades are the simplest and easiest to scale to different sizes. Another advantage is that a blade can be combined with a hopper to have the recoating system and powder reservoir in a single unit. A simple blade, however, applies relatively high shear forces on the surface of the layer, which can damage the previous layers, and has little effect on the compaction of the powder. More complex designs of the blade's edge have been proposed based on numerical simulations, for example, by Haeri et al. [91].

Rollers are also widely used and can be forward-rotating or counter-rotating. Both can potentially increase the compaction of the layers and the density of the powder bed. However, forward-rotation leads to a greater risk of powder sticking to the surface of the roller and damaging the previous layers. Counter-rotation can instead "activate" the flow of the powder in front of the roller and distribute it more homogeneously. A disadvantage of roller recoaters is that they can be supported only at two points (the two extremities). This can be problematic for large powder beds, because the midpoint deflection increases with increasing length L according to $\delta_{\max} = \frac{5 \cdot w \cdot L^4}{384 \cdot E \cdot I}$.

where w is the own weight load, E is the elastic modulus, and I is the moment of inertia.

Vibration can be applied to hoppers, as well as to blades and rollers, to improve the flow and packing of powders. Furthermore, advanced recoater designs have been developed combining multiple elements, as shown in Figure 3.21b. Using a blade and a roller or two rollers, for example, offers the possibility to combine the spreading of smooth and flat layers with an increased compaction of the powder. Similarly, the "triple act system" by company ExOne combines a hopper with vibration, a first roller for distributing the powder, and a tailing roller for defining the final layer height and for compacting the powder.

Discrete element methods (DEMs) can be used to simulate various effects on the powder spreading, both considering the setup design (shape or the recoater, recoating speed, layer thickness, etc.) and the powder characteristics (particle size, shape, material, etc.). Despite being computationally expensive, by limiting the simulation to a section of the powder bed and taking advantage of symmetry to set the boundary conditions, the simulation of systems with thousands of particles is nowadays possible at an increasing level of complexity.

3.2.3.3 Process Parameters and Binder Saturation

The mechanical stability and the geometrical accuracy of BJT printed parts depend on the interplay of many different parameters.

Figure 3.22 shows an Ishikawa (fishbone) diagram of the main printer parameters that are accessible to the user, although the actual parameters will depend on the specific manufacturer. Each parameter needs to be not only individually optimized but also optimized alongside the other feedstock, printer, and post-processing parameters.

For example, a smaller layer thickness may be chosen to obtain a less stepped surface of the part. The new layer thickness, however, needs to be compatible with the powder in use; otherwise, it may result in defective or rough layers. Design rules or simple rules of thumb have been developed for some of these parameters, such as the layer thickness should be at least three times larger than the average particle size [92]. In most cases, for each new material, a new set of parameters is individually optimized, which can be done with the help of design of experiment (DOE), such as the Taguchi method [93].

A parameter specific to BJT that will be discussed in more detail in the following is the binder saturation, *S*. The binder saturation is defined as the fraction of open pores in the part that is filled with liquid bonding agent. In AM, a part can be represented as a repetition of voxels (three-dimensional pixels), and a voxel in BJT can usually be identified as the smallest volume into which one printed drop is deposited.

In raster scanning, which is commercially the most common type of printing configuration, the printhead traverses the powder bed along the *x*-axis and activates multiple nozzles spaced in the *y*-direction (Δy) to print the layer image. If needed, the printhead then moves on the *y*-axis and scans a new swath along the *x*-direction.

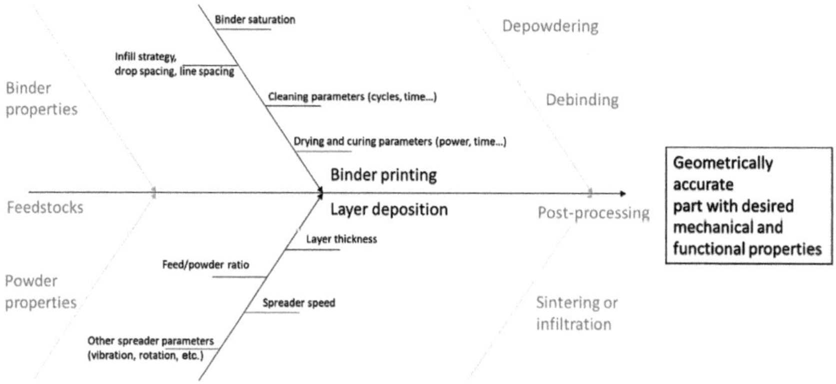

Figure 3.22 Ishikawa diagram of typical printer parameters in BJT.

For this configuration, S can be defined at voxel level as Eq. (3.9):

$$SS = \frac{V_L}{V_{OP}} = \frac{V_{drop}}{V_{voxel} \cdot (1-f)} = \frac{V_{drop}}{(\Delta x \cdot \Delta y \cdot \Delta z) \cdot (1-f)} = \frac{V_{drop}}{(\Delta x \cdot \Delta y \cdot LT) \cdot (1-f)} \tag{3.9}$$

V_L is the volume of printed liquid in the part.
V_{OP} is the volume of open pores in the part.
V_{drop} is the volume of a single printed drop.
V_{voxel} is the volume of a voxel.
f is the packing fraction of powder in a voxel (assumed to be homogeneous in the part).
Δx, Δy is the printing resolution (drop-to-drop distance) in x and y direction.
$\Delta z = LT$ is the layer thickness.

This expression offers a convenient relationship between several machine parameters and a single process parameter S. In fact, V_{drop} depends mostly on the printhead and inkjet parameters used, and its exact value can be measured; Δx is the distance between drops along the x-axis, which can be set in the software; Δy is the distance between printed lines. This is the physical y-distance between nozzles in a single-pass system, or it can be varied in a multi-pass system by interpolating the y position of the head between passes; finally, the packing fraction f of powder in the powder bed can be measured. Given a value of layer thickness LT, which is typically selected by the user, the printer software calculates Δx and Δy to obtain the desired saturation.

Figure 3.23 shows exemplarily how the contour of a printed part changes at increasing saturation [94]. Notice how the contours are sharp at low saturation and become blurry and wider at the bottom at increasing saturation. This effect, due to excess binder, is also known as "binder bleeding."

On the other hand, if the binder saturation is too low, the printed part may be too weak to be handled. The green strength must be sufficient to remove the part from the powder bed, depowder (clean) it and transfer it to post-processing.

The binder saturation does not necessarily need to be homogeneous in the part. There are strategies to print, for example, a higher saturation in the outer shell of the part to make it strong and a lower saturation in the inside. A lower saturation in the bulk of the part is advantageous to reduce the overall amount of binder used, to reduce costs, and to simplify the successive post-processing. Variations in binder saturation can be achieved in different ways. In the simplest configuration, each layer is processed as a black and white image, in which each black pixel is printed with a

Figure 3.23 Example of contour changes of a BJT part printed at increasing saturation. Reproduced from [94] / with permission of Elsevier.

drop of binder and each white pixel is not printed. Transforming some black pixels to white in a black area (e.g. in the core) is a mechanism known as "dithering" and can be used to reduce the average saturation in that area. In other cases, some printheads have the possibility to print gray-scale gradients by varying the drop size, for example, with larger drops (higher saturation) in darker gray pixels and the opposite in lighter gray pixels.

The effect of the single printing parameters and of the interaction between binder drops and powder is quite complex and cannot be reduced to a value of saturation. Detailed studies revealed that droplet velocity, spacing, and inter-arrival time play a role in the formation of continuous printed lines [95]. At a microscopic level, the interaction between binder and powder is a complex process for which the result (i.e. the distribution of binder in the printed part) depends on impact dynamics, wetting, spreading, and, finally, on the drainage and imbibition equilibrium.

Despite these complexities, the concept of saturation is a simplified way to optimize the printing process, which has been shown to be a useful starting point even without a full understanding of the complex physics behind it.

3.3 Material Jetting

3.3.1 Definition and Classification of Material Jetting Technologies

MJT is defined in ISO/ASTM 52900:2021 as a "process in which droplets of feedstock material are selectively deposited." In the initial years of MJT, the term "solid free forming" (SFF) was employed to delineate the aforementioned definition. Figure 3.24 displays schematically the principle of MJT. The feedstock/ink is applied droplet-by-droplet onto a substrate or previous printed layer through either (a) a nozzle array (Figure 3.25a) or (b) a single-nozzle (Figure 3.25b).

In the past three decades, a variety of MJ technologies have evolved, employing the aforementioned configurations as well as different approaches of droplet generation and ejection, and different feedstocks. Therefore, two distinctions are made: On the one side (a) the printing approach (see Figure 3.26a), CIP or

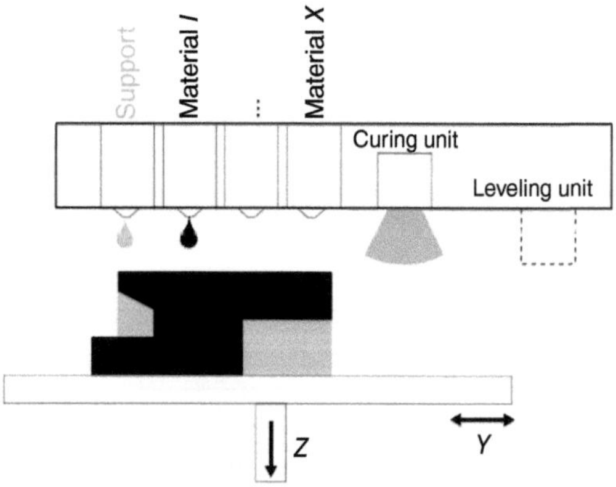

Figure 3.24 Schematic of the material jetting (MJT) process.

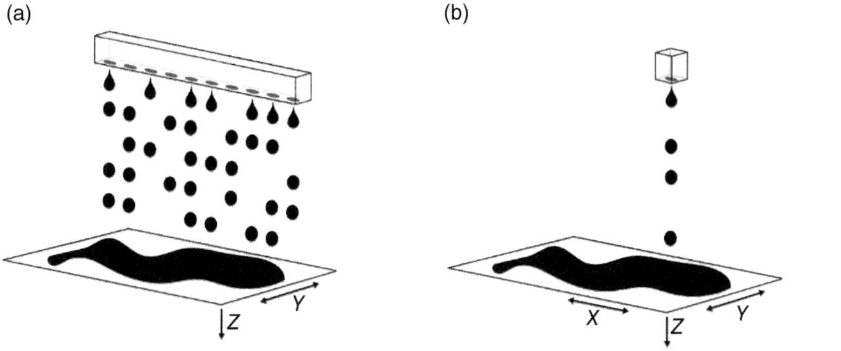

Figure 3.25 Schematic of a printhead configuration with (a) a nozzle array or (b) with a single nozzle.

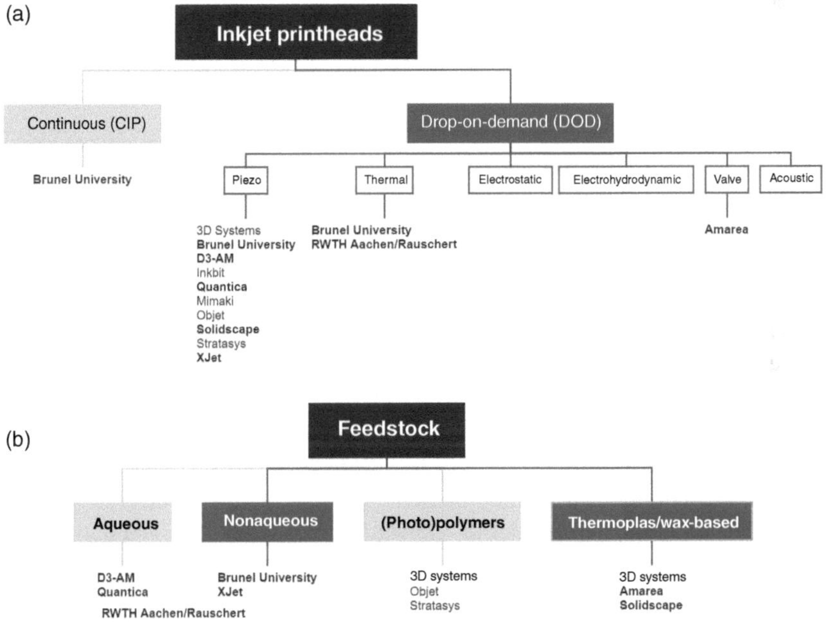

Figure 3.26 (a) Organigram providing a comprehensive list of the various research departments and companies that have engaged or are engaged in material jetting. These entities are classified based on the type of **inkjet printhead** they have utilized or are currently utilizing. (b) Organigram, classifying the research groups and companies working in material jetting according to the applied **feedstock**. The names printed in bold provide work on the material jetting of **ceramics**.

DOD, where it is further differentiated between the approaches of piezo, thermal, electrostatic, electrohydrodynamic, valve, and acoustic, as in Section 3.1. IJP. The second distinction is according to the feedstock, as organized in the organigram of Figure 3.26b, whether aqueous, nonaqueous, polymeric, or thermoplasts/wax-based feedstocks are used. Some of the established research groups and companies that have contributed to and/or currently apply MJT technologies are listed below, with those working with ceramics printed in bold. The distinct MJ technologies have the potential to outperform other AM technologies due to their ability to combine different materials in a straightforward manner. This capability provides a significant degree of freedom in design, as the support material can be utilized and effortlessly removed. The simultaneous printing of multiple inks also provides the ability to combine multiple materials in a droplet-by-droplet manner, enabling the fabrication of multifunctional components by merging materials of distinct properties. In addition, MJT confers a notable degree of freedom regarding material selection, exhibiting independence from color, shape, and particle size constraints, depending on the specific technology employed. The feedstock used for MJ is slurry-based, where powder particles are dispersed in a viscous medium. High solids loadings can be achieved. Ensuring adequate dispersion of powder particles is imperative, as it dictates the packing of particles within a dried layer. This, in turn, determines the green density and, consequently, the sintered density.

3.3.2 Historical Development

MJT's origins can be traced to the domain of investment casting, where the concept was devised to circumvent the initial step of mold preparation, thereby enabling direct wax model shaping through three-dimensional printing. The development of the "3D Model Maker" pioneered by Royden Sanders utilized single nozzles with piezoelectric pulse generation to print wax-like materials, marking a significant milestone in the field [96]. This innovation subsequently led to the foundation of "Solidscape Inc." [96]. Another significant advancement in the field of MJ was achieved through the pioneering work of Rami Bonen, Gerskon Miller, and Hanan Gothait. Their innovation involved the application of photopolymeric inks through a piezoelectric printhead (nozzle array), which were then cured through a UV lamp post-printing process. Consequently, they founded the company "Objet Geometries Inc." with the commercial printer named "PolyJet." Subsequently, Objet was merged with "Stratasys," a provider of AM solutions [97]. The merged entity now offers PolyJet machines, which enable multi-material printing of photopolymers by combining materials with distinct properties. The "MultiJet" printer manufactured by "3D Systems" employs the same working principle and is capable of printing photopolymers and wax-based materials [98].

3.3.2.1 Material Jetting of Ceramics

As is the case with conventional IJP, the initial MJT of ceramics was executed using the working principle of CIP [99–101]. The research group led by Julian R.C. Evans at Brunel University in the United Kingdom conducted investigations into the application of multiple layers of ZrO_2 and subsequently TiO_2 onto a substrate through CIP. In subsequent experiments, they also applied thermal and piezoelectric DOD printheads to free-form ZrO_2 as well as Si_3N_4 materials. These investigations primarily utilized a feedstock based on a nonaqueous solvent [102]. Subsequent attempts to manufacture ceramics through AM were undertaken by Solidscape Inc. [103, 104].

They incorporated Al_2O_3 powders into a wax-based feedstock and utilized the aforementioned 3D Model Maker for printing. A third research group, led by Telle at RWTH Aachen University in Germany, focused on the IJP of ZrO_2 for dental applications using thermal IJP with a commercially available printhead from HP. This group utilized both aqueous and nonaqueous solvents as feedstock [105–109]. The working principle of the aforementioned thermal IJP and the ZrO_2 ink system was subsequently adopted by Rauschert GmbH.

The MJT of ceramics has historically been the focus of various research and development projects, although it has barely led to expanded investigations. Only a small number of companies currently engage in commercial activities related to MJT technologies for ceramic materials. These enterprises have diversified their offerings, distinguishing themselves through unique product names that denote their specialization, in an effort to differentiate themselves in the market. These include Nano-Particle Jetting™ (NPJ), Multi-Material Jetting (MMJ), and Micro-Particle Jetting™ (MPJ).

3.3.2.2 Nano-Particle Jetting (NPJ)TM

The MJT technology of NPJ™ was developed by XJet Ltd. [110], a company founded by the same individual who established Objet. In NPJ, powder particles of ceramics or metals are dispersed in nonaqueous solvents and printed using a commercially available piezoelectric printhead. A water-soluble support material is employed for shape stabilization and the ability to print overhangs [111]. The material ink is printed, followed by the printing of the support ink. Subsequently, a roller is utilized to level the applied material, and a drying step is performed after every printed layer. Post-printing, the support material is removed, and a thermal post-processing step is initiated. The NPJ technology affords high precision with a resolution of ~20 μm, a minimum feature size of ~100 μm, and layer heights ranging from 7 to 10 μm for ceramic materials. In their portfolio, they offer the following metallic and ceramic materials: stainless steel 17-4 PH, stainless steel 316L, zirconia, and alumina.

3.3.2.3 Multi-Material Jetting (MMJ)

MMJ is a relatively new MJT technology that allows the combination of multiple materials. This technology was originally developed at the Fraunhofer IKTS, where it was called "Thermoplastic 3D Printing" (T3DP) [112]. The MMJ technology was outsourced and is now offered by AMAREA technology GmbH. In the case of MMJ, powder particles are dispersed in a wax-based thermoplastic. The feedstock is solid at room temperature and must be heated to eject droplets. The ink is applied at a temperature of 70–100 °C through a Vermes micro-dispensing system [113, 114]. The micro-dispensing system is predicated on the DOD valve approach with piezoelectric motion control, wherein nozzles of various diameters can be mounted, thereby enabling the attainment of disparate droplet volumes and layer heights. The MMJ printing system includes three to six dispensing systems, providing the opportunity to increase productivity and combine different materials. MMJ has demonstrated the ability to combine, for example, ceramic and metallic materials or materials of different colors. The use of a thermoplastic as feedstock requires a separate debinding step after printing prior to sintering. A broad portfolio of materials has been demonstrated to be printable with MMJ, including ceramics (Al_2O_3, AlN, ATZ, SiO_2, porcelain, SiSiC, SSiC, Si_3N_4, TiO_2, and ZrO_2), metals (Ag, SS17-4PH, and SS316L), and hardmetals (WC-Co and WC-Ni) [114].

3.3.2.4 Micro Particle Jetting™ (MPJ)

MPJ has been developed by D3-AM GmbH and launched at the end of the year of 2023 [115–117]. It has the capability to manufacture parts of comparably large dimensions and offers considerable freedom in material selection. MPJ is based on an innovative printhead that has been developed for over a period of 10 years in the Durst Group AG, Brixen, Italy. This printhead is capable of handling abrasive glaze materials with coarse particles. The development of MPJ™ has involved a transformation of this printhead, now enabling the printing of ceramics in three dimensions. It is possible to print powder particles with D99 <60 μm from a technical perspective. An aqueous suspension, filled with ceramic powder particles, is used for MPJ as feedstock. A second ink is used as a support structure, especially for inner channels and overhangs. The material ink is applied droplet-by-droplet, forming a layer. The water is evaporated in a layerwise manner, resulting in dried green parts. The materials are subjected to a thermal post-processing stage, which is carried out in accordance with standard sintering protocols. A rather uncomplicated and expeditious debinding step is enabled due to the low organic content (<10 vol%) that is used. This also facilitates the fabrication of wall thicknesses up to 30 mm. As previously mentioned, MPJ technology offers a high degree of flexibility in material selection, particularly with regard to particle size and color. The printing of coarse powder particles eliminates the need for additional grinding, as is often required in other technologies. Furthermore, MPJ allows for the processing of both dark and bright materials. Ceramic materials, including sintered silicon carbide (SSiC) and alumina (Al_2O_3), have been successfully manufactured using MPJ. The technology's potential extends beyond ceramic materials, encompassing a diverse range of ceramic materials and other material classes, such as hardmetals. A notable advantage of MPJ by D3-AM is its substantial build area of 480 × 400 mm, which facilitates high productivity and enables the fabrication of ceramic components of considerable size, a capability that is exclusive to limited AM technologies.

3.3.3 Process Variations

To date, the aforementioned three MJT technologies are available for commercial use: NPJ™, MMJ, and MPJ™. Therefore, the comparison of process variations will be on those three technologies. Table 3.2 presents a comparative analysis of the process specifications for the technologies of NPJ, MMJ, and MPJ. Upon consideration of the organigrams in Figure 3.26 and the aforementioned historical evolution of MJT, it becomes evident that the majority of prevalent MJT technologies utilize disparate printheads and feedstocks. With respect to the printhead, it has been demonstrated that the nozzle diameter exerts a significant influence on the particle size that can be

Table 3.2 Process specifications of the different material jetting technologies, NPJ, MMJ, and MPJ.

Technology	Feedstock	Particle size (μm)	Printhead	Droplet volume (pL)	Layer height (μm)	Build area (mm)	Accuracy $x-y$ (μm)	Printable materials
NPJ Nano-Particle Jetting™	Nonaqueous	n.a.	Piezo	<100	7–10	530 × 280	50	Metals, ceramics
MMJ Multi-Material Jetting	Thermoplasts	*a)	Valve	500–45,000	70–300	530 × 300	200–1000	Metals, hardmetals, ceramics
MPJ Micro Particle Jetting™	Aqueous	<60	Piezo	>1,000	100–180	480 × 400	200	ceramics

a) Depending on the selected nozzle size.

printed. A commonly accepted guideline suggests that the particle size (D99) should not exceed one-tenth of the nozzle diameter to ensure printability. In the case of NPJ, standard printheads are utilized, accompanied by nozzle diameters of approximately 50 μm. This implements the utilization of particles of a few micrometers (<5 μm), or, as the name of the technology suggests, nano-particles. In the context of ceramic processing, nano-particles are often defined as powder particles of size smaller than 100 nm. The acquisition of nano-particles necessitates a sophisticated preparation process, frequently involving supplementary milling steps, which consequently engenders elevated material expenditures. In contrast, MMJ and MPJ permit the use of standard, commercially available powders.

NPJ and MPJ employ volatile solvents in the preparation of their feedstock. For curing, energy is induced, resulting in solidification of the layer. In contrast, MMJ utilizes a thermoplastic polymer as its feedstock, which is printed in its liquid state and subsequently solidifies due to cooling. The thermoplastic contributes to the green body's stability, forming a matrix between the powder particles. Consequently, a comparable time-consuming debinding step is required for the thermoplastic removal.

Taking the layer height of the aforementioned technologies into consideration, it reveals that NPJ possesses relatively thin layer heights of less than or equal to ten micrometers. This observation provides a basis for assessing the precision or resolution of the respective technologies. The components that are printed using NPJ demonstrate a high degree of resolution. For MMJ, the layer height depends on the selected nozzle diameter and may range between 70 and 300 μm. In the case of MPJ, the layer thickness results in a resolution of approximately 160 μm.

The size of the build area for the three different technologies may be similar, allowing for the simultaneous printing of a high number of components, resulting in high productivity. However, MPJ may be the only technology capable of producing large parts (>300 × 300 mm) with a wall thickness of up to 30 mm due to its feedstock system and the use of commercially available powders.

3.3.4 Challenges in Material Jetting

MJT offers a number of advantages, including the ability to deposit materials contactless and the potential to combine multiple materials in a straightforward manner, droplet by droplet. However, the challenges associated with each MJ technology have a significant impact on the resulting final component and may be the reason that MJ in ceramic AM may still be in its initial stage compared to other ceramic AM processes. The predominant challenges are discussed in the next paragraphs.

3.3.4.1 Nozzle Failures

The ability to obtain free nozzles throughout the printing process may represent the most significant challenge in MJT, particularly when printheads with a nozzle array are used. The phenomenon of nozzle clogging, which refers to the full plug of a nozzle, has been demonstrated to result in a defective layer and, consequently, a component of substandard quality. However, beyond the occurrence of full nozzle

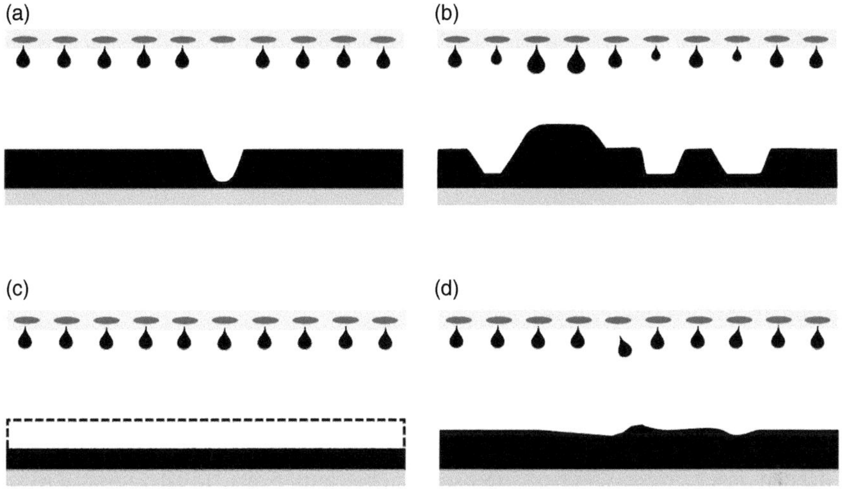

Figure 3.27 Schematics of nozzle failure for a nozzle array, which are commonly observed in inkjet printing: (a) nozzle clogging, (b) droplet volume variation, (c) nozzle shrinkage, and (d) position error.

clocking, additional nozzle failures may arise, as illustrated in Figure 3.27a–d [47], yielding subsequent printing outcomes. Figure 3.27a illustrates nozzle clogging of multiple nozzles within a nozzle array, accompanied by the representative print. Figure 3.27b demonstrates the initial phase of nozzle clocking, characterized by the shrinkage of the nozzle diameter. The deposition of ink material on the nozzle inner walls leads to a reduction in nozzle diameter, resulting in a smaller droplet size and reduced layer height. Figure 3.27c illustrates the issue of variation in droplet volume, which may be attributable to inhomogeneities in droplet ejection control or selective nozzle shrinkage. The straight trajectory of a droplet from the orifice onto the substrate suggests proper nozzle adjustment, while misorientation of the nozzle can result in droplets moving crosswise, as demonstrated and labeled as "position error" in Figure 3.27d. This issue may also be attributed to the pollution of the orifice or the presence of entrapped air bubbles within the ink, which can impede the efficient ejection of droplets. In the event of nozzle failure, it is often necessary to replace both the printhead and the nozzles. However, the incorporation of a circulation system serves to prevent such failures by facilitating continuous ink movement and ensuring the systematic cleaning of the entire system and printhead after printing. Such a circulation system has been incorporated in both the NPJ and MPJ technologies. The design of the printhead used in MPJ allows the circulation of material at elevated flow rates. The utilization of commercially available printheads may encounter limitations in terms of flow rates.

A multitude of factors may contribute to the failure of nozzles, including elevated viscosity, excessive solid loading, an imbalanced particle size to nozzle diameter ratio, inadequate dispersion of powder particles or entrapped air bubbles. It is imperative that all these parameters are taken into account during the development and preparation of the ink.

3.3.4.2 Droplet Ejection, Impingement, and Fusion

The formation of a droplet at the nozzle, its ejection, and its final characteristics are strongly influenced by ink parameters and the aforementioned condition of the nozzle. The droplet's characteristics also depend on the force with which it is ejected from the orifice. Inadequate printing parameters, especially for ink of low viscosity, have been observed to result in the formation of satellites, as described in Section 3.1 and illustrated in Figure 3.28. Figure 3.28 presents the print of two distinct printing configurations. Configuration (a) illustrates the print of a properly formed droplet, while configuration (b) depicts the formation of a droplet accompanied by satellites [114].

The quality of the individual droplets exerts a substantial influence on the quality of the print. In the event of proper droplet formation, a certain degree of droplet homogeneity can be attained. Conversely, this is not the case in configuration (b).

To record the quality and volume of individual droplets and enable adjustments if needed, laser profile scanners are used. Such scanners are utilized in both MMJ [118] technology and other MJT technologies, including those developed by Inkbit [119]. In the case of MMJ, the scanner records the properties of printed test droplets prior to component printing. After analysis, automatic adjustments regarding droplet volume may be made. Inkbit uses the scanner during printing and may provide real-time adjustments.

The quality of the individual droplets is a critical factor in the subsequent fusion process, which results in the formation of a homogeneous layer. In the context of inks comprising both aqueous and nonaqueous components, where the viscosity is minimal, the fusion of droplets can occur with remarkable rapidity and efficiency. The incorporation of thermoplasts or waxes results in the immediate solidification of the droplets upon contact with the substrate. Consequently, a droplet fusion factor (*dff*) has been developed to quantify the extent of overlap between adjacent droplets [120]. The droplet fusion factor is given by the following equation (3.10):

$$dff = \frac{a_{red}}{d_{char}} \tag{3.10}$$

where d_{char} is the characteristic diameter in (m) representing the distance between the center points of two adjacent identical droplets that are in contact but do not overlap, as illustrated in Figure 3.29. a_{red} is the reduced distance of the center points of two adjacent droplets as they overlap in (m), see Figure 3.29.

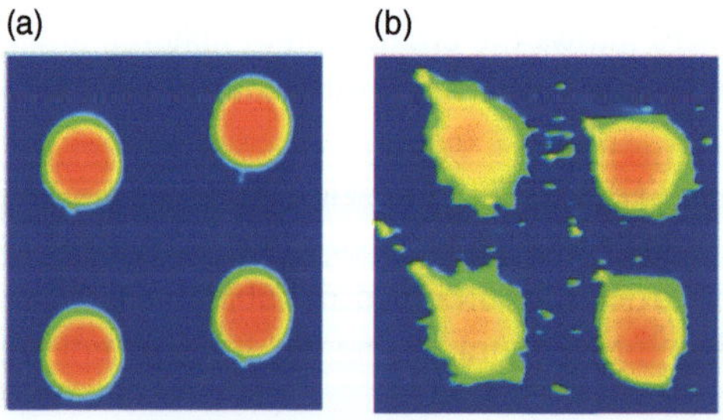

Figure 3.28 Images of (a) a properly formed droplet and of (b) a droplet of poor quality with accompanying satellites. [114] / MDPI / CC BY 4.0.

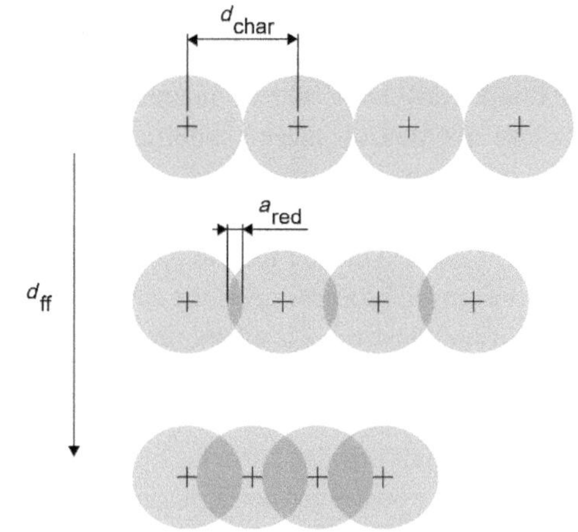

Figure 3.29 Schematic defining the characteristic diameter (d_{char}) and reduced distance (a_{red}) for adjacent droplets [114, 120].

Figure 3.29 also shows droplets at different degrees of fusion, where the *dff* increases from top to bottom [114, 120].

The extent of the overlap, the dimensions (height, width) of the fused material, and the resulting *dff* are contingent on ink properties, particularly viscosity and surface tension, the printing distance between droplets, and the droplet volume.

3.3.4.3 Leveling

Nozzle failure and insufficient droplet fusion can lead to variations in layer height, resulting in inhomogeneities within a single layer. A planar layer is imperative for an adequate build-up process. As demonstrated in Figure 3.30, an error in a single layer can propagate and compromise the integrity of the entire printed component.

Various technologies employ distinct strategies to ensure the planarity of the printed layer. In instances where low-viscosity inks, doctor blades may be employed to level the material layer in its viscous state prior to curing. Post-curing techniques may employ a miller to remove and level the material layer in the solid state.

In the case of thermoplastic/wax-based feedstock, it has been demonstrated that the application of heat onto the printed surface induces melting and, consequently, leveling of the printed layers [121].

3.3.4.4 Curing and Crack Formation

The process of curing can be defined as the phase during which the printed ink transitions from a viscous state to a solid state. In the case of wax-based feedstocks, this transition occurs as a result of cooling. In the context of MMJ, the wax-based ink is

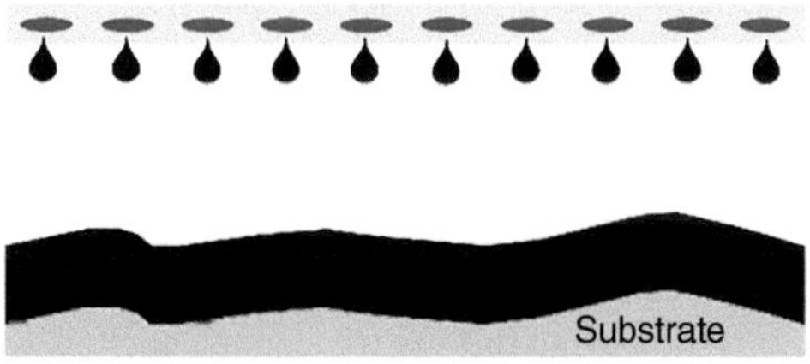

Figure 3.30 Schematic illustrating the effect of an uneven surface, indicating the need of leveling step.

ejected at a temperature of max. 100 °C [114]. During the trajectory and subsequent impingement on the substrate, the droplets undergo a process of solidification due to cooling. Conversely, when nonaqueous or aqueous slurries are used as feedstock, as in NPJ and MPJ, the layer is solidified as the solvent is removed through evaporation. The initiation of the process of evaporation may be facilitated by the implementation of heaters in conjunction with air circulation, as well as infrared emitters.

In the context of MJT technologies, where the solvent undergoes evaporation during the curing process—as is the case in the NPJ and MPJ—the prevention of crack formation constitutes a substantial challenge. Crack formation is a result of internal stresses (σ) that evolve during the curing process. The stresses increase with the inverse capillary radius, thus resulting in the tendency of crack formation for layers with small powder particles (<1 μm). In order to facilitate the printing of inks comprising powder particles measuring less than 1 μm, it is possible to reduce the layer height. Eq. (3.11) delineates the maximum permissible layer height for curing without the occurrence of cracking [122]:

$$h_{\text{max}} = \left(\frac{K_c}{1.4 \cdot \sigma}\right)^2 \tag{3.11}$$

where K_c is the stress intensity factor in [MPa·m$^{1/2}$], and σ the stresses evolving during curing in (MPa). Stresses may be controlled primarily by the capillary radius as a result of particle size. However, Kc may be tailored by the binder system used. A binder system that is flexible and exhibits some plasticity may be able to achieve higher stress intensity factors, resulting in the ability to increase the maximum printable layer height.

3.3.5 Applications

A comparison of various MJT technologies reveals significant differences in their working principles, as well as in the properties of the raw, printed, and sintered materials. These differences lead to notable variations in the applications produced by these distinct technologies. As a result, the range of applications covers a wide spectrum of industrial, medical, automotive, aerospace, and luxury sectors. These technologies can produce feature sizes as small as 100 μm or with wall thicknesses reaching up to 30 mm. Additionally, due to the droplet-by-droplet deposition process, multiple materials can be combined at the droplet level, enabling the creation of parts with integrated materials that offer optical structuring or multifunctional properties.

The ability to apply multiple materials in MJT technologies is complemented by the use of a support material, which is removed after the printing process. The support material facilitates the production of internal structures, intricate geometries, and fine details, which are often difficult or impossible to achieve with other AM technologies.

Figure 3.31 presents a representative illustration of a ZrO_2 component manufactured through NPJ™. This advanced manufacturing technique has been applied in both industrial and luxury sectors, such as in the production of multi-way nozzles,

Figure 3.31 Exemplary ZrO_2 part manufactured with Nano-Particle Jetting™ demonstrating the ability of this technology to obtain fine features and high resolution. [123] / Elsevier / CC BY 4.0.

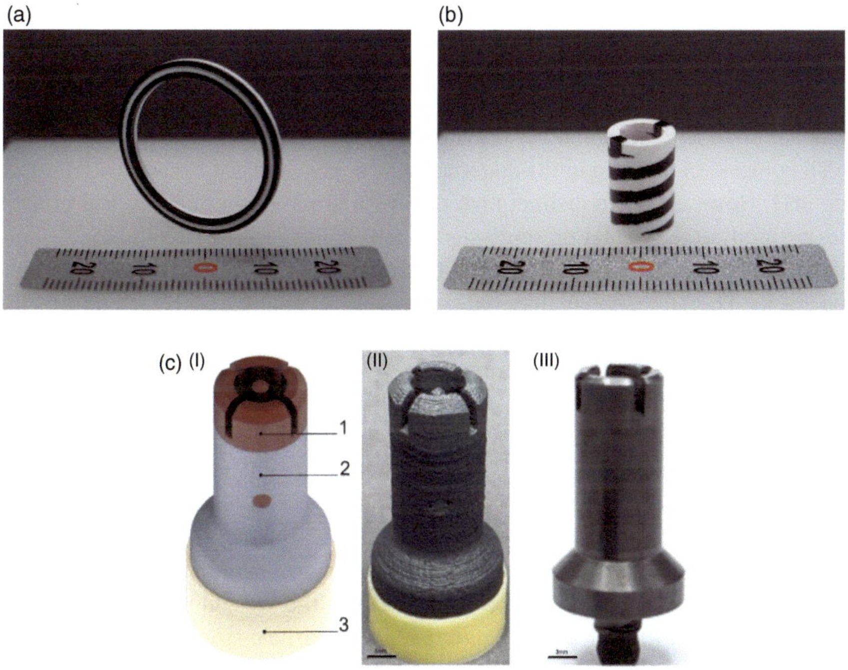

Figure 3.32 Exemplary parts manufactured with multi-material jetting. (a) and (b) are components made of black and white ZrO_2. (c) Plasma plug consisting of two distinct materials with different electrical properties. (I) Represents a model of this multi-material plasma plug, where material (1) is semiconductive ($SiC + MoSi_2$), material (2) is insulating (Si_3N_4), and material (3) is a support material. (II) Multi-material plasma plug in the green state and (III) in the sintered state. Reference [124] / Elsevier / CC BY 4.0.

watch components, and customized earbuds. Figure 3.32 also highlights the versatility of MJT by showcasing the combination of multiple materials. The first two examples (Figure 3.32a, b) depict visual representations of white and black ZrO_2, demonstrating a multi-material approach that has gained considerable attention in the luxury sector, particularly for jewelry applications. The subsequent example in Figure 3.32c illustrates an industrial use case of a plasma plug, where two materials with distinct electrical properties are integrated. Figure 3.32c (**I**) depicts a schematic of the component, where material (i), a combination of SiC and $MoSi_2$, exhibits semiconductive properties, while (ii), Si_3N_4, is an insulating material. Material (iii) serves as a support material to enable overhang structures. The printed component in its green state and in its sintered state are depicted in Figure 3.32c (**II**) and Figure 3.32c (**III**), respectively. The examples presented have been realized through MMJ technology. Amarea GmbH, distributing the MMJ technology, offers an extensive material portfolio, enabling both monolithic and multi-material compositions. The resolution and layer height of the valve micro-dispensing system vary, depending on the nozzle diameters employed.

The ability to jet multiple materials with distinct functional or optical properties into a single component is a unique feature of MJT technology. This capability is particularly valuable in the field of dentistry, where the jetting of materials with varying colors holds significant potential. More accurate aesthetic matches include smoother transitions. Such advancements could lead to improved quality and reduced costs in the production of customized tooth implants, offering more accurate aesthetic matches and including smoother transitions.

The process of printing multi-materials may appear straightforward when utilizing MJT. However, when distinct materials with different thermo-physical properties are used, sintering may present a significant challenge. Mismatching coefficients of thermal expansion can lead to high stresses during the thermal treatment, causing cracks. Therefore, not all material systems are compatible with each other. This must be considered when manufacturing multi-material components.

The examples presented in Figure 3.33a, b have potential applications in both the semiconductor industry and aerospace. Figure 3.33a showcases an aerospike nozzle with internal channels composed of Al_2O_3, highlighting the capability to engineer intricate internal channels, such as those designed for cooling purposes. This capability underscores a substantial advantage of MJT over other AM technologies. Figure 3.33b illustrates impellers made from sintered silicon carbide (SSiC), showcasing another application of this advanced manufacturing technique. To date, only limited reports have demonstrated the feasibility of AM of SSiC. The material's dark color and the required small particle size, to reach sufficient density, present challenges for AM technologies, such as vat photopolymerization and BJT. The third component, illustrated in Figure 3.33c, is a "big ring" with a diameter of 350 mm, composed of silicon carbide (SiC) in the green state, which is intended for utilization in the semiconductor industry. The material used is silicon carbide (SiC). While the component's geometry may seem straightforward, its large diameter, narrow wall thickness, and the dark color of the powder pose challenges for both AM and conventional shaping techniques. These three applications have been manufactured using MPJ™ by D3-AM GmbH, a technology that stands out for its ability to produce advanced ceramics in large formats.

3.3.6 Potential of Material Jetting

This chapter has shown that MJT presents promising opportunities in the field of AM for ceramics and offers a range of unique advantages. Unlike powder-based

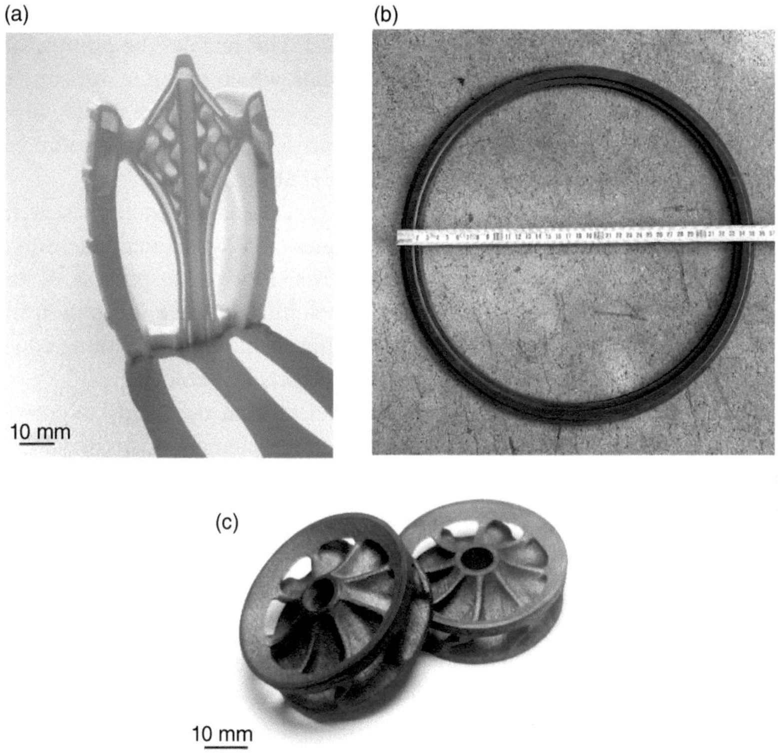

Figure 3.33 Exemplary components manufactured with Micro-Particle Jetting™. (a) An aerospike (Al_2O_3), demonstrating the feasibility of inducing internal channels. (b) A ring with a diameter of 350 mm of SiC in the green state that finds its application in the semiconductor industry. (c) Impellers made of dense sintered silicon carbide (SSiC).

processes, MJ operates with ceramic suspensions, making the process powderless and eliminating limitations related to powder flowability. Due to the drop-by-drop deposition of material, properties can be precisely tailored, enabling the production of customized material characteristics.

Another key advantage of MJT is the capability for multi-material printing, allowing the fabrication of ceramic components with functionally graded properties or combined material systems. In dental restorations, this enables the production of customized crowns and bridges with optimized mechanical and aesthetic properties, including smooth transitions between hard and soft ceramics for improved strength and translucency.

Beyond dental applications, MJT allows the combination of conductive and nonconductive ceramics, opening up new possibilities in electronic components, sensors, and circuit-integrated ceramics. The ability to precisely deposit different materials within a single build enables embedded electrical pathways and customized insulation structures, which are crucial for advanced energy storage devices and high-performance electronics.

Additionally, MJT facilitates the production of complex internal structures, as it utilizes a support material that can be easily removed. This enables the printing of intricate geometries, internal channels, and fine details, which are often difficult or impossible to achieve with other AM technologies. Such features are particularly advantageous in heat exchangers, fluid flow components, and biomedical scaffolds, where internal architecture plays a crucial role in overall performance.

While challenges such as optimizing ink formulations remain, ongoing research in suspension chemistry, advanced sintering strategies, and hybrid manufacturing approaches continues to enhance the industrial feasibility of the process. With these technological advancements, MJT has the potential to drive innovation in high-performance ceramic manufacturing, providing solutions for cutting-edge applications in aerospace, energy, electronics, and medical technology.

By combining material versatility, geometric freedom, and the ability to create highly intricate structures, MJT is ideally positioned to complement existing ceramic AM technologies and enable the next generation of advanced ceramic components.

3.4 Post-Processes

Post-processing depends strictly on the ceramic material in use; however, some general steps are common for most materials and BJT or MJT technologies.

The evolution of a printed part's microstructure along a typical process chain for BJT and MJT is schematized in Figure 3.34.

1) **Drying and/or curing:** To increase productivity, drying and curing are often not fully completed at each layer during the part's build-up. If necessary, the part is thus subject to a thermal process at low temperatures (around 50–200 °C) to complete drying and/or curing.
2) **Cleaning:** In BJT, the printed part at this stage is surrounded by loose powder and has to be cleaned with a brush or by pressurized air. The leftover powder can, in most cases, be sieved and reused. In MJT, the support material has to be removed from the printed part. This is achieved either through a difference in solubility (e.g. the support is water-soluble, but not the part) or mechanically.
3) **Debinding:** Debinding entails the removal of organic components from the part before high-temperature processing. For most BJT and MJT processes, debinding is achieved thermally at temperatures between 100 °C and 600 °C, at which the organic compounds used fully decompose. During this decomposition, it is greatly advantageous to have a network of open pores in the green part, so that the gases developed in the bulk can escape to the surface and be released without causing internal pressure and defects. In BJT and in solvent-based (aqueous or nonaqueous) MJT, there is sufficient open porosity available in the green parts, thus debinding can be performed without issues at relatively high heating rates. More care should be taken in the case of wax-based MJT, because all the volume between particles in the green part is initially filled with wax/thermoplastic polymer.

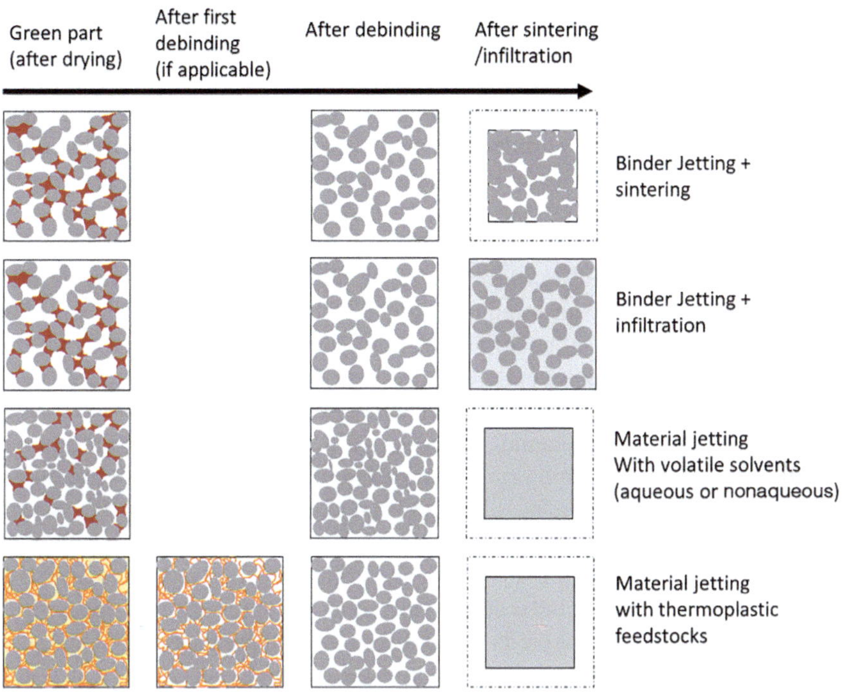

Figure 3.34 Evolution of a printed part's microstructure along a typical process chain for BJT and MJT. [125] / with permission of Elsevier.

4) **Sintering:** Most advanced ceramics after debinding are sintered in a high-temperature furnace, if necessary, with the assistance of a specific atmosphere (vacuum, inert, reactive) and gas pressure. Sintering defines a process of bonding, densification, and/or recrystallization of powder compacts, that is, a treatment in which a green body is converted to a stronger monolith [126]. Sintering is driven by the reduction of free energy of the powder system and is activated by diffusion of atoms across the boundaries between particles, in the solid state or through a liquid phase, but well below the melting point of the material.

   Sintering densification is associated with a geometrical shrinkage (Eq. (3.12)), which needs to be compensated for at the design stage. A common way to compensate for shrinkage is to use the linear shrinkage factors, which are first calibrated on a series of runs with representative specimens of simple geometry. The linear shrinkage factors are defined as Eq. (3.12):

$$Sf,x = \frac{L_{\text{green},X} - L_{\sin,X}}{L_{\text{green},X}} \qquad Sf,y = \frac{L_{\text{green},Y} - L_{\sin,Y}}{L_{\text{green},Y}} \qquad Sf,z = \frac{L_{\text{green},Z} - L_{\sin,Z}}{L_{\text{green},Z}} \tag{3.12}$$

   where $L_{\text{green},X}$ is the measurement of a characteristic linear geometrical feature of the specimen (e.g. side length of a cube) in the green state, measured along the

X printing direction. $L_{\text{sin},X}$ is the measurement of the same geometrical feature on the same specimen after sintering. The same for Y and Z directions.
The flexible production of specimens by AM allows to easily run a series of prints to iteratively adjust the shrinkage factors (Sf_X, Sf_Y, Sf_Z). However, it should be considered that the shrinkage of AM parts is generally anisotropic, thus $Sf_{,X} \neq Sf_{,Y} \neq Sf_{,Z}$. Especially in BJT, shrinkage along the layer stacking direction (Z) is significantly higher than in the layer plane. If the aim is to achieve full densification, a low green density leads to a high shrinkage. As discussed in the BJT chapter, the difficulties in handling fine powders in BJT often lead to low green density for advanced ceramics, which, in turn, leads to insufficient densification or very large shrinkage (>20% linear). On the other hand, the green density and consequently the shrinkage factors in MJT will be, in most cases, closer to those of the traditionally processed materials.

5) For specific advanced ceramics, the green part can be densified by infiltration with a second material, following a processing route that is alternative to sintering. Infiltration has the main advantage that almost full densification (close to 100 % relative density) can be obtained at moderate temperatures and with minimal (sometimes close to 0 %) shrinkage. Infiltration can be pressure-assisted, which is rather complex and costly, or more often pressureless, relying on wetting of the liquid phase on the preform and driven by capillary pressure.
The limitation is obviously that a multiphase material is obtained, which will have different properties compared to the starting ceramic. As already discussed in Section 3.2.2.4, the infiltration of liquid silicon in silicon carbide preforms is a major example of the infiltration approach in BJT, to produce complex parts of SiSiC material (also known as reaction-bonded silicon carbide, RBSiC). It should be mentioned that other infiltration-based approaches have been investigated as ways to densify BJT parts, such as polymer infiltration and pyrolysis (PIP) [127] and chemical vapor infiltration (CVI) [128]. Due to the need for multiple infiltration cycles, these methods are rather laborious and have not yet found major commercial applications in ceramic BJT; however, they may provide interesting solutions in the future as research progresses.
MJT, on the other hand, has mostly focused on sintered ceramics, especially due to the ability to handle very fine powders (down to nanopowders). New technologies on the market such as MMJ and MPJ™ have, however, enabled the printing of coarser particles, opening the door to the infiltration route also for MJT.

References

1 Heilmann, J. (2005). The application of inkjet for the printing of PDLC display cells. In: *NIP & Digital Fabrication Conference*, vol. 21, 48–51. Society of Imaging Science and Technology.

2 Knobloch, A. (2005). Printed RFID labels based on polymer electronics. In: *NIP & Digital Fabrication Conference*, vol. 21, 121–123. Society of Imaging Science and Technology.

3 Lee, H.-H., Chou, K.-S., and Huang, K.-C. (2005). Inkjet printing of nanosized silver colloids. *Nanotechnology* 16 (10): 2436.

4 Szczech, J., Zhang, J., and Gamota, D. (2005). Printed electronic using traditional graphic arts printing technologies. In: *NIP & Digital Fabrication Conference*, vol. 21, 213–213. Society of Imaging Science and Technology.

5 Bennett, R., Edwards, C., Lee, J., and Silz, K. (2003). Precision industrial ink-jet printing technology for full color PLED display and TFT-LCD manufacturing. In Proc. of the 3rd International Display Manufacturing Conf.

6 Haskal, E., Buechel, M., Dijksman, J.F. et al. (2002). 21.1: ink jet printing of passive-matrix polymer light emitting displays. In: *SID Symposium Digest of Technical Papers*, vol. 33, no. 1, 776–779. Wiley Online Library.

7 Basaran, O.A. (2002). Small-scale free surface flows with breakup: drop formation and emerging applications. *Am. Inst. Chem. Eng. AIChE J.* 48 (9): 1842.

8 Bietsch, A., Zhang, J., Hegner, M. et al. (2004). Rapid functionalization of cantilever array sensors by inkjet printing. *Nanotechnology* 15 (8): 873.

9 Carter, J.C., Alvis, R.M., Brown, S.B. et al. (2006). Fabricating optical fiber imaging sensors using inkjet printing technology: a pH sensor proof-of-concept. *Biosens. Bioelectron.* 21 (7): 1359–1364.

10 Chee, M., Yang, R., Hubbell, E. et al. (1996). Accessing genetic information with high-density DNA arrays. *Science* 274 (5287): 610–614.

11 Chen, T. (2006). Piezoelectric inkjet print head technology for precision dispensing application. In: *NIP & Digital Fabrication Conference*, vol. 22 (ed. IS&T), 66–69. Society of Imaging Science and Technology.

12 Cooley, P.W., Wallace, D.B., and Antohe, B.V. (2001). Applications of ink-jet printing technology to BioMEMS and microfluidic systems. *Microfluid. BioMEMS* 4560: SPIE: 177–188.

13 James, P. and Papen, R. (1998). Nanolitre dispensing—a new innovation in robotic liquid handling. *Drug Discov. Today* 3 (9): 429–430.

14 Koo, H.-S., Chen, M., and Pan, P.-C. (2006). LCD-based color filter films fabricated by a pigment-based colorant photo resist inks and printing technology. *Thin Solid Films* 515 (3): 896–901.

15 Darhuber, A.A. and Troian, S.M. (2005). Principles of microfluidic actuation by modulation of surface stresses. *Annu. Rev. Fluid Mech.* 37 (1): 425–455.

16 Nakanishi, T., Ohtsu, I., Furuta, M. et al. (2005). Direct MS/MS analysis of proteins blotted on membranes by a matrix-assisted laser desorption/ionization-quadrupole ion trap-time-of-flight tandem mass spectrometer. *J. Proteome Res.* 4 (3): 743–747.

17 Ohtsu, I., Nakanisi, T., Furuta, M. et al. (2005). Direct matrix-assisted laser desorption/ionization time-of-flight mass spectrometric identification of proteins on membrane detected by western blotting and lectin blotting. *J. Proteome Res.* 4 (4): 1391–1396.

18 Thomson, W. (1867). Improvements in receiving or recording instruments for electric telegraphs. GB Patent 2147.

19 Plateau, M. (1856). XXXVII. On the recent theories of the constitution of jets of liquid issuing from circular orifices. *Lond. Edinb. Dubl. Philos. Mag. J. Sci.* 12 (79): 286–297.
20 Rayleigh, L. (1878). On the instability of jets. *Proc. Lond. Math. Soc.* 1 (1): 4–13.
21 Eggers, J. (1997). Nonlinear dynamics and breakup of free-surface flows. *Rev. Mod. Phys.* 69 (3): 865.
22 Sweet, R.G. (1965). High frequency recording with electrostatically deflected ink jets. *Rev. Sci. Instrum.* 36 (2): 131–136.
23 Hertz, C. and Simonsson, S. (1968). Ink jet recorder. US Patent US3416153 A.
24 Kobayashi, H., Koumura, N., and Ohno, S. (1981). Liquid recording medium. US Patent US4243994 A.
25 Yamada, T., Matsuda, Y., Yoshino, M., and Sakata, M. (1988). Micro-dot ink jet recorder. US Patent US4746928 A.
26 Wijshoff, H. (2010). The dynamics of the piezo inkjet printhead operation. *Phys. Rep.* 491 (4-5): 77–177.
27 Hansell, C.W. (1946). Jet sprayer actuated by supersonic waves. US Patent US2512743A.
28 Zoltan, S. (1970). Pulsed droplet ejecting system. US Patent US3683212A.
29 Stemme, N. (1972). Arrangement of writing mechanisms for writing on paper with a colored liquid. US Patent US3747120A.
30 Kyser, E.L. and Sears, S.B. (1970). Method and apparatus for recording with writing fluids and drop projection means therefor. US Patent US3946398A.
31 Li, J., Rossignol, F., and Macdonald, J. (2015). Inkjet printing for biosensor fabrication: combining chemistry and technology for advanced manufacturing. *Lab Chip* 15 (12): 2538–2558.
32 Le, H.P. (1998). Progress and trends in ink-jet printing technology. *J. Imag. Sci. Technol.* 42 (1): 49–62.
33 Endo, I., Sato, Y., Saito, S., Nakagiri, T., and Ohno, S. (1979). Liquid jet recording process and apparatus therefor. GB Patent GB2007162.
34 Vaught, J.L., Cloutier, F.L., Donald, D.K. et al. (1984). Thermal ink jet printer. US Patent US4490728A.
35 Kamisuki, S., Hagata, T., Tezuka, C. et al. (1998). A low power, small, electrostatically-driven commercial inkjet head. In: *Proceedings MEMS 98. IEEE. Eleventh Annual International Workshop on Micro Electro Mechanical Systems. An Investigation of Micro Structures, Sensors, Actuators, Machines and Systems (Cat No. 98CH36176)*, 63–68. IEEE.
36 Lepowsky, E., Muradoglu, M., and Tasoglu, S. (2018). Towards preserving post-printing cell viability and improving the resolution: past, present, and future of 3D bioprinting theory. *Bioprinting* 11: e00034.
37 Umezu, S., Suzuki, H., and Kawamoto, H. (2005). Droplet formation and dropping position control in electrostatic inkjet phenomena. In: *NIP & Digital Fabrication conference*, vol. 21, 283–286. Society of Imaging Science and Technology.

38 Umezu, S., Nakazawa, R., and Kawamoto, H. (2006). Cross-talk of multi-nozzle in electrostatic inkjet system. In: *NIP & Digital Fabrication Conference*, vol. 22, 66–68. Society of Imaging Science and Technology.

39 Galliker, P. (2021). Electrohydrodynamic print head with structured feed layer. WO Patent WO2021008699A1.

40 Hoath, S.D. (2016). *Fundamentals of Inkjet Printing: The Science of Inkjet and Droplets*. John Wiley & Sons.

41 Achrainer, J. (2010). Valve device of an application device for applying fluid to a substrate, and applicator. US Patent US7837071B2.

42 Dawson, T. and Ellis, H. (1994). Will ink jets ever replace screens for textile printing? *J. Soc. Dye. Colour.* 110 (10): 331–337.

43 Nguyen, Q.-H., Choi, S.-B., and Kim, J.-D. (2008). The design and control of a jetting dispenser for semiconductor electronic packaging driven by a piezo stack and a flexible beam. *Smart Mater. Struct.* 17 (6): 065028.

44 Wang, J., Li, M., and Nguyen, T. (2023). Mechanisms of droplet formation and deposition in drop-on-demand needle-valve inkjets for precision 3D microprinting. *Int. J. Precis. Eng. Manuf.-Green Technol.* 10 (6): 1433–1446.

45 Parashkov, R., Becker, E., Riedl, T. et al. (2005). Large area electronics using printing methods. *Proc. IEEE* 93 (7): 1321–1329.

46 Derby, B. (2010). Inkjet printing of functional and structural materials: fluid property requirements, feature stability, and resolution. *Annu. Rev. Mater. Res.* 40 (1): 395–414.

47 Elkaseer, A., Chen, K.J., Janhsen, J.C. et al. (2022). Material jetting for advanced applications: a state-of-the-art review, gaps and future directions. *Addit. Manuf.* 60: 103270.

48 Reis, N. and Derby, B. (2000). Ink jet deposition of ceramic suspensions: modeling and experiments of droplet formation. *MRS Online Proc. Libr. (OPL)* 625: 117.

49 Du, Z., Lin, Y., Xing, R. et al. (2018). Controlling the polymer ink's rheological properties and viscoelasticity to suppress satellite droplets. *Polymer* 138: 75–82.

50 Zhang, Z., Jin, Y., Yin, J. et al. (2018). Evaluation of bioink printability for bioprinting applications. *Appl. Phys. Rev.* 5 (4): 041304.

51 McKinley, G.H. (2005). Dimensionless groups for understanding free surface flows of complex fluids.

52 Sachs, E.M., Haggerty, J.S., Cima, M.J., and Williams, P.A. (1993). Three-dimensional printing techniques. US Patent US5204055A.

53 Sachs, E., Cima, M., Williams, P., Brancazio, D., and Cornie, J. (1992). Three dimensional printing: rapid tooling and prototypes directly from a CAD model.

54 Zhou, K. and Han, C. (2023). Metal binder jetting. In: *Metal Powder-Based Additive Manufacturing*, 237–290. Wiley-VCH GmbH.

55 The ExOne Company. (2025). https://www.exone.com/en-US/X1-Tooling/X1-ThermoForm. https://www.exone.com/en-US/X1-Tooling/X1-Layup. https://www.exone.com/en-US/X1-Tooling/X1-Washout (accessed 25 February 2025).

56 W.E. Forum (2024). High-emitting sectors: challenges and opportunities for low-carbon suppliers.

57 Lowke, D., Dini, E., Perrot, A. et al. (2018). Particle-bed 3D printing in concrete construction–possibilities and challenges. *Cem. Concr. Res.* 112: 50–65.

58 Michael Hansmeyer (2025). Digital Grotesque. https://www.michael-hansmeyer.com/digital-grotesque-I (accessed 25 February 2025).

59 Beau Jackson (2025). https://3dprintingindustry.com/news/behind-3d-printed-bridge-exclusive-interview-catalonian-designers-iaac-101391/ (accessed 25 February 2025).

60 D-Shape (2025). Una casa tutta di un pezzo. https://d-shape.com/Prodotti/unacasatuttadiunpezzo/ (accessed 25 February 2025).

61 European Space Agency (2025). Building a lunar base with 3D printing. https://www.esa.int/Enabling_Support/Space_Engineering_Technology/Building_a_lunar_base_with_3D_printing (accessed 25 February 2025).

62 Ma, G., Hu, T., and Li, Z. (2024). Binder jetting 3D printing rock analogs using magnesium phosphate cement. *Constr. Build. Mater.* 420: 135620.

63 Walling, S.A. and Provis, J.L. (2016). Magnesia-based cements: a journey of 150 years, and cements for the future? *Chem. Rev.* 116 (7): 4170–4204.

64 Salari, F., Zocca, A., Bosetti, P. et al. (2024). Powder-bed 3D printing by selective activation of magnesium phosphate cement: determining significant processing parameters and their effect on mechanical strength. *Open Ceram.* 18: 100609.

65 Salari, F., Bosetti, P., and Sglavo, V.M. (2022). Binder jetting 3D printing of magnesium oxychloride cement-based materials: parametric analysis of manufacturing factors. *J. Manuf. Mater. Process.* 6 (4): 86.

66 Weger, D. and Gehlen, C. (2021). Particle-bed binding by selective paste intrusion—strength and durability of printed fine-grain concrete members. *Materials* 14 (3): 586.

67 Davidovits, J. (1991). Geopolymers: inorganic polymeric new materials. *J. Therm. Anal. Calorim.* 37 (8): 1633–1656.

68 Turner, L.K. and Collins, F.G. (2013). Carbon dioxide equivalent (CO_2-e) emissions: a comparison between geopolymer and OPC cement concrete. *Constr. Build. Mater.* 43: 125–130.

69 Duxson, P., Provis, J.L., Lukey, G.C., and Van Deventer, J.S. (2007). The role of inorganic polymer technology in the development of 'green concrete'. *Cem. Concr. Res.* 37 (12): 1590–1597.

70 Xia, M. and Sanjayan, J. (2016). Method of formulating geopolymer for 3D printing for construction applications. *Mater. Des.* 110: 382–390.

71 Xia, M., Nematollahi, B., and Sanjayan, J. (2018). Influence of binder saturation level on compressive strength and dimensional accuracy of powder-based 3D printed geopolymer. In: *Materials Science Forum*, vol. 939 (ed. U. Kazuo, H. Justnes, K.-B. Kim, and T. Akitsu), 177–183. Trans Tech Publ.

72 Nematollahi, B., Xia, M., and Sanjayan, J. (2019). Post-processing methods to improve strength of particle-bed 3D printed geopolymer for digital construction applications. *Front. Mater.* 6: 160.

73 Voney, V., Odaglia, P., Brumaud, C. et al. (2020). Geopolymer formulation for binder jet 3D printing. In: *Second RILEM International Conference on Concrete and Digital Fabrication: Digital Concrete 2020 2*, 153–161. Springer.

74 Ur Rehman, A. and Sglavo, V.M. (2020). 3D printing of geopolymer-based concrete for building applications. *Rapid Prototyp. J.* 26 (10): 1783–1788.

75 Elsayed, H., Gobbin, F., Picicco, M. et al. (2022). Additive manufacturing of inorganic components using a geopolymer and binder jetting. *Addit. Manuf.* 56: 102909.

76 Lowke, D., Talke, D., Dressler, I. et al. (2020). Particle bed 3D printing by selective cement activation—applications, material and process technology. *Cem. Concr. Res.* 134: 106077.

77 Wilbig, J. (2024). Metrology for additively manufactured medical implants. Doctoral dissertation, Technical University of Clausthal.

78 Wang, P., Robertson, G., Gibson, B.T. et al. (2024). Improved productivity with multilaser rotary powder bed fusion additive manufacturing. *3D Print. Addit. Manuf.* 11 (1): 231–241.

79 Hauser, C., Lewis, D., Dunschen, M., Egan, M., and Sutcliffe, C. (2007).Image transformations and printing of plaster layers in spiral growth manufacturing.

80 Du, W., Ren, X., Pei, Z., and Ma, C. (2020). Ceramic binder jetting additive manufacturing: a literature review on density. *J. Manuf. Sci. Eng.* 142 (4): 040801.

81 Lv, X., Ye, F., Cheng, L. et al. (2019). Binder jetting of ceramics: powders, binders, printing parameters, equipment, and post-treatment. *Ceram. Int.* 45 (10): 12609–12624.

82 Ziaee, M. and Crane, N.B. (2019). Binder jetting: a review of process, materials, and methods. *Addit. Manuf.* 28: 781–801.

83 Kunchala, P. and Kappagantula, K. (2018). 3D printing high density ceramics using binder jetting with nanoparticle densifiers. *Mater. Des.* 155: 443–450.

84 Fatah, N. (2009). Study and comparison of micronic and nanometric powders: analysis of physical, flow and interparticle properties of powders. *Powder Technol.* 190 (1-2): 41–47.

85 Molerus, O. (2002). The role of science in particle technology. *Powder Technol.* 122 (2-3): 156–167.

86 Krupp, H. (1967). Particle adhesion theory and experiment. *Adv. Colloid Interf. Sci.* 1: 111–239.

87 Yang, J., Sliva, A., Banerjee, A. et al. (2005). Dry particle coating for improving the flowability of cohesive powders. *Powder Technol.* 158 (1-3): 21–33.

88 German, R.M. (1992). Prediction of sintered density for bimodal powder mixtures. *Metall. Trans. A.* 23: 1455–1465.

89 Du, W., Singh, M., and Singh, D. (2020). Binder jetting additive manufacturing of silicon carbide ceramics: development of bimodal powder feedstocks by modeling and experimental methods. *Ceram. Int.* 46 (12): 19701–19707.

90 Diener, S., Zocca, A., and Günster, J. (2021). Literature review: methods for achieving high powder bed densities in ceramic powder bed based additive manufacturing. *Open Ceram.* 8: 100191.

91 Haeri, S., Wang, Y., Ghita, O., and Sun, J. (2017). Discrete element simulation and experimental study of powder spreading process in additive manufacturing. *Powder Technol.* 306: 45–54.

92 Utela, B., Storti, D., Anderson, R., and Ganter, M. (2008). A review of process development steps for new material systems in three dimensional printing (3DP). *J. Manuf. Process.* 10 (2): 96–104.

93 Shrestha, S. and Manogharan, G. (2017). Optimization of binder jetting using Taguchi method. *JOM* 69: 491–497.

94 Gilmer, D., Han, L., Hong, E. et al. (2020). An in-situ crosslinking binder for binder jet additive manufacturing. *Addit. Manuf.* 35: 101341.

95 Colton, T. and Crane, N.B. (2021). Influence of droplet velocity, spacing, and inter-arrival time on line formation and saturation in binder jet additive manufacturing. *Addit. Manuf.* 37: 101711.

96 (2025). Solidscape. https://solidscape.com/ (accessed 26 February 2025).

97 (2025). Stratasys. https://www.stratasys.com/en/ (accessed 26 February 2025).

98 3D Systems. https://www.3dsystems.com/ (accessed 26 February 2025).

99 Blazdell, P., Evans, J., Edirisinghe, M. et al. (1995). The computer aided manufacture of ceramics using multilayer jet printing. *J. Mater. Sci. Lett.* 14: 1562–1565.

100 Blazdell, P. (2003). Solid free-forming of ceramics using a continuous jet printer. *J. Mater. Process. Technol.* 137 (1–3): 49–54.

101 Song, J.H., Edirisinghe, M.J., and Evans, J.R. (1999). Formulation and multilayer jet printing of ceramic inks. *J. Am. Ceram. Soc.* 82 (12): 3374–3380.

102 Slade, C. (1998). Freeforming ceramics using a thermal jet printer. *J. Mater. Sci. Lett.* 17: 1669–1671.

103 Zhao, X., Evans, J., Edirisinghe, M., and Song, J. (2003). Formulation of a ceramic ink for a wide-array drop-on-demand ink-jet printer. *Ceram. Int.* 29 (8): 887–892.

104 Zhao, X., Evans, J.R., Edirisinghe, M.J., and Song, J.H. (2002). Direct ink-jet printing of vertical walls. *J. Am. Ceram. Soc.* 85 (8): 2113–2115.

105 Fischer, H., Telle, R., Ebert, J., and Özkol, E. (2008). Method for producing three-dimensional ceramic mouldings. EU Patent EP2007698B1.

106 Cappi, B., Özkol, E., Ebert, J., and Telle, R. (2008). Direct inkjet printing of Si3N4: characterization of ink, green bodies and microstructure. *J. Eur. Ceram. Soc.* 28 (13): 2625–2628.

107 Gingter, P., Wätjen, A., Kramer, M., and Telle, R. (2015). Functionally graded ceramic structures by direct thermal inkjet printing. *J. Ceram. Sci. Technol.* 6 (6): 119–124.

108 Özkol, E., Ebert, J., Uibel, K. et al. (2009). Development of high solid content aqueous 3Y-TZP suspensions for direct inkjet printing using a thermal inkjet printer. *J. Eur. Ceram. Soc.* 29 (3): 403–409.

109 Özkol, E., Zhang, W., Ebert, J., and Telle, R. (2012). Potentials of the "Direct inkjet printing" method for manufacturing 3Y-TZP based dental restorations. *J. Eur. Ceram. Soc.* 32 (10): 2193–2201.

110 (2025). XJET. https://xjet3d.com/ (accessed 26 February 2025).

111 Dayagi, Y., Benichou, A., and Kritchman, E. (2016). Support ink for three dimensional (3D) printing. US Patent US11000897B2.

112 Scheithauer, U., Schwarzer, E., Poitzsch, C. et al. (2015). Method for producing ceramic and/or metal components. WO Patent WO2015177128A1.

113 Scheithauer, U., Schwarzer, E., Richter, H.J., and Moritz, T. (2015). Thermoplastic 3D printing—an additive manufacturing method for producing dense ceramics. *Int. J. Appl. Ceram. Technol.* 12 (1): 26–31.

114 Reckert, A., Lang, V., Weingarten, S. et al. (2024). Quality prediction and classification of process parameterization for multi-material jetting by means of computer vision and machine learning. *Journal of Manufacturing and Materials Processing* 8 (1): 8.

115 (2025). D3-AM. https://d3-am.com (accessed 27 February 2025).

116 (2025). https://www.voxelmatters.com/d3-am-to-present-new-lab-ii-ceramic-3d-printer-at-formnext/ (accessed 27 February 2025).

117 Kety S., (2024). Technical ceramics: current advancements & limitations in multi-material 3D printing. 2024-09/10 *3D ADEPT MAG.*

118 Scheithauer, U., Weingarten, S., Schwarzer, E. et al. (2019). Process for the layerwise additive production of three-dimensionally formed components. DE Patent DE102017217682A1.

119 Inkbit. (2025). https://inkbit3d.com/ (accessed 27 February 2025).

120 Scheithauer, U., Johne, R., Weingarten, S. et al. (2018). Investigation of droplet deposition for suspensions usable for thermoplastic 3D printing (T3DP). *J. Mater. Eng. Perform.* 27: 44–51.

121 Scheithauer, U., Weingarten, S., Lang, V., Glausch, F., and Wiemer, H. (2024). Method for the additive manufacture of components, in which a component is produced by depositing at least one material in the form of droplets. Patent US20240262032A1.

122 Hu, M., Thouless, M.D., and Evans, A.G. (1988). The decohesion of thin films from brittle substrates. *Acta Metall.* 36 (5): 1301–1307.

123 Oh, Y., Bharambe, V., Mummareddy, B. et al. (2019). Microwave dielectric properties of zirconia fabricated using NanoParticle Jetting™. *Addit. Manuf.* 27: 586–594.

124 Ziener, J., Scheithauer, U., Gottlieb, L. et al. (2024). Additive manufacturing of ceramic multi-material heating and ignition elements for a sustainable space access. *Acta Astronaut.* 219: 609–617.

125 Zocca, A. and Günster, J. (2024). Towards a debinding-free additive manufacturing of ceramics: a development perspective of water-based LSD and LIS technologies. *Open Ceram.* 19: 100632.

126 Vakifahmetoglu, C. and Karacasulu, L. (2020). Cold sintering of ceramics and glasses: a review. *Curr. Opinion Solid State Mater. Sci.* 24 (1): 100807.

127 Cramer, C.L., Armstrong, H., Flores-Betancourt, A. et al. (2020). Processing and properties of SiC composites made via binder jet 3D printing and infiltration and pyrolysis of preceramic polymer. *Int. J. Ceram. Eng. Sci.* 2 (6): 320–331.

128 Terrani, K., Jolly, B., and Trammell, M. (2020). 3D printing of high-purity silicon carbide. *J. Am. Ceram. Soc.* 103 (3): 1575–1581.

4

Material Extrusion (MEX)

Frank Clemens[1] *and Esther García-Tuñón*[2]

[1] *High Performance Ceramics, Department of Advanced Materials and Surfaces, Empa, Überlandstrasse 129, 8600 Dübendorf, Switzerland*
[2] *University of Liverpool, School of Engineering and Materials Innovation Factory, Harrison Hughes Building, The Quadrangle, University of Liverpool, L69 3GH, Liverpool, UK*

4.1 Introduction

Material extrusion processes can be divided into direct ink writing (DIW) and fused filament fabrication (FFF) processes, each with well-differentiated underlying physical processes and material requirements. In the context of DIW feedstocks, rheological properties are of paramount importance. Ceramic feedstocks must exhibit a yield stress behavior and structural recovery to ensure printability and shape fidelity. Conventional FFF involves the utilization of a filament that is driven between two gears into the printhead. In the printhead's heating zone, the polymer is melted and subsequently extruded through a nozzle and cooled down below the melting point of the polymer. Therefore, a thermoplastic binder system has been employed for the FFF printing of ceramics. The following chapter will describe the DIW and FFF processes for ceramic powders in detail. Post-processing after printing will be discussed in Section 4.5. This chapter ends with an overview of the state-of-the-art on the fabrication of ceramic matrix composites using MEX techniques. The diverse research and the extensive literature already published in MEX of ceramics cannot be presented in detail within this chapter. The authors strongly encourage the reader to consult the resources signposted, while bearing in mind that this chapter will primarily focus on the fundamental aspects.

4.2 Direct Ink Writing

DIW is an AM technique, originally named "robocasting" or robotic-assisted deposition [1], that relies on the continuous extrusion of highly concentrated ceramic suspensions directly onto a substrate. The breadth and scope of the research in DIW of ceramics have consistently grown [1–3] and are currently being applied in diverse ceramics applications from biomedical devices and scaffolds [4, 5], bone implants [6, 7], optical materials [8], piezoelectric components [9, 10],

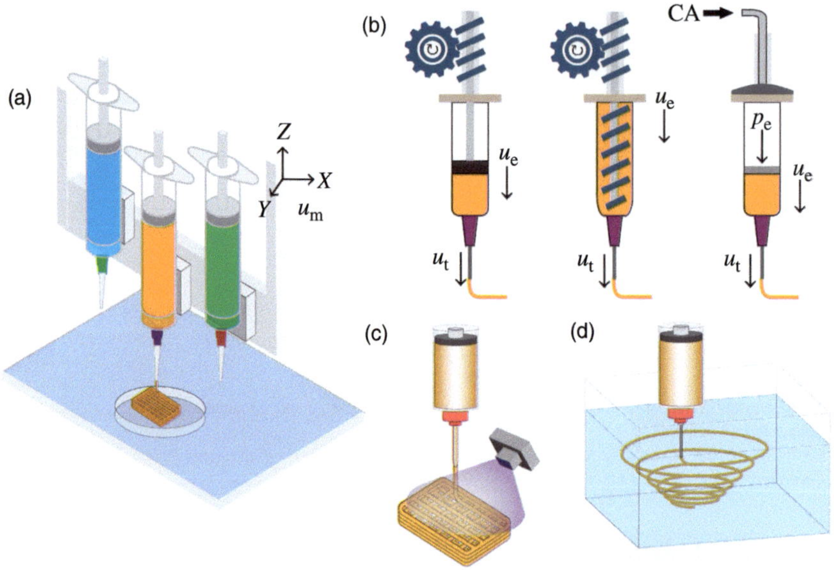

Figure 4.1 Schematic representations of DIW systems. (a) A DIW Cartesian system equipped with three syringe pumps. Where u_m corresponds to the motion of the $X-Y-Z$ Cartesian robot. (b) Extrusion systems: plunger displacement, screw/mixer, and pressure-controlled plunger displacement. Where u_e corresponds to the velocity of the plunger, u_t corresponds to the velocity at the tip of the nozzle, CA stands for the compressed air, and p_e is the pressure acting on the plunger using pressure-dispensing accessories. DIW variants illustrating in situ curing (c) and embedded printing (d).

energy [11] and catalysis [12–14], lightweight structures [15], extreme environments [16–18], multi-material sensors [19], composites [20–22], and bioinspired materials [23–25]. Achieving complex ceramic structures and composites via DIW requires first gaining a good understanding of the relevant DIW system and technical constraints (Section 4.2.1), and the design of printable ceramic feedstocks (Section 4.2.2) with specific rheological properties (Section 4.2.3). This chapter also introduces concepts in non-Newtonian fluid mechanics to provide a theoretical basis for selecting process parameters (Section 4.2.4).

4.2.1 DIW Systems

The fundamental working principle is common for any DIW system: an extrusion device (or multiple, Figure 4.1a) synchronized with a controlled 3D moving platform. The specifications and working limits of different DIW systems vary; however, these are rarely reported in enough detail. Comparison of systems and technical specifications is therefore challenging, particularly given the number of studies that use self-constructed DIW printers (examples can be found in [12–14, 26]). This section provides a general overview of different DIW setups and variants.

The 3D motion can be provided by moving parts connected to the extrusion device (Figure 4.1a) or a fixed system extruding on a moving platform. $X - Y - Z$ Cartesian gantry systems with servo motors are designed for precise control with micron-level positioning accuracy and repeatability [7, 24, 27, 28]. However, not every DIW system will be equipped with this technology, and therefore, the precision that can be achieved will vary. Feedstock extrusion can be driven through different means. Using rod-type actuators to control linear motion with varying forces enables precise syringe plunger movement [7, 26, 29]. Modifications of this extrusion system (Figure 4.1b) include mixers, screw extruders, or a combination of pressure and screw systems [30]. Alternatively, the plunger can be driven by pressure-dispensing tools, which in some cases might restrict working limits due to the high force required to initiate the flow (when working with stiff ceramic feedstocks [31]) and to achieve the precision required to maintain a continuous steady state flow without pressure fluctuations. In short, the specifications of the DIW systems (and selected process parameters, Section 4.2.4) can impose a technical limit on the "printability" of ceramic feedstocks or on the part resolution that can be practically achieved. Unfortunately, equipment specifications and working limits are rarely explicitly detailed in DIW literature, thus adding to the challenge of establishing standards across the field.

Variants to DIW systems include add-on systems to facilitate in situ colloidal assembly, temperature control, or UV curing (Figure 4.1c). Commercial pressure-controlled extrusion systems designed for bioprinting are being used in DIW of ceramics [21], enabling hybrid processes, combining extrusion with in situ curing. Combining direct extrusion with laser writing of photoresponsive silica colloids enables the fabrication of structures with controlled features ~10 μm [32], while UV-curing assisted DIW has recently been used to create zirconia-based composites with aligned alumina platelets [21]. The UV-irradiation triggers the photopolymerization process to avoid common issues due to drying (cracks, warping, or uneven shrinkage) [33]. Instead of UV-curing, some variants combine embedded 3D printing with microwave-activated curing to create complex ceramic structures with controlled composition [34]. Embedded DIW is a versatile variant enabling the fabrication of more complex shapes and multi-material composites [35]. For example, embedded-DIW of preceramic polymers (Figure 4.1d) using a pressure-controlled DIW system enables the fabrication of complex free-form ceramic components through the optimization of printing parameters (p_e and u_m) for a given tip diameter [36].

Each DIW system imposes specific considerations on feedstock design, rheological requirements and process parameters selection. Therefore, covering the fundamentals of each of these variants is beyond the scope of this chapter. Sections 4.2.2–4.2.4 primarily focus on a DIW setup without supporting fluids or "in situ" consolidation.

4.2.2 Feedstock Preparation: Additives and Processes

Feedstocks in DIW of ceramics consist of highly concentrated suspensions (called ceramic inks, pastes, slurries, ceramic gels, and, more broadly, formulations) that require a careful design process (Figure 4.2). Successful performance requires a systematic consideration and adjustment of inter-related aspects. Each material

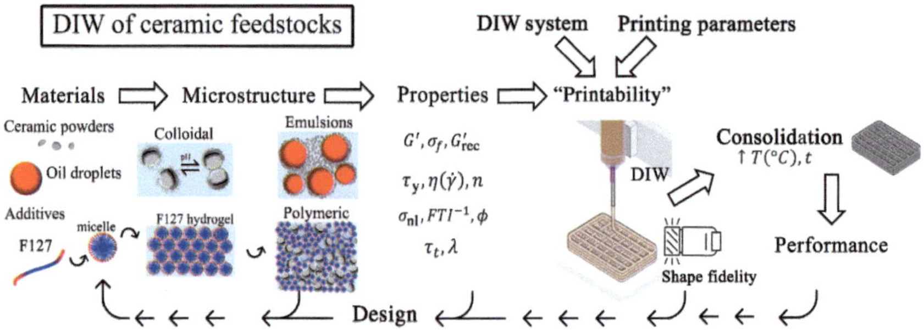

Figure 4.2 Scheme illustrating the design process to achieve "printable" feedstocks for DIW of ceramics. The structure–property relationship in ceramic feedstocks, the DIW system, and process parameters, each play a role in the shape fidelity that can be achieved. Typical microstructures (colloidal, emulsions, and polymeric (F127) structures) illustrate the diversity of feedstocks used in DIW of ceramics. Printed ceramic "green" parts must undergo a consolidation process. DIW of ceramics and composites requires balancing the different stages in the manufacturing process. The structural or functional performance after consolidation also plays a role in the design process.

of interest (with specific properties: particle size, distribution, surface chemistry, specific surface area, etc.) will require suitable additives, and the adjustment of their formulation (e.g. concentrations, solvent/s, pH, or temperature) to achieve performance considering both the DIW process (Figure 4.2) and the final use of the part (structural and functional properties). DIW of ceramics brings an added level of complexity compared to other DIW applications (e.g. bioprinting) due to the need to compromise the processing and post-processing stages. The design of ceramic feedstocks for DIW aims to minimize the concentration of additives (so they can be eliminated during heat treatment before consolidation at high temperature, Section 4.4) to ensure the maximum density of final parts. There exist reviews [4, 11, 37, 38] and recent meta-analysis [39] of ceramic "inks" for DIW that provide different perspectives on the state-of-the-art and a comprehensive compilation of specific formulations and their applications (e.g. in energy [11] and biomedical applications [4]). In this section, we emphasize the importance of the structure–rheology relationship [40, 41] in the design of printable ceramic feedstocks.

Advanced ceramics and glasses (e.g. alumina (Al_2O_3 [13, 42–44]), titania (TiO_2), zinc oxide (ZnO [31]), calcium phosphate [7], strontium titanate, $SrTiO_3$, yttria-stabilized zirconia (YSZ [45]), silicon carbide (SiC [46]), boron carbide (B_4C [26]), quartz or lanthanum magnesium titanate ($La(Mg_{0.5}, Ti_{0.5})O_3$, LMT [47], and bioactive glasses [48] or 45S5 [49]) do not exhibit the intrinsic plasticity of traditional ceramics, thus requiring careful formulation design to enable dispersion, controlled assembly and shaping [50]. DIW feedstocks often contain high concentrations of ceramic particles with sizes ranging from nm to μm, which require the use of additives to aid dispersion (e.g. citric or ascorbic acids) and binding (e.g. polyvinyl acetate, PVA). In some cases, anti-foaming agents (1-octanol [7]) and rheology modifiers (different cellulose derivatives, collagen, gelatine, or 2D colloids) might be needed to achieve the required properties and rheological behaviors (Figure 4.2). When processing a ceramic feedstock for DIW, the ceramic powders, additives, and solvents will form different microstructures depending on the intrinsic properties of the powders, and the nature of the stabilization mechanism. DIW feedstocks can be classified according to the resulting microstructures in: colloidal systems, emulsions, foams, and polymeric systems (or colloidal gels, including hydrogels and organogels [9, 38, 39]). Although this general classification encompasses many of the existing DIW feedstocks, other approaches might not require powder processing, for example when using preceramic polymer-based inks [51, 52] or a blend of ceramic colloids and polymeric (boro-poly-carbo-siloxane, BPCS) ceramic precursors [46]. There are also examples of flexible feedstocks that can be shaped after the DIW process to enable the fabrication of even more complex shapes [19, 53, 54].

Colloidal "sol–gel" systems rely on a gel transition driven by interparticle or intermolecular interactions. Note that the concept of sol applied here to a ceramic suspension can be argued. This can be achieved through different means, using polyelectrolytes that respond to pH (poly-ethylenimine, PEI [26, 55], and branched copolymer surfactants), or through the combination of different additives (citric acid and ascorbic acid to aid dispersion, polyvinyl alcohol, PVA, for binding, and polyethylene glycol, PEG, as a rheology modifier [45]). "Sol"–gel systems exhibit tunable interactions that enable controlled particle assembly, yield stress and shear-thinning behavior [41]. Carboxymethyl cellulose (CMC [49]) and 2D colloids of graphene oxide (GO [43]) are examples of multifunctional additives used in feedstocks for DIW. They are multifunctional owing to their roles as dispersants and binders to provide the required rheological behavior (Figure 4.2 and Section 4.2.3). Ceramic emulsions [44] and foams [56] are metastable colloidal systems with more complex microstructures, which require the stabilization of the interface between two immiscible solvents (e.g. oil and water) or a solvent and air (in foams).

Colloidal gels or polymeric systems are described as those in which the rheological response is assumed to be governed by a polymer gel matrix containing the ceramic particles [38, 39]. These systems provide gel-embedded suspensions that can be subclassified according to the solvents used in hydrogels (water [16, 24, 27, 29]) and organogels (e.g. ethanol [48] for glass particles using additives: ethyl cellulose and PEG, and propanol-2 using polyvinyl butyral (PVB) and PEG [47]). A common formulation base to design feedstocks for DIW is Pluronic F127 [7, 14, 16, 24, 27, 29], which is also widely used in drug delivery and nanomaterial design. F127 is a polyethylene oxide (PEO) and polyethylene oxide (PPO) triblock copolymer (PEO-PPO-PEO) that forms lyotropic liquid crystalline phases (LLC) in aqueous solutions at certain concentrations and temperatures, exhibiting unique structural transitions [57]. F127 with concentrations of 25 wt% in water forms a "hard" gel above its transition temperature (~18 °C) due to the formation of micelles that arrange in a cubic structure [58]. The bulk rheological properties show a type III strain thinning behavior [59] during an amplitude sweep (oscillatory shear test). Structural studies demonstrate that the hydrogel microstructure evolves when subjected to shear [60, 61]. Due to its common use in DIW, the link between structure, properties, flow behavior, and "printability" of DIW feedstocks based on F127 is the subject of current studies. The behavior of these gel feedstocks can change with particle concentration and composition, sometimes in ways that negatively affect their printability [27, 38, 62] due to changes in the microstructure with the addition of powders. This example emphasizes the importance of the interplay between structure, properties, flow behavior, and "printability," involved in the design process (Figure 4.2) of ceramic feedstocks for DIW.

Feedstocks for DIW variants. Designing formulations with tailored flow behaviors for each specific DIW process is essential to meet the requirements of the chosen method. For example, DIW in air requires the design of yield stress fluids (YSFs) with rheological properties within certain thresholds to ensure good shape fidelity (Section 4.2.3 [27–29, 43]). However, the rheological requirements change in embedded printing where the behavior of two different feedstocks has to be matched (and in this circumstance, thixotropy might be desired [35]). The design process for embedded DIW is the same (Figure 4.2), albeit it has to be applied to both the supporting and extruded feedstocks [35]. Likewise, DIW-UV curing hybrid systems require complex fluids that solidify in situ. Achieving defect-free structures via a photopolymerization-DIW hybrid system requires additional considerations in the structure–rheology design process (Figure 4.2) to achieve ceramic pastes with controlled "reactivity" [21, 33, 63]. Their performance will also be affected by UV-light intensity, curing time, the refractive index of the ceramic material, and other critical factors in the wet processing of ceramics (e.g. concentration, surface chemistry, surface area, particle size, and distribution [50]).

4.2.3 Feedstock Characterization

DIW feedstocks are subjected to complex flows and transitions throughout the printing process; therefore, their rheological study is an essential tool to characterize their behavior and to adapt process parameters (Section 4.2.4) as required. This section introduces YSFs and rheological experiments to characterize them, followed by a critical discussion on the quantification of "printability" and the state-of-the-art in the field linking rheology and "printability."

Introduction to yield stress fluids (YSFs). The rheological response of concentrated ceramic suspensions depends on the timescale (rate) and the level (magnitude) of force imposed. Ceramic feedstocks for DIW are best described as thixotropic elastic-visco-plastic (or TEVP) materials [64]. TEVP fluids show a time-varying or thixotropic response even at a fixed amplitude of forcing [64], viscoelasticity (viscous and elastic behaviors), and plasticity (or ability to deform permanently under stress). These complex materials show a transition (yielding) from solid-like to liquid-like behavior under deformation and are commonly referred to as YSFs. Below the yield stress (τ_y), YSFs behave like a solid, and at stresses above, they "flow". However, not all "yielding" is the same, soft materials can also exhibit brittle and ductile behaviors. These soft materials are at the center of ever-growing research from experimental and theoretical approaches [64–69].

Modeling complex (YSFs): Generalized Newtonian fluids (GNFs) models (that follow the general form: $\tau = -\eta(\dot{\gamma})\dot{\gamma}$ [70]) can represent the flow behavior of YSFs fairly well. For example, by fitting experimental data to a power-law model,

$$\eta(\dot{\gamma}) = K\dot{\gamma}^{n-1}, \tag{4.1}$$

where K, the consistency index, and n, the flow index, are the fitting parameters and descriptors of the liquid regime [65, 71]. A flow index, $n = 1$ corresponds to a Newtonian fluid, when $n < 1$ the material is shear thinning, and $n > 1$ corresponds to shear-thickening behavior. However, GNF models do not consider the possible deformations in the solid regime and do not capture transient behaviours. The Oldroyd model was the first perspective that considered the different features of the material behavior above and below the yield condition [72]. These considerations remain in more recent models, for example, Saramito [73] and de Souza Mendes and Thompson [74]. The latter can also account for time-dependent thixotropic effects. The Kamani–Donley–Rogers (KDR [75]) model unifies the physics above and below the yield stress, accounting for the nonlinear rheology of YSF. A modification of the KDR model includes the brittleness factor (Bt) to differentiate brittle and ductile behaviors [67]. These models offer a theoretical base to advance fundamental understanding of DIW feedstocks.

Measuring the rheological properties of DIW feedstocks: There are many different experiments that can be performed in a rotational rheometer, using different shear conditions (continuous or oscillatory), geometries (e.g. parallel plate, cone plate, cup, and bob), and settings (e.g. shear rate, strain, or stress controlled [70, 76]). Here, we describe four selected protocols (Figure 4.3) using steady and oscillatory shear experiments to provide a set of comprehensive rheological metrics linked to "printability." The reader should also be aware that DIW feedstocks can also be subjected to elongational flows in extrusion processes [41, 77]. Extensional properties can be quantified using different methods to determine the extensional yield stress [77], and the strain-to-break (ϵ_{STB}, the engineering strain at which the filament breaks) [78].

In **steady shear experiments,** a shear force is applied continuously in the same direction. An increasing shear rate is applied to the sample during a flow sweep, allowing time at each point in the sweep to reach a steady state. This is a commonly reported test in DIW literature to analyze the shear-rate-dependent viscosity and provide evidence of the shear-thinning behavior. The experimental data can be fitted (Figure 4.3a) to the Herschel-Buckley model (HB),

$$\tau = \tau_y + K\dot{\gamma}^{n-1}, \tag{4.2}$$

where τ_y is the yield stress [79], which corresponds to the stress value above which the material flows, and that needs to be overcome during extrusion in the DIW process. Feedstocks for DIW show $n < 0$ values and often $n \to 0$. Combining up and down flow sweeps, it is possible to determine the shear history dependence (thixotropy) and the τ_y^{stat} and τ_y^{dyn} values [80]. Thixotropy can also be assessed

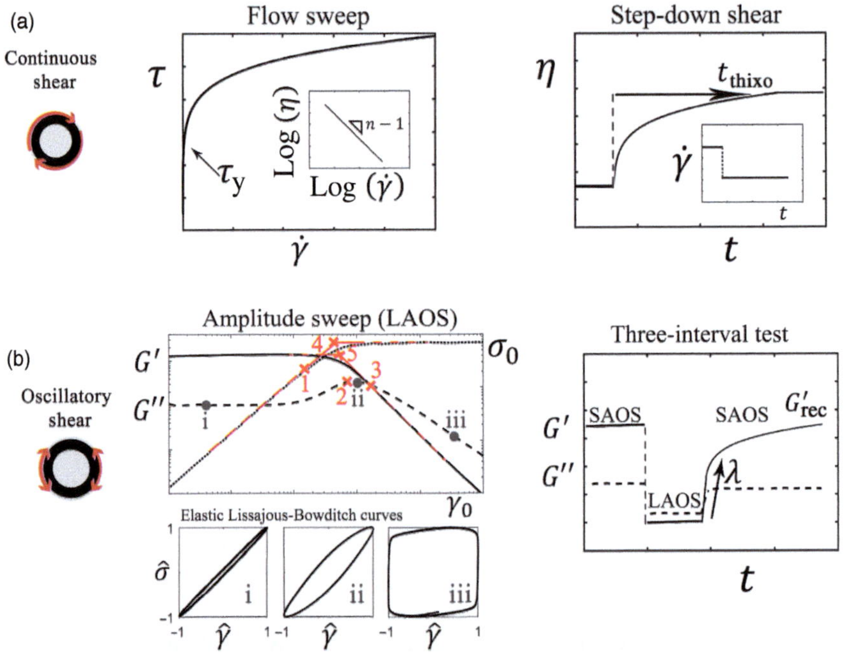

Figure 4.3 Rheological experiments to quantify rheological indicators for ceramic feedstocks. (a) Steady shear tests: flow sweep and step-down steady shear ($\dot{\gamma}_1 > \dot{\gamma}_2$). (b) Oscillatory shear experiments: amplitude sweep (LAOS), and three interval oscillatory test (3ITT, or SAOS-LAOS-SAOS). The points highlighted with red crossed symbols, in the LAOS experiment (labels 1–5), mark the points that have been associated with a "yield stress" value at the: (1) nonlinear onset, (2) G'' maximum, (3) moduli crossover (3, $G' = G''$), (4) σ_0 trend crossover, and (5) G' trend crossover. The insets underneath show the Elastic Lissajous-Bowditch plots for three points in the amplitude sweep (i, ii, and iii). Adapted from Ref. [65].

using step-down steady shear, by first applying a high-shear rate on the sample ($\dot{\gamma}_1$) and monitoring the stress when reducing the shear rate to a smaller value ($\dot{\gamma}_2$ Figure 4.3a). A thixotropic timescale can be determined to assess restructuring using the "thixotropic" stress (τ_{thixo}) and time [41]. Continuous shear experiments are widely used to quantify flow behavior; however, it can often be practically impossible or challenging to collect reliable data using continuous shear when working with very stiff ($G' \sim 1$ MPa or above) printable feedstocks [28]. When the zero-shear viscosities of DIW feedstocks are very high (which can reach up to 250 kPa s), the experimental data can show inconsistent trends (e.g. sudden changes in viscosity or stress, and even negative n values). These signs can flag flow instabilities, banding, wall slip, or fracture [28]. Transient effects at low shear rates and flow instabilities at high-shear rates result in a very narrow window that can be used to determine fitting HB parameters reliably, although data are reported between ~0.1 and 100 s^{-1} [38]; in many cases, data above ~1 s^{-1} can be flawed. Ensuring rigorous data collection and careful analysis are essential to "avoid bad data" [81].

In **strain-controlled oscillatory shear experiments,** a sinusoidal strain waveform is applied to the material, and the stress response is recorded and analyzed. The input signal can be manipulated with varying frequency (f, Hz) and amplitude (γ_0, %), thus providing a versatile and powerful tool to study the viscoelastic, linear and nonlinear fingerprints of ceramic feedstocks [82]. For example, structure evolution tests [43], amplitude sweeps (Figure 4.3b) [16, 62], and oscillatory intervals at varying strains (named 3ITT [38] or SAOS-LAOS-SAOS sequence [27]). Amplitude sweeps probe the material under a wide range of deformations by applying a sinusoidal strain input with increasing amplitude from small to large (SAOS to LAOS), commonly known as LAOS tests. LAOS experiments are widely used within the ceramics DIW field [38] to determine the storage modulus (G'_{LVR}) in the linear viscoelastic region (LVR), and the "yield stress" at the moduli crossover ($G' = G''$). In this chapter, we denote the crossover point as the flow stress, σ_{f}, since it corresponds with the solid-to-liquid or "flow" transition. The LVR (or LVER) is the region at small deformations where the storage and loss moduli are independent of the strain amplitude, the relationship between the stress and strain amplitudes is linear, and the sinusoidal raw signals do not present any distortion.

A note on determining the yield stress. Although the yield stress has historically been the focus of controversial debates and discussions [83], there currently exists a consensus within the rheology and fluid mechanics communities to describe "yielding" as a process, a flow transition from solid-to-liquid like behavior [40, 41, 65, 69]. There is also an agreement on the importance of this yielding transition and its energy transitions (recoverable and unrecoverable). Constitutive models and rheological features of the yielding process can enable the differentiation of brittle and ductile behaviors in soft materials [67, 84]. Despite the positive consensus, some confusion remains since there is not a settled or standard approach to quantify the "yield stress," which also varies with the type of experiment (oscillatory or continuous shear). For example, in an oscillatory amplitude sweep (Figure 4.3b), there are up to five different points (marked with red crossed symbols and labels 1–5) that have been described as the "yield stress" [65]. Using continuous shear, an up-flow sweep provides the static "yield stress" ($\tau_{\mathrm{y}}^{\mathrm{stat}}$) and the reverse downflow sweep provides the dynamic "yield stress" ($\tau_{\mathrm{y}}^{\mathrm{dyn}}$). The reader is encouraged to carefully note the points of reference used in every resource consulted to overcome the confusion caused by the inconsistencies and the nuance of the terminology used.

Advanced LAOS analyses: LAOS results are often reported to qualitatively compare different formulations within a DIW study. However, an analysis of the results based only on the storage and loss moduli (G', G'') with increasing strain amplitude (γ_0) cannot capture distinctive features in the solid-to-liquid (or "flow") transition. Performing analyses on oscillatory raw data signals (stress, strain and time) using different mathematical frameworks, for example, Fourier-transform (FT) rheology [85], the sequence of physical processes (SPP) [68], and recovery rheology [86], to study nonlinear behaviors can provide additional measures to compare the yielding process of DIW feedstocks. When the response of a material becomes nonlinear (outside the LVR), the sinusoidal stress signals become distorted; however, these distortions are harmonic [85, 87]. FT analyses enable the mapping of these higher harmonics $\left({}^{I_n}/_{I_1}\right)$ [62], to accurately identify the onset of nonlinearities (σ_{nl}, γ_{nl}) and other nonlinear material measures [87]. These metrics can be used to quantify the extension (strain space) of the flow transition region (e.g. using inverse of the flow transition index, FTI^{-1} [28]). The Lissajous–Bowditch (L–B) curves [62] provide a visual representation of the raw data and qualitative information of nonlinearities. Examples of elastic L–B curves (plotting, $\hat{\sigma} = {}^{\sigma}/_{\sigma_0}$ versus $\hat{\gamma} = {}^{\gamma}/_{\gamma_0}$) are included with the amplitude sweep to show these curves at three different points in the experiment (i, ii, and iii, Figure 4.3b). The *L–B* curve for point "i" (within the LVR) shows almost a straight line associated with purely elastic behavior. The distorted ellipse in point "ii" (within the flow transition region) means that the response is viscoelastic and nonlinear, while the shape close to a square in point "iii" (in the liquid regime under large deformations) has been associated with "perfect plastic" flow [88]. The visual information provided by the elastic curves can be quantified using the dissipation ratio, ϕ [27, 62, 88]. Non-monotonic ϕ and σ_0 trends with γ_0 associated with structural disruptions during yielding can compromise printability [62, 89]. The SPP analysis enables the quantification of the yielding process in each oscillatory cycle within the sweep, through intra-cycle trajectories of instantaneous moduli, G_t' and G_t'', in Cole-Cole plots [90]. The deltoid shapes, location, and evolution within these Cole-Cole plots for a series of F127 DIW feedstocks provide insights into the different transitions that each formulation undergoes [27].

Oscillatory tests can also be used to determine the extent and timescale of the structural recovery (liquid-to-solid transition) after a large deformation. For example, by applying a three-interval series of oscillatory experiments with small-large-small amplitude strains. This test can be called a SAOS-LAOS-SAOS sequence [27, 28], or a three-interval thixotropic test, 3ITT [38]. The extent of recovery is calculated through the comparison of the storage modulus (G') in the first interval (that should be within the LVR) and the average value in the plateau of the third interval (G'_{rec}, $\frac{G'_{\text{rec}}}{G'_{\text{LVR}}} \times 100$ [27]). While the mutation time, λ [28, 62, 91], enables the quantification of the timescales of the recovery between the second and third intervals. When $\lambda \to 0$, the structural recovery is fast (ideal for DIW), while increasing λ values can compromise printability due to slow restructuring.

"Printability" or DIW performance from a structure–rheology perspective: Structure and rheology are entwined, both determine the feedstock's performance or "printability," a concept that is widely used but rarely quantitatively defined. A single and universally agreed-upon definition of "printability" is still not settled in the field. However, the literature [27, 38, 39, 42, 89] converges on the idea that "printability" is an intricate set of metrics that characterize the ability of the material to: (i) be extruded (ceramic feedstocks must flow through narrow nozzle tips, in a continuous and steady state), (ii) retain its shape, support its weight (without slumping or collapsing), and (iii) achieve dimensional accuracy in the final printed structure (shape fidelity).

Some studies in DIW of ceramics assess "printability" based on rheological parameters that describe the flowing material (e.g. HB fitting parameters, τ_y, K and n), while others use rheological measures of strength associated with the solid-like behavior (e.g. G'_{LVR}, σ_f, and the figure of merit G'_{LVR}/σ_f [42]). Smay et al. proposed the first model to predict the printability of ceramic feedstocks to print structures with unsupported spans [92]. Smay's model can be rearranged to define a dimensionless parameter, S (Eq. (4.3) [39]), which is calculated based on rheological metrics associated with the "strength" of the equilibrium structure.

$$S = \frac{G'_{eq}}{1.4\rho g\left(\frac{L}{D_t}\right)^4 D_t} \geq 1, \tag{4.3}$$

where G'_{eq} is the equilibrium storage modulus, ρ is the density of the ink, g is the gravitational acceleration, L is the span length, and D_t is the nozzle tip diameter, and 1.4 is the shape factor for circular filaments. $S > 1$ is indicative of no deflection and therefore that the feedstock is printable. An alternative "printability" criteria was proposed by Barki et al., through the printability number, Ξ [80]. According to this model, an optimal yield stress is required to retain the shape of printed parts to counter the slumping and deformation caused by gravity and surface tension. The outcome of a recent meta-analysis on DIW "inks" [39] is a new printability criterion based on two rheological properties, G'_{eq} and σ_f (equivalent to $\tau_{y/FP}$ [39]) and their product (from Chan's criterion [44]), that provide specific thresholds to mark the boundaries of a "printability" region around existing data.

These models and criteria provide useful guidelines in feedstock design to determine a "strength" threshold required to achieve self-supporting structures. However, they cannot account for the multiple features described in the "printability" definition. When used in isolation, "strength" rheological indicators cannot capture: (i) the characteristics of the material before and after "yielding," (ii) distinctive yielding processes (transient behaviours), (iii) the timescales of the "post-yield" restructuring, (iv) the sensitivity to process parameters, and (v) the loss of printing accuracy due to microstructure disruption at the nozzle tip. A quote from a recent meta-analysis articulates this reflection: "*the 'printability' criteria… may not fully capture the complex rheological behaviors of the inks and their relation to printing*" [39], thus pointing out a growing research area [27, 78, 86, 93]. Understanding transient nonlinear rheology [86], the interplay between structure–rheology and flow [27], and their link to "printability" (including shape fidelity) is key to DIW processes. Different states of the material [64] in the solid state, liquid state (flow), and the nonlinear transitions in between [65, 67, 86], yielding and restructuring [27, 61], must be better understood. Extensional properties should also be studied [77].

Recent progress linking rheology and "printability": Recent studies [27, 40, 41, 93] convey that the DIW process can be associated with at least three material characteristics and rheological criteria: ability to "flow" (yielding and flow through the nozzle and tip), "recover" (restructuring, transition from liquid-to-solid), and achieve strength to build a free-standing unsupported object. Another aspect associated with printability is "drawability," or the ability to stretch the feedstock to retain the filament shape [78]. We adopt emerging terminology for these criteria: "flowability," "recoverability," "material strength," and "drawability." Alternative terms for "flowability" currently used within ceramics DIW are: "pumpability," and "extrudability" [38]. The term "buildability" is also used in some studies instead of strength [78]. The "printability" description can thus be articulated using four inter-related criteria, each associated with relevant rheological metrics for the DIW process:

1) "Flowability" represents the solid-to-liquid transition, which is the ability of the material to yield and steadily flow under an applied shear force without disruptions (no phase separation, clogging at the tip or discontinuous flow) [27, 93]. "Flowability" rheological descriptors are: σ_{nl}, σ_{f}, $\mathrm{FTI}^{-1} = \frac{\sigma_{\mathrm{f}}}{\sigma_{\mathrm{nl}}}$ determined from LAOS [27]; and τ_{y} and n from steady shear [38, 39]. However, these metrics should be interpreted in context with the trends in material functions: $G'(\gamma_0)$, $G''(\gamma_0)$, $\sigma_0(\gamma_0)$, and $\phi(\gamma_0)$ that can flag signs of microstructural disruption, for example, is there a stress overshoot in the solid-to-liquid transition before the crossover [62]? Does ϕ show a non-monotonic trend [27, 62]? Does the feedstock exhibit type I or type III LAOS behavior [85]? FT and SPP frameworks provide further insight into nonlinear transient yielding. An Ashby-type property map based on ϕ and FTI^{-1} values has been proposed to compare the "flowability" of different formulations, showing abrupt or gradual yielding [27].
2) "Recoverability" is the ability of the material's microstructure to reform, rebuild, or "recover" enough elasticity (e.g. G'_{rec} or G'_{eq}) in short timescales (e.g. λ, mutation time) to become solid-like upon deposition and retain the filament shape accurately (no spreading). "Recoverability" [27, 93] can be quantified using G'_{rec}, and λ from oscillatory shear tests (three intervals, Figure 4.3b), or τ_{thixo} and the thixotropic time, t_{thixo} [41] from steady shear. Ashby-type property maps have been proposed using $\%G'_{\mathrm{rec}}$ versus λ_{I} [27], and reduced metrics from thixotropy spectra analyses [94].
3) "Material strength" is the ability of the material to withstand the forces imposed by gravity and surface tension, the weight of upper layers in the 3D part, and retain the features of unsupported features (spans and overhangs). "Strength" can be quantified through rheological descriptors: G'_{rec} and σ_{f}, which are already considered in existing printability criteria (e.g. printability number, Ξ by Barki et al. [80], the figure of merit by Feilden $G'_{\mathrm{LVR}}/\sigma_{\mathrm{f}}$ [42], G'_{eq} by Smay et al. [92] and an inverse gravitational Bingham number, $\frac{\rho g h}{2\sigma_{\mathrm{f}}} < 1$ [86]). Some of these criteria include the diameter of the nozzle tip (e.g. Ξ [80], and G'_{eq} [92]), assuming that the printed filament retains this dimension. This aspect depends on shape fidelity, often qualitatively assessed through geometrical deviation from the computer-aided design (CAD).
4) "Drawability" (gap-spanning ability) is the ability of a material to be stretched into a straight, cylindrical filament, while maintaining high shape fidelity [78]. This criterion is linked to extensional properties, or the "extensibility" of YSFs. "Drawability" or "stretchability" can be quantified using the extensional strain-to-break (ϵ_{STB}), critical yield stress in elongation ($\sigma_{\mathrm{c,elong}}$), and draw ratio ($DR$). An Ashby-type "printability" map is proposed to plot the ϵ_{STB} versus σ_{f}, or alternatively, $\sigma_{\mathrm{c,elong}}$ versus σ_{f}.

"Printability" quantification through shape fidelity analyses: "Printability" has also been defined as a print quality criterion (or shape fidelity [95]) without consideration of rheological descriptors. Some studies assess shape fidelity through qualitative assessments of the geometrical deviation [38], comparing the printing dimension (D_p) and the predicted dimension (D_{pre} [63]), or the relative loss of predesigned macropores [27]. The printed dimensions of the filament diameter can be quantified through imaging techniques, using optical [96], scanning electron microscopy [29], or X-ray tomography [22]. Performing image analysis on a pattern or structure with specific features (pores or spaced lines at a certain pitch) can help avoid subjective measurements and experimental errors [97]. A recent study on DIW of cements points out that the alignment method (between the scanned print and model) is a source of biases when quantifying print quality, which can be addressed through automated image processing and alignment methods [95].

Without dismissing the importance of accurately quantifying shape fidelity or print quality, there seems to be a common "qualitative" or high-tolerance approach in the ceramic DIW context to assess geometric deviation of printed parts. More emphasis has (so far) been put on feedstock design and characterization, perhaps owing to the need to compromise shape fidelity with structural and functional performance. It must also be noted that feedstocks for DIW can also behave differently and result in varying shape fidelity "scores" (subjective or quantitative) depending on the process parameters (Section 4.2.4) and the technical specifications of the DIW system used (Section 4.2.2). Although different studies diverge on the chosen criteria (rheological or geometrical) to assess printability, overall, the literature converges on the idea that there is not a single criterion, but rather, a combined set of inter-related rheological criteria (ability to "flow/extrude," "recover/rebuild," "build/strength," and "draw/stretch"), shape fidelity, and process parameters.

4.2.4 Printing Parameters

Systematic studies and explicit rationale on the choice of parameters in DIW of ceramics are important aspects that require further and more careful study within the field. Here, we introduce fundamental concepts underpinning the DIW process and signpost specific sources in relevant areas: extrusion, fluid mechanics, and YSFs. Choosing the DIW settings or process parameters should be done with careful consideration of non-Newtonian fluid mechanics concepts [70, 98]. DIW is, in essence, the flow of a complex fluid through a contraction obeying the continuity equation (mass conservation principle). The volumetric flow, Q, is constant through the contraction, which results in Eq. (4.4) for an incompressible fluid.

$$Q = u_e A = \text{constant}, \tag{4.4}$$

where u_e is the extrusion velocity (m/s, velocity of the moving syringe plunger with a cross section, A (m^2)). From Eq. (4.4), the average velocity at the nozzle tip (u_t) is calculated using Eq. (4.5),

$$u_t = u_e \left(\frac{D_e}{D_t} \right)^2, \tag{4.5}$$

where D_e and D_t are the diameters of the cross section of the plunger (before the contraction) and of the nozzle tip (after the contraction). Considering ideal flow conditions (likely not realistic in practice) through two infinite parallel plates (Figure 4.4a), we assume that the extrusion flow during the DIW process reaches steady state is fully developed and laminar. Steady state means that the velocity profiles do not change with time $\left(\frac{du}{dt} = 0\right)$; fully developed means that the velocity profiles do not change along the "pipe," in the x direction, $\left(\frac{du}{dx} = 0\right)$; and laminar flows are expected given the small diameter of the nozzles and high viscosity of DIW ceramic feedstocks (which results in very small Reynolds numbers, $Re = \frac{\rho u D}{\eta} \ll 1$). This idealization enables the derivation of a simple dimensionless equation to assess the impact of printing settings (e.g. flow rate and nozzle tip diameter) on the shearing profile acting on the material (feedstock) as it passes through a narrow nozzle. Considering Newton's laws of motion and applying a force balance for this idealized pressure-driven flow, we obtain that the driving forces due to the pressure difference must be equal to the resistance forces due to friction on the wall (shear stress, τ, Pa). Assuming that when the feedstock is flowing, it can be represented by the power-law GNF model (Eq. (4.1) can be rewritten as $\tau = K\left(\frac{du}{dy}\right)^n$, where n is 1 for Newtonian fluids, <1 for shear thinning, and >1 for shear-thickening materials [99]), a dimensionless equation to calculate the velocity profiles can be derived (4.6):

$$\hat{u} = 1 - \hat{y}^{\frac{n+1}{n}}, \tag{4.6}$$

where $\hat{u}$ is the dimensionless velocity (u/u_0, and u_0 is the maximum velocity at $y = 0$), n is the flow index, and $\hat{y}$ is the dimensionless height (y/r, Figure 4.4a). The dimensionless velocity and shear profiles (Figure 4.4b) strongly depend on the flow index value, showing parabolic velocity profiles for Newtonian fluids ($n = 1$). When $n \to 0$, for example, for DIW feedstocks with $n = 0.1$ [28], the velocity profiles correspond to plug flow. Small n values, reducing the tip diameter and increasing the extrusion velocity, lead to increasing shear rates acting on the material (Figure 4.4c).

The reader is advised that this is an idealized and oversimplified approach to estimate the shear profiles with the only purpose of qualitatively illustrating the impact of process parameters on the shear forces acting on the feedstock (Figure 4.4). Understanding the flow of non-Newtonian (complex) fluids is a vast and active research field; many of these studies are relevant for DIW of ceramic feedstocks. A study combining magnetic resonance imaging (MRI) and finite element simulations using a generalized constitutive equation [100] provides a detailed view of paste extrusion and the characteristic flow of a model YSF (Carbopol) through a contraction. The challenge is that ceramic DIW formulations do not always behave as a "model" YSF and often exhibit TVEP behaviors [64]. Many efforts are dedicated to advancing fundamental understanding of non-Newtonian flows with huge industrial relevance. For example, studying complex flows of GNFs [100] and the diverse yielding behaviors that soft materials can exhibit [65, 67–69, 102, 103].

Additional process parameters that must be considered in DIW are: the height of the nozzle tip from the substrate (H, Figure 4.4d), the velocity of the $X-Y-Z$ Cartesian robot or moving platform (u_m), the extrusion velocity at the plunger (u_e),

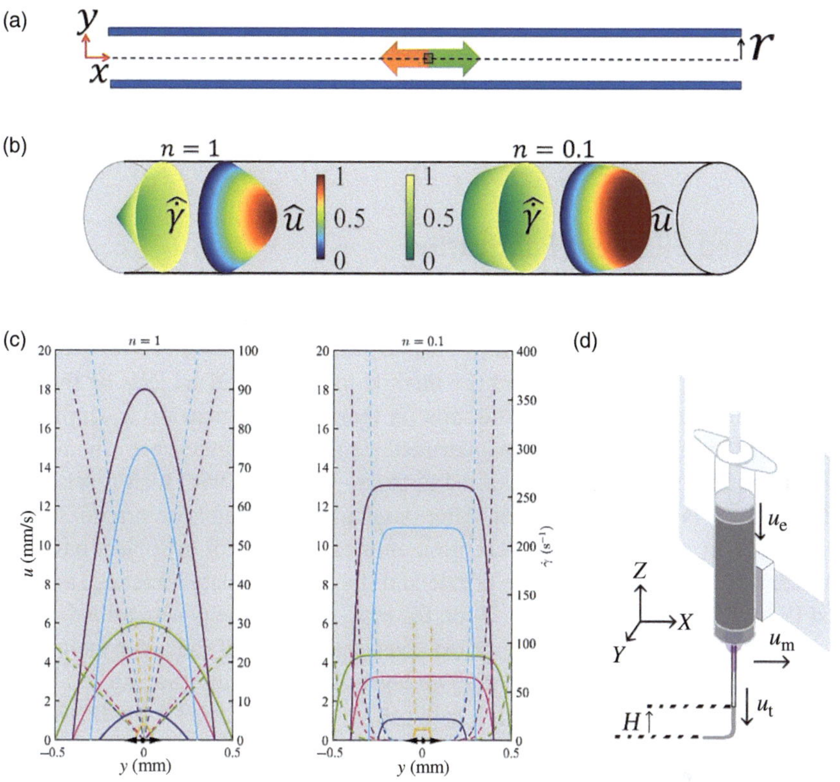

Figure 4.4 Selection of process parameters in DIW. (a) Schematic of a simplified 2D flow through parallel planes showing the force balance used to derive Eq. (4.6). (b) Dimensionless 3D shear rate ($\hat{\dot{\gamma}}$) and velocity ($\hat{u}$) profiles for flow index (n) values of 1 (Newtonian) and 0.1 (plug flow). (c) Dimensional examples of shear rate ($\dot{\gamma}$, dashed lines) and velocity profiles (u, continuous lines) for different tip sizes (with diameters ranging from 100 to 800 μm) and printing velocities (u_t, varying between 1 and 12 mm/s). (d) Scheme showing the process parameters that determine printing patterns (u_e, u_t, u_m, and H).

and at the nozzle tip (u_t) [104]. The ratio of Cartesian velocity (u_m, or plane velocity of a moving platform [104]) to volumetric extrusion rate (Q) and the height rescaled by the nozzle diameter ($\frac{H}{D_t}$) leads to different patterns that can be predicted theoretically: droplets, discontinuous lines, continuous lines, meanders, and alternated loops [104]. These parameters also play a role in the origin of corner defects [105] due to constant extrusion parameters during acceleration and deceleration. Although u_p and u_e are dependent through the continuity equation, printing settings within DIW software (e.g. the "pumping scalar" [27, 89]) can alter this relationship "on the fly" to adjust the printing pattern of the extruded filament. This is, of course, practically useful; however, we encourage those working on DIW of ceramics to perform a thorough, systematic study of process parameters and the feedstock's rheological properties from an experimental and theoretical perspective.

Despite the admitted idealized considerations explicitly stated, the content here presented shows that the selection of process parameters in DIW (e.g. flow rate, nozzle tip diameter and length, print velocity and height) will have an impact on two aspects: (i) the shearing conditions and state of the material (Figure 4.4b) and (ii) printing patterns of the filament. Process parameters also play a role in "printability" and shape fidelity, and consequently on the performance of ceramic parts.

4.3 Fused Filament Fabrication

Fused deposition modeling (FDM) goes back to S. Scott Crump [106–108]. To avoid trademark issues, RepRap community created in 2005 the term FFF. Over the years, the process was still renamed, and the method to fabricate 3D objects by thermoplastic material extrusion is still the same. A thermoplastic material is supplied to a printhead, where it is melted and extruded via a nozzle. Underneath the nozzle, the extruded filament is then deposited in the x–y direction with high precision to generate a 2D layer of a 3D object. After passing the orifice of the nozzle, the filament cools and solidifies. Once the programmed 2D layer of the object has been printed, the distance between the nozzle and the printing plate increases, and the next layer of the object is printed. Upon the placement of the new extruded strand (e.g. filament) on the previously printed layer, the surface of the latter melts once more, thus causing the fusing of the two layers [109]. During the late 1990s and early 2000s, researchers initiated the application of this process for the fabrication of metal and ceramic components [110]. Similar to the preceding section, the present section is organized in accordance with the discussion of the various printing heads (Section 4.3.1), the feedstock preparation (Section 4.3.2), the feedstock characterization by looking on the rheological aspect (Section 4.3.3), and, finally, the printing parameters that are required to achieve higher mechanical strength by eliminating printing defects (Section 4.3.4).

4.3.1 Printhead Systems

Over the years, different printheads have been developed for the thermoplastic MEX AM process. At the beginning of the FFF process, Bowden (Figure 4.5a) and direct dive extruders (Figure 4.1b) were developed for the purpose of feeding the thermoplastic-ceramic filament into the heating zone of the printhead. Ceramic filaments with a diameter of 1.75 or 2.85 mm are still fabricated by the thermoplastic extrusion process [111]. Analogue to the fabrication of polymer filaments, the precise diameter can be achieved in a post-processing stage by adjusting the speed of the winding unit with an optical sensor that measures the diameter of the filament. In a filament printer based on a Bowden extruder, the driving wheels that transport the filament into the printhead are close to the filament spool. A plastic tube between the driving wheels and the printhead end guides the ceramic filament. Based on this design concept, the printheads are light, and a higher print speed can be achieved. However, the filament brittleness makes it difficult to print 1.75 mm filaments with such a printer [112]. Today, direct drive printheads are more favored

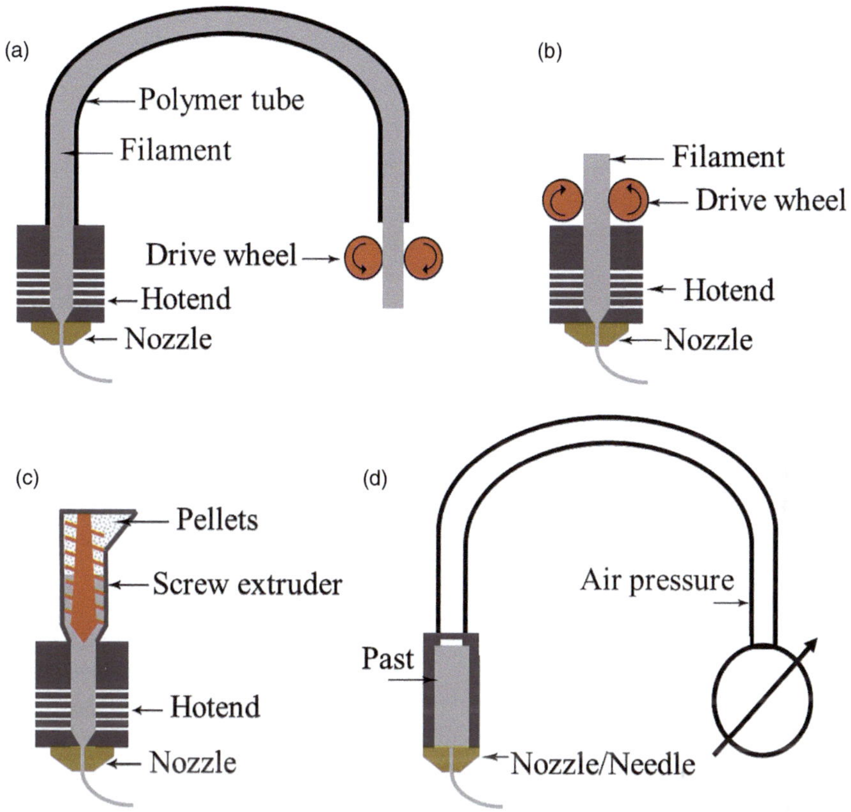

Figure 4.5 The scheme illustrates the various printheads utilized in the thermoplastic MEX process. The feeding of the ceramic feedstocks, which is based on a mixture of thermoplastic additives and ceramic powder, can be fed into the hot zone of the printhead by multiple methods: (a) Bowden, (b) direct drive, (c) auger, and (d) piston extruder.

for ceramic filament printing [113–116]. The drive wheels are located on the printhead above the heat sink. Filament printers are user-friendly devices that allow straightforward material changeovers.

The cost of ceramic filaments remains significantly higher, at almost double the price of ceramic pellets for ceramic injection molding. Consequently, printheads with an auger extruder become a more attractive option for the thermoplastic-based MEX AM process [117–121]. The extruder head is designed similarly to extruders inside an injection molding machine, incorporating a hopper and a screw (Figure 4.5c). In this process, pellets are used to feed the printhead, rather than filaments. The pellets are filled in a hopper and then transported to the orifice of the printing nozzle by an auger. Bellini et al. already demonstrated in 2005 the use of an auger extruder for the 3D printing of ceramics [122].

The working principle of piston extruder heads is similar to DIW, where a pressurized gas or plunger will push the material through a nozzle or needle (Figure 4.5d). Scheithauer et al. introduced a printhead utilizing a heated cartridge designed for printing a ceramic paste [123].

4.3.2 Feedstock Preparation: Additives and Processes

In previous years, terms like fused granular fabrication (FGF) [121], fused "deposition of ceramics" (FDC) [124, 125], FDM [126–128], FFF [115, 129–135], composite extrusion modeling (CEM) [136, 137], mini extruder deposition (MED) [122], extrusion deposition additive manufacturing (EDAM) [138], and fused feedstock deposition (FFD) [139] have been used for the thermoplastic based ceramic MEX AM process. However, the process from powder to thermoplastic-ceramic pellet or filament for use in the MEX AM process remains unchanged.

In a first step, a mixture of thermoplastic materials is mixed with ceramic powder. The mixture of thermoplastic-ceramic composite is known as ceramic feedstock. Due to the inherent properties of thermoplastics, the mixing process is conducted at a temperature above the melting point of these polymers. As shown by Clemens, different compounding methods have been used to achieve ceramic feedstocks [111]. An internal mixer or torque rheometer is a process that is carried out in batches. First, the thermoplastic materials are melted in the mixing chamber before the ceramic powder is added. The mixing process can be monitored and used to identify the right combination of organic additives and ceramic powder.

Roll mills and shear roller mixing are also two devices often used to prepare the ceramic feedstock [111]. In both equipment, the mixing process takes place in the gap (continuously adjustable) between two counter-rotating heated rolls. Due to the high-shear force and the continuous process involving the shear roller mixing system, this device is nowadays almost everywhere to compound feedstocks for the ceramic industry. The material is transported from one end of the rolls to the other, and at the end, a granulator element takes place to pelletize the homogenized feedstock. In order to feed a roll mill or a shear roller mixing system, a premix of the thermoplastic additives and the ceramic powder is required.

Alternative double or single screw extruders are used for the compounding of ceramic feedstock. A premixture of the ceramic feedstock is placed in the hopper above the feeding zone of the extruder. An alternative option is to use a gravimetric or volumetric dosing system to bring the raw materials into the feeding zone of the extruder. Subsequently, the screw draws the mixture into the heated mixing zone where the thermoplastic components melt, and the feedstock is homogenized under high-shear conditions. A variety of screw part elements (e.g. feed, transition and dispersive mixing elements, kneading blocks, and gear mixers) are used to achieve optimal homogenization of the feedstock. Granulation of the feedstock occurs at the end of the extruder or by using strand pelletizers behind the die orifice.

Empa has developed a low-shear mixing process for use with small quantities of thermoplastic feedstocks [140]. In this process, a blend of thermoplastics is dissolved in a suitable solvent and then mechanically mixed with ceramic powder. Following the evaporation of the solvent, the feedstock is sufficiently homogeneous to enable fiber extrusion with a diameter below 500 μm.

As previously stated, a ceramic feedstock is typically composed of a blend of thermoplastics and a ceramic powder [141]. Quackenbush et al. classified these thermoplastic binder additives into four categories [142]. The major binder component, now called the backbone polymer, determines the general range of properties of the final binder; the minor binder component is easily removed, and its removal creates the pore structure during the first debinding step. Due to its low molecular weight, this component also reduces the viscosity of the raw material. Often, the minor binder component can reach up to 60 wt% of the total binder content to achieve an open interconnected pore structure after the first debinding step. Thereby, nowadays, the component is called the primary binder; the plasticizer reduces the viscosity of the raw material, allowing the final product to flow easily; the surfactant promotes the wetting of the ceramic powder with the various thermoplastic components. It is well known that a thermoplastic additive can fall into more than one of these categories. A good example of this is stearic acid. Stearic acid is a surfactant, but it can also act as a plasticizer when a polyolefin is used as the main binder component. Countless binder systems have been reported for the thermoplastic MEX AM process. Therefore, Table 4.1 summarizes commonly used thermoplastic additives from the literature and groups them into categories with specific functions.

In addition to the binder additives summarized in Table 4.1, Sarraf et al. investigated crosslinking agents to prevent deformation of the printed ceramic parts during the thermal debinding process [143]. The use of a crosslinking additive for the preceramic silicon resin and the thermoplastic binder ethylene vinyl acetate (EVA) significantly reduced deformation during thermal treatment.

Table 4.1 Summary of commonly used thermoplastic additives for thermoplastic MEX AM feedstock formulations in literature.

Category	Function	Examples from literature
Major binder component (backbone component)	• Mechanical strength • Shape stability	LDPE, HDPE, PP, APO, PA, EVA, PLA, PCL, ABS, PEEK, PS, PVB, EEA, PEO, PPO, polyepoxides, POM, PCP, PU
Minor binder component (primary binder component)	• Decrease viscosity • Improve debinding process	Paraffin waxes, Carnauba wax, PE-waxes, PEG, PVA, PEGMA
Plasticizer	• Increases plasticity • Increases flexibility • Decreases viscosity	• DBP • Mineral oil
Surfactant	• Increases the wetting • Increase the interface between powder and polymer	• Fatty acids (e.g. stearic acid, oleic acid, lauric acid) • Oleylamine • Octadecylamine
Elastomer	• Increases flexibility	• TPEs, SEBS, TPU
Tackifier	• Increases adhesion during printing	• Resins (e.g. terpenes, rosins)
Commercial feedstocks and thermoplastic binders	• Easy to use • Mixing with powder	• Inmafeed • Inmaflow • Embemold CC • Embemold K83 • Kcmix 3.3

4.3.3 Feedstock Characterization

Similar to DIW, rheological analysis is used to characterize the flow behavior of the thermoplastic feedstock. One method of analyzing flow behavior is to use a torque rheometer, also known as an internal mixer [144]. Given that the measured torque (M) is related to the applied shear stress (τ) and the roller speed (N) is proportional to the shear rate ($\dot{\gamma}$), the apparent viscosity can be calculated by:

$$\eta_{\text{app}} = \frac{\tau_{\text{app}}}{\dot{\gamma}_{\text{app}}} = \frac{K_1 M}{K_2 N}, \tag{4.7}$$

where η is the apparent viscosity, and K_1 and K_2 are constants that depend on the dimensions of the rheometer [145]. A torque rheometer can be fitted with sigma, roller, Banbury, delta, or cam rotors. With roller rotors, the apparent shear rate can be calculated by

$$\dot{\gamma}_{\text{app}} = \frac{2\pi r N}{60y}, \tag{4.8}$$

where r is the roller diameter, and y is the gap between the roller and the chamber wall. It can be concluded that the smallest gap between the roller and the chamber wall provides the maximum shear rate, and that the widest gap provides the lowest rate.

Another method of characterizing the viscosity of a feedstock for the thermoplastic MEX process involves the use of a capillary rheometer [146–148]. To investigate the apparent viscosity of a thermoplastic-ceramic feedstock, a capillary with a specific diameter-to-length (l/d) ratio is utilized [146]. In order to calculate the real viscosity in accordance with the Cogswell theory [149], a short (e.g. zero length) die of the same diameter is required. The inlet pressure drop can be determined from the short die, and the true viscosity can be calculated. Alternatively, the inlet pressure drop can be extrapolated using the Bagley method [150]. The inlet pressure is extrapolated by plotting the pressure against the l/d ratio. The measured pressure is used to calculate the shear stress and, later, the viscosity.

$$\tau = \frac{(P_{\text{L}} - P_0)r}{2L}, \tag{4.9}$$

$$\dot{\gamma} = \frac{4Q}{\pi r^3}\left(\frac{1}{4}\left(3 + \frac{d\ln\dot{\gamma}_{\text{a}}}{d\ln\tau_{\text{R}}}\right)\right), \tag{4.10}$$

where P_{L} and P_0 are the capillary pressures measured at a given length (L) and zero length, respectively. Q denotes the volumetric flow of the feedstock, while r signifies the radius of the capillary. The term enclosed in the bracket in Eq. (4.10) is known as the Weissenberg-Rabinowitsch correction, which is used to determine the true shear rate.

As an alternative approach, the flow behavior of the thermoplastic feedstock can be determined using a micro compounder [119, 151]. This micro compounder comprises a conical co-rotating twin screw that functions to transfer the feedstock material from the feeding zone to a bypass pneumatic valve [152]. In this configuration, the material can either be extruded or directed to an integrated back channel for circulation. In the circulation mode, the rheological properties of the feedstock can be recorded. Unfortunately, this device can only determine the apparent viscosity and apparent shear rate.

Finally, the rheology can be analyzed by a rotational and oscillatory parallel plate rheometer, as outlined in Section 4.2.3 [128, 153]. Gorjan et al. demonstrated how the content of surfactant affected the flow behavior of the MEX AM alumina feedstock [154]. It was evident that an alteration in rheological behavior occurred when the stearic acid content exceeded the threshold required for the complete saturation of a monolayer of stearic acid on the powder surface. Below the threshold, the viscosity of the feedstock could be better described by the Cross and Carreau/Gahleitner model. Conversely, above the threshold, the Herschel-Bulkley and power-law models provided a more accurate description of the flow behavior.

As shown in Section 4.2.3, the power-law rheological model is a simple yet frequently effective two-parameter model. In contrast, the Herschel-Bulkley model is a widely utilized three-parameter model that is employed to describe the viscosity of fluids with a yield point as a function of shear rate. It is particularly effective for suspensions with a high concentration of particles, which often exhibit shear-thinning and yield point behavior. The Cross and Carreau/Gahleitner models are typically employed for polymers with low to moderate concentrations of particles. In both models, a parameter η_0, which is associated with the viscosity plateau at low shear rates, can be calculated. A surfactant at a concentration that falls below the threshold can lead to undesirable flow of the printed components during the thermal debinding process, as shown by Gorjan et al. [113].

Hanemann et al. used oscillation measurements to investigate the rheological behavior of zirconia feedstocks with different fatty acids [155]. A significant correlation was identified between the various fatty acids and the resultant viscoelastic flow behavior of the feedstocks. Nötzel et al. used the same method to verify the optimal concentration of stearic acid for their zirconia filaments [148]. In all of the materials used, a crossover of the storage and loss moduli was observed. Therefore, the initial printed layers were able to resist the applied pressure by overprinting the subsequent layers, thereby maintaining their position and shape. The results are in good agreement with Gorjan et al. [154].

4.3.4 Printing Parameters

Rane et al. investigated the effect of printing parameters on the mechanical properties of Al_2O_3 and ZrO_2 ceramics [156]. They investigated the printing orientation and the printing speed on the Vickers hardness, flexural strength, and fracture toughness of the sintered Al_2O_3 and ZrO_2 parts. The vertical printed parts had a slightly higher hardness value than the horizontal parts. However, due to the high deviation, the differences were not significant. The flexural strength of the samples built in horizontal orientation was found to be higher than that of vertical orientation. Similar results have been observed by Cano et al. for ZrO_2 ceramics [157]. However, it is worthwhile to mention that the density after sintering was strongly affected by the printing orientation. ZrO_2 ceramic parts, printed with 0° orientation, resulted in relative densities of 98.73% and 98.23% after printing and sintering, respectively. The parts printed with an orientation of 90° resulted in a significantly lower relative density, namely 96.35 and 95.82%. Therefore, it can be assumed that this effect is mainly caused by defects during the printing process.

Smirnov et al. investigated the statistical relationship between the 3D printer nozzle diameter, layer height, and filling on the mechanical strength of ZrO_2, as determined by a biaxial bending test in accordance with the ISO 6872:2019 [158]. Statistical analysis methods were used to establish the relationship between printing parameters and calculated strength values. They conclude that a higher layer height is preferred to achieve higher mechanical strength because fewer printed horizontal plies result in fewer voids between them. The use of a "line" infill type resulted in the highest mechanical values due to better fusion. In their studies, a print height of 0.4 mm with a nozzle diameter of 0.6 mm gave the best mechanical results.

Several researchers have used the design of experiment (DoE) or machine learning algorithms to predict a relationship between the printing parameters and the mechanical strength.

There are many challenges during MEX AM fabrication, which depend on the ceramic feedstock properties and printing parameters. Anitha et al. focused on optimizing the surface quality of the FDM process [159]. Therefore, they used Taguchi's analysis to investigate three parameters, namely the road width, build layer thickness, and speed of deposition. The material flow on the extrusion nozzle has been analytically modeled by Bellini et al. and Venkataraman et al. [124, 160]. A compendium of variables that must be given full consideration when undertaking optimization of printing parameters is provided in Figure 4.6.

As previously stated, the nozzle diameter, layer thickness, infill type (e.g. line, zigzag, and concentric), infill density, printing orientation (e.g. raster angle) and speed all have a significant impact on the final mechanical properties of the ceramic parts. However, there are several other printing parameters that must be taken into consideration in order to achieve an optimal printing result.

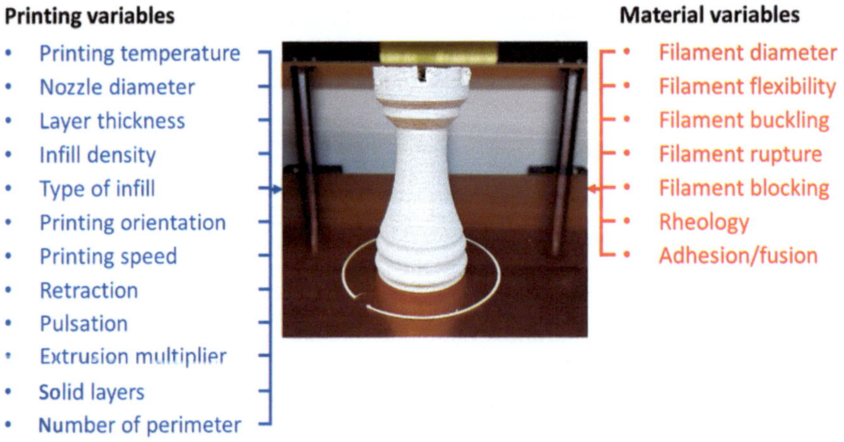

Figure 4.6 Overview of printing and material variables that affect the quality of the thermoplastic MEX printing process.

Printing temperature is a parameter that can be easily optimized to achieve a reliable printed object [154]. In the context of printing heads that are equipped with an auger extruder, the issues of retraction and pulsation are of particular significance. These phenomena have the capacity to compromise the resolution of the printed object and its mechanical strength. The extrusion multiplier is a key component in the analysis of extrusion processes, as it serves to define the extrusion flow rate. This parameter is critical in understanding the output of the material through the nozzle over a given time unit. In the process of manufacturing lightweight ceramic structures by means of the MEX technique, the application of a solid layer to the uppermost and nethermost surfaces is imperative [114]. The utilization of an infill pattern characterized by an infill density below 100% is furthermore required between these two aforementioned layers. An increase in the thickness of the solid layer is directly proportional to an enhancement in the mechanical properties of the lightweight ceramic component. Three-dimensional prints possess an outer shell, known as a perimeter. The number of perimeters has a direct impact on the mechanical strength of the ceramic components. Frequently, disparities in shrinkage between the perimeter and the infill, attributable to particle orientation, give rise to internal stresses that diminish the mechanical strength of the ceramic components.

The process of printing with a MEX AM printer is contingent on several material properties. In order to avoid the occurrence of air gaps between the printed lines, it is imperative that the diameter of the filament is maintained with the utmost precision. Deviations in filament diameter result in the formation of uncontrolled voids between printed lines. The ability to transport filament through the printing machine is a significant consideration; therefore, flexibility of the ceramic filaments, even with higher powder concentration, is required. Buckling happens where the feeding system meets the inlet of the hotend on the printer, so the feedstock shouldn't melt all of a sudden. If the filament is too brittle or if platelet-like particles are used, the high friction in the gear motor can cause it to break (filament rupture). MEX AM printers have an electrical powder limit. This means that the material can only be fed into the hotend if the right viscosity is adjusted. The thread that is pressed out and the layer that was printed underneath must stick together well so that the two layers can fuse during the remelting of the printed layer.

4.4 Heat Treatment and Post-Processing

AM of ceramics involves multistep processing from 3D shaping to consolidation. The MEX AM via DIW or FFF produces "green parts" which must undergo subsequent post-processing for debinding and sintering. Thermal treatments can lead to substantial changes in composition and properties, for example, through phase transitions, liquid phase formation, and uncontrolled grain growth. The theoretical basis for the selection of sintering conditions involves considerations of densification mechanisms, microstructural evolution and the thermodynamics of phase transformations [4].

4.4.1 Post-Processing for DIW-Printed Ceramic Parts

Post-processing of DIW-printed parts includes drying, debinding, and sintering conditions. The microstructure of HA scaffolds made via DIW can be tailored through the selection of starting powders, drying, debinding, and sintering conditions [7]. Another example of microstructure control is partial sintering of zirconium diboride, ZrB_2, to create porosity at the microscale [18]. Using carbon as a sintering aid and firing temperatures at 2275 °C, boron carbide complex parts with high densities can be achieved [17]. Phase transformations and reactions play a critical role, for example, using preceramic polymers for SiC requires specific temperature holds during the sintering process to facilitate low-temperature thermal crosslinking [161]. Constant heating rates at 5 °C/min are used for complex Al_2O_3 structures to prevent defects due to gas by-products during debinding [14]. Sintering temperatures and heating rate for Al_2O_3 vary between 1000 and 2000 °C, depending on the study. For example 1600 °C for two hours with specific heating rates; 3 °C/min between 25 and 600 °C; 5 °C/min between 600 and 1600 °C [30]. Another study sintered alumina at 1550 °C for 6 hours in air after drying the printed samples [12]. YSZ structures are subjected to a two-step process to achieve full densification: heating up to 150 and 250 °C for plasticizer evaporation, followed by debinding at 500 °C and sintering at 1450–1500 °C [53]. A controlled atmosphere in a tube furnace prevents the oxidation of SiC parts subjected to a three-stage temperature ramp: at 5 °C/min from room temperature up to 1800 °C, with two-hour holds at 300 °C (curing), 1250 °C (pyrolysis), and 1800 °C (sintering) [22, 46]. Lead zirconate titanate (PZT) ceramics also required a controlled atmosphere (Pb-rich), and temperature holds at 325 and 500 °C for 1 hour for organic decomposition, prior to sintering at 1250 °C for 2 hours [9]. A thermal cycle for bioglass 45S5 to burn out CMC additives was established based on TGA analyses [49].

4.4.2 Post-Processing for FFF-Printed Ceramic Parts

One of the major disadvantages of the thermoplastic MEX process is the debinding process. The formation of an interconnected pore structure is of high importance to avoid blister and crack formation during the debinding process. Louis et al. investigated the physicochemical processes during binder decomposition of porous ceramic bodies [162]. In the study, they examined the binder–vapor interface, which has been demonstrated to penetrate toward the interior of the green body in either a planar or nonplanar front. The description of the debinding process is even more intricate because thermoplastic binder systems are based on different components, as discussed in Section 4.3.3 [142]. The initial debinding step is focused on the removal of a primary binder component, earlier called a minor binder component, that exhibits an interconnected pore structure and can be readily extracted. It is noteworthy that the current focus on the so-called minor binder component is typically between 50 and 60 wt% of the total thermoplastic binder system, with the objective of achieving sufficient interconnected pore structure for the subsequent thermal debinding step. Therefore, the naming has been changed into primary binder component.

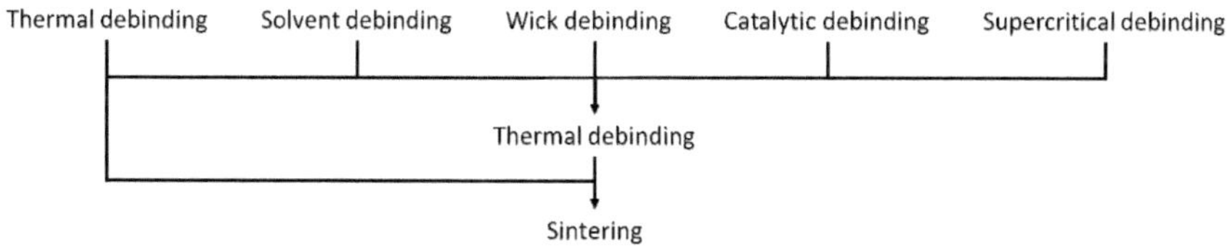

Figure 4.7 Summary of commonly used debinding methods. The debinding process is typically comprised of two distinct stages. In the initial stage, the primary binder components are eliminated through thermal, solvent, wick, catalytic, or supercritical debinding processes. Once a sufficient interconnected pore structure has been obtained, the thermal debinding step is initiated. This step is crucial for the complete decomposition of the backbone binder components before the sintering step.

In the field of thermoplastic MEX AM, the two-step debinding process is a frequently employed industrial procedure [163]. As illustrated in Figure 4.7, a comprehensive overview of the various debinding process methods that have been developed historically is provided.

In addition, the implementation of a two-stage debinding process is advised. The elimination of the thermoplastic binder can be accomplished through single-stage thermal heat treatment. Unlikely, there is any open porosity at the beginning of the debinding process in order to prevent the formation of blisters and cracks during the debinding process, it is essential to ensure the diffusion of oxygen into the green body, as well as the diffusion of evaporated, decomposed, or combusted binder products from the body. This process has been successfully utilized in the literature for the debinding of thin-wall ceramic structures [146, 164–167]. Salehi et al. used the single-step debinding process by means of free kinetic modeling to investigate the debinding process in more detail [165]. It has been established that the debinding process can be subdivided into three distinct stages (thermo-oxidation, scissoring, and evaporation). The single-step thermal debinding process is acutely sensitive to alterations in heating rates, dwell times, and debinding atmospheres [168]. Gorjan et al. have demonstrated that this debinding method has been employed with success in the fabrication of 3D-printed bioceramic scaffold structures [113]. Pure thermal debinding has been mainly used for thin parts like fiber, tubes, or scaffold structures.

In the context of green ceramic bodies with substantial wall thickness, the initial creation of open interconnected porosity assumes critical importance. The process of evaporation and decomposition of the backbone binder components generates internal mechanical stress due to gas formation if the pores are not open or the interconnected pore structure is not sufficient to channel the gas to the surface of the ceramic body. As the pressure within the ceramic body increases, this can result in the formation of blisters and cracks during the process of debinding. Consequently, thermoplastic binder components exhibiting divergent decomposition characteristics (e.g. melting points and partial vapor pressures) are utilized with greater frequency.

In the two-stage debinding process, the solvent, wick, supercritical or catalytic debinding method is combined with a subsequent thermal debinding step.

The majority of commercial thermoplastic binder systems for ceramic processing employ solvent debinding methods to achieve a porous interconnected network for the subsequent thermal debinding step. The removal of thermoplastic binder components by solvent extraction results in an interconnected porous ceramic structure [169, 170]. The solvent debinding process has been employed for the purpose of removing binder components of the thermoplastic binder system, and this method has been found to be successful in the context of thermoplastic MEX AM described in Section 4.3.1 [171, 172].

In the field of injection molding, researchers have developed thermoplastic binder systems comprising a component that can be dissolved in water. Rak et al. gave a summary of the different water-soluble thermoplastics that have been studied over the years, including polysaccharides and polyethylene glycol (PEG) [163]. Yang et al. presented a water-soluble plastic binder system based on PEG polymers with different molecular weights [173]. By this, blister and crack formation during the solvent debinding could be avoided successfully. One of the newer studies on porous network evolution in ceramic parts by solvent debinding has been published by Rolere et al. [174]. The relationship between the creation of open porosity and the efficiency of subsequent thermal debinding processes was presented with the objective of avoiding structural defects and reducing the duration of the entire debinding procedure.

Another method of generating a porous network is the wicking method. In this process, the green ceramic part is embedded in a powder bed (e.g. fine ceramic powder). The binder removal can be described by fluid wicking, diffusion control, and permeation control [175–178]. At temperatures above the melting point, the binder components are known to desorb from the ceramic body by means of capillary forces into the surrounding porous material. In order to guarantee an effective wicking process, it is essential to ensure that the average pore size of the powder bed is lower than that of the ceramic body. Following the draining of the melted binder components into the powder bed, the thermal debinding stage (diffusion and permeation control) can be employed, due to the interconnected porosity that has been achieved. As demonstrated by Gorjan et al., the wicking process has been optimized in such a way that 95 wt% of thermoplastic binder can be removed below 240 °C [179]. The final debinding process is undertaken using a thermal treatment with a constant heating rate of 10 °C/min. The two-stage wicking and thermal debinding process has been successfully validated for utilization in the thermoplastic MEX AM of different ceramics by several research groups [180, 181].

Instead of the solvent debinding step, the employment of a catalytic method has been demonstrated to be a viable alternative for the generation of an interconnected porous ceramic body. Ter Maat et al. developed a binder system known as Catamold, which involves the use of a catalyst to crack the polymeric components at the surface of the green ceramic body into monomers, causing them to evaporate [182–184]. It can thus be concluded that porosity is generated at the surface of the ceramic part and continues in a worm-like structure into the center. This technique has been demonstrated to facilitate the achievement of reduced debinding times without causing warping. This is due to the fact that the binder removal takes place at temperatures below the softening or melting temperature of the polymers. Following the expiration of BASF's patents, further investigation has been conducted into a catalytic thermoplastic binder system for use in the ceramic shaping process, with a basis in polyoxymethylene (POM) [185–187].

Finally, debinding of thermoplastic binder components can be achieved by supercritical extraction. In this process, gas is converted into a fluid under supercritical conditions. The diffusion coefficient of the supercritical fluid is analogous to that of the gas, and the solubility is similar to that of the liquid phase. The primary benefit of this fluid lies in its absence of surface tension, thereby negating the effects of capillary forces. Consequently, inhomogeneities within the body, such as agglomerates, microdefects, and density variations, are rendered less consequential. The prevalent utilization of carbon dioxide as a gas is attributable to its safety and cost-effectiveness. Nevertheless, the employment of alternative gases remains a possibility [188–191]. The supercritical debinding method has been the subject of investigation in the context of injection molding. However, this method is not considered to be a viable option due to the high cost and complexity of the necessary machinery.

Hadian et al. described a debinding process for large 3D-printed ceramic parts, comprising three debinding steps [172]: solvent debinding, wick debinding (partial debinding), and thermal debinding. This process facilitates the debinding of components with a wall thickness of 30 mm or greater. The investigation employed free kinetic modeling to examine the implementation of the final thermal debinding step, both prior to and following solvent debinding, as well as after the solvent and wick debinding. A constant mass-loss rate of 0.1%/min was selected to generate the final thermal debinding program.

Following the thermal debinding stage, the ceramic is subjected to a sintering process that results in its densification. The observation of different sintering mechanisms is possible, depending on the type of ceramic material. These mechanisms include solid phase, liquid phase, viscous flow sintering, or reactive phase sintering [192]. In all cases, the driving force for densification is the reduction of surface energy. The green ceramic density is typically in the range of 45–60 vol%, which corresponds to a volume shrinkage of between 35 and 50%. This is equivalent to a linear shrinkage of 18–25%. As outlined in Section 4.4.1, a number of case studies on the sintering of ceramics have previously been discussed. It is evident that there is no distinction in the sintering mechanisms between DIW and FFF samples after the debinding stage.

4.5 AM of Fiber-Reinforced Composites

4.5.1 Fiber-Reinforced Composites by DIW Printing

The flexibility of DIW systems, feedstocks,and process parameters enables customization of the technology to create ceramic matrix composites via different means. A common approach relies on the extrusion of reinforced feedstocks by including fibers or platelets within the formulation. The adjustment of process parameters (tip length, diameter, and extrusion velocity) enables the alignment of reinforcing fibers in the flow direction. Using platelets instead of fibers expands the ability to tune the microstructure of bioinspired ceramic composites. Bouligand-type structures can be achieved by rotating the raster direction of each layer 30° (Figure 4.8). Using a composite feedstock containing Al_2O_3 platelets and powders in a F127 hydrogel matrix is possible to create these nature-inspired structures able to guide crack propagation and twisting in three dimensions [24]. These platelet-reinforced ceramic matrix composites can exhibit higher strength and stiffness than traditional short carbon fiber-reinforced epoxy composites.

DIW of SiOC ceramic precursors and SiC powders containing a relatively high amount of short carbon fibers (>30 vol%, <1 mm) enabled the fabrication of complex CMC structures with ~75% porosity and compressive strength of ~4 MPa [20]. Another study follows a similar approach to fabricate boron nitride-reinforced polysilazane-derived ceramic composites [193]. DIW feedstocks containing a silicon carbide (SiC) precursor polymer and zirconium diboride (ZrB2), reinforced with silicon carbide fibers (SiC_f), provide a route to fabricate ultra-high temperature ceramic matrix composites (UHTCMCs) suitable for extreme environments [22].

A higher microstructural control can be achieved through coextrusion using concentrical tips or customized cartridges. Coextrusion of core–shell composites combining carbon and alumina can create a network of continuous weak interlayers to increase fracture resistance (Figure 4.8b) [194]. The study also investigates the correlation between the microstructure and idealized velocity profiles in the nozzle, which depend on the flow behavior of the feedstocks and process parameters (Section 4.2.4). Coextrusion is also exploited in other studies to produce ceramic filaments with ductile metallic cores [195] and carbon-alumina core–shell composites [196]. Combining both strategies, reinforcement and coextrusion, enables the fabrication of carbon fiber-reinforced core and SiC shell composites with strengths of ~123 MPa and porosities of ~31% [197].

DIW variants are also exploited to fabricate ceramic composites, for example through UV-assisted DIW, or embedded DIW in a supporting fluid. Alumina-toughened zirconia (ATZ) composites with 3 wt% Al_2O_3 platelets were fabricated using UV-curing-assisted DIW. The aligned Al_2O_3 platelets enhanced crack deflection, improving fracture toughness [21]. Steel-reinforced alumina composites can be fabricated via embedded-DIW using F127 hydrogels for both the flowing and supporting fluid [35]. Steel frameworks with different architectures can be printed within the Al_2O_3 matrix (Figure 4.8b). Auxetic steel structures result in work of fracture 50% greater than that obtained with simple fiber arrays and orders of magnitude above the fracture energy of the matrix [35].

These examples of fiber-reinforced ceramic composites demonstrate the flexibility of the DIW process in customization of feedstocks, systems, variants, and process parameters.

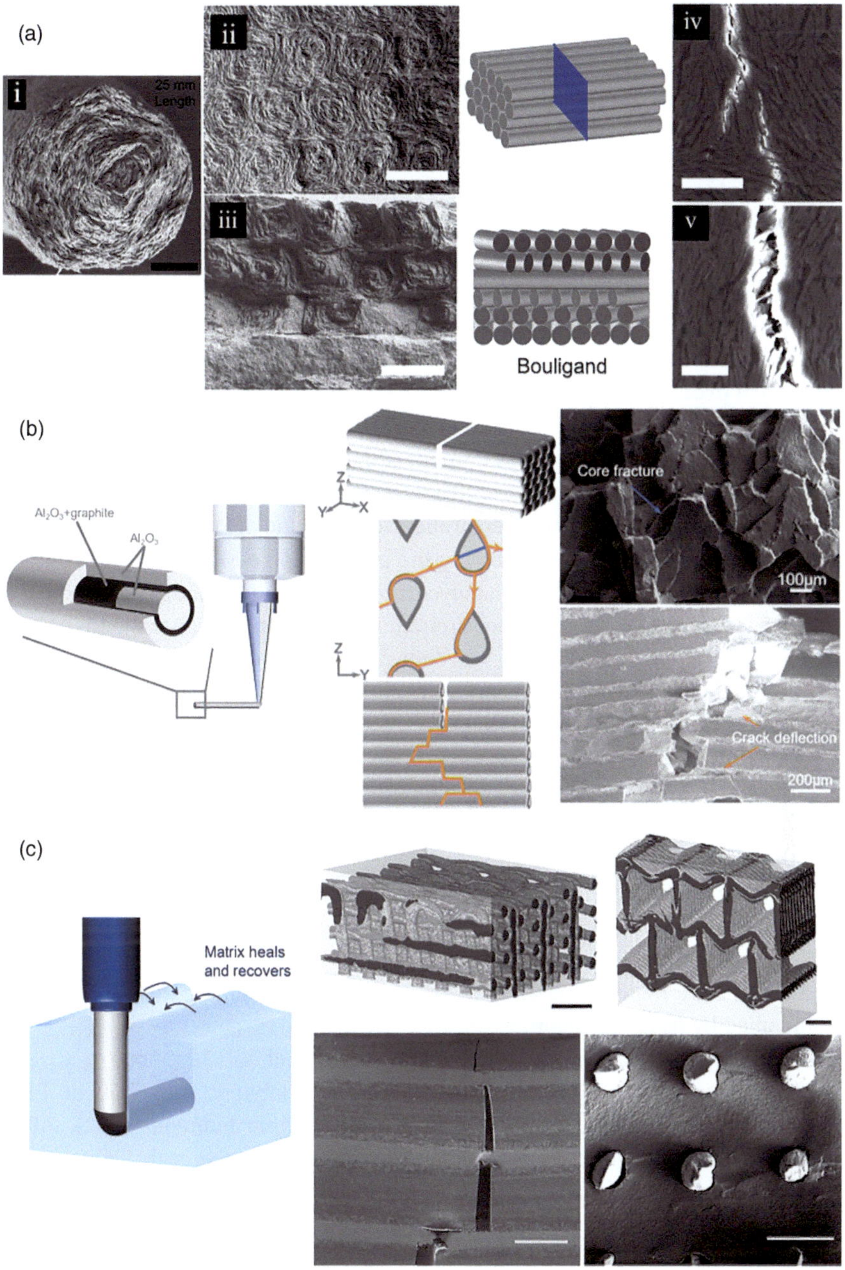

Figure 4.8 Examples of composites fabricated with DIW adaptations and variants. (a) DIW of alumina-F127 ceramic gel reinforced with alumina platelets. [24] / Springer Nature / CC BY 4.0. (b) Coextrusion of carbon-alumina core–shell composites. [25] / Elsevier / CC BY 4.0. (c) Embedded printing of steel-alumina composites. [25] / Elsevier / CC BY 4.0.

4.5.2 Fiber-Reinforced Composites by FFF Printing

The manufacture of fiber-reinforced ceramic composites using thermoplastic MEX AM was initiated a few years ago. To date, research has focused primarily on non-oxide ceramics. However, in the last two years, the scope has been extended to include oxide ceramic matrix composites (O-CMC).

FFF of carbon/carbon composites: A study on C_f/C composites to print rocket nozzles for reentry vehicles was conducted by investigating the use of different commercial carbon fiber-reinforced polymer filaments [138]. Following the pyrolysis process, a variety of carbon residues were obtained. PEI and PEEK were determined to be more effective polymer precursors for carbon/carbon composite processing when compared to PPS, PESU, and PSU, owing to their notably superior char yield.

FFF with preceramic polymer impregnation: Abel et al. developed a SiC_f/SiC filament with a fiber content of up to 30 vol% for the thermoplastic MEX process [198]. It was determined that a fiber content of 5% was sufficient to achieve damage tolerance and that to achieve a dense structure, the printed parts were impregnated with a preceramic polymer. Other researchers used a blend of PCL, ZrB_2, and carbon fibers for filament fabrication. Following the pyrolysis process, the printed components underwent impregnation with polycarbosilane, resulting in the formation of a C_f/ZrB_2-SiC composite [199]. In order to circumvent potential reactions with the carbon fibers, a flexible carbon coating was deposited on the surface of the carbon fiber monofilament via a hydrothermal carbonization reaction. The resultant composites demonstrated a mechanical strength of 218 MPa in three-point bending.

FFF with liquid silicon impregnation: Freudenberg et al. developed a C_f/SiC composite with a three-point flexural strength of 60 MPa when printed in the 0° direction [200]. In the 45° and 90° directions, the samples achieved mechanical strengths of 51 and 10 MPa, respectively. In a separate study, a mechanical strength of 235 MPa for C/SiC composite was attained by employing PEEK filament comprising 60 vol% carbon fibers in a four-point-bending test [201]. The pyrolysis of the PEEK polymer and the subsequent impregnation with liquid silicon resulted in the formation of a SiC matrix. The samples exhibited porosity ranging from 10% to 20%, accompanied by residual silicon.

FFF with chemical vapor impregnation: Mei et al. developed a C_f/SiC composite by combining carbon fibers, preceramic polymer, and PCL or PLA as a thermoplastic matrix for the fabrication of ceramic filaments [202]. After the pyrolysis process, the porous structures were impregnated with SiC through chemical vapor impregnation (CVI). This resulted in a 7.6-fold increase in mechanical strength when compared with parts without carbon fibers. In a subsequent study, the group optimized the fiber matrix interface during the CVI process [203]. Prior to the CVI with SiC, a pyrolytic carbon layer was formed around the fiber. Consequently, a 203% increase in mechanical strength was attained.

FFF of fiber oxide ceramic matrix composites: Hadian et al. developed an Al_2O_{3f}/Al_2O_3-ZrO_2 composite [204]. The commercial binder kcmix 3.3 from Krahn Ceramics (Germany) was used for the filament fabrication. Despite a fiber length that diminished to 450–250 μm after the compounding step, a four-point bending strength of 110 MPa was attained on filaments with a fiber content of 15 vol%. As demonstrated by micro-CT analysis, crack deflection, bridging, and fiber pullout were observed. The printed samples exhibited a modestly diminished degree of mechanical strength, a phenomenon that can be attributed to the volume effect of the tested sample. A comparison of the same test volume revealed that the mechanical strength of the filaments and the printed part was equivalent. Despite the presence of defects in the printed components, which were likely caused by the printing process. The mechanical strength of printed parts that lacked fibers was found to be 6.4 times lower than that of their counterparts. Stiller et al. investigated 3D printing of Al_2O_{3f}/Al_2O_3, using a thermoplastic binder based on PVB, PEG and stearic acid [205]. With a fiber content of 40 vol%, the researchers achieved a four-point bending strength of 30 MPa.

4.6 Summary

In this chapter, we have provided an overview of DIW and FFF processes. Owing to the versatility of DIW systems and the flexibility to formulate bespoke feedstocks, DIW has become an accessible and exciting research area within the ceramics field. DIW has expanded beyond the ceramic context, being currently exploited for other advanced functional materials. Designing and characterizing DIW feedstocks is attracting researchers from diverse scientific communities, with the common goal of better understanding printability and performance in DIW. Using complementary rheological studies enables achieving a holistic understanding of DIW feedstocks, including yielding, material strength, flow, restructuring timescales, and extensional properties. Further investigating the impact of process parameters and quantifying shape fidelity are also key to the success of DIW processes. The versatility and flexibility of DIW systems and feedstocks have been demonstrated with specific examples of ceramic composites with controlled microstructures and improved mechanical performance. For the FFF process, several extrusion heads are now available, but due to the brittleness of ceramic filaments, the trend is toward pellet extrusion, even though the resolutions are lower and the extrusion speed of the Bowden extruder cannot yet be achieved. It is now possible to obtain commercial ceramic filaments or to use commercial ceramic feedstocks designed for ceramic injection molding. Rheological characterization of MEx AM thermoplastic feedstocks is well established and can be used for quality assurance, research, and development. The process strongly depends on the binder composition selected and the size of the printed parts. Debinding using water-soluble binder additives has been commercialized. However, the reuse of these feedstocks is still a challenge because the flow behavior can be drastically altered after the heat treatment during the thermoplastic shaping process. Composite materials are not only achievable by multi-material printing; in recent years, researchers have also started to print fiber-reinforced ceramics.

Acknowledgments

Dr Esther García-Tuñón would like to acknowledge Dr Bin Ling for his sketching contributions in some of the schemes in Section 4.2, and to Dr David Dennis for the many interesting discussions on fluid mechanics. This work was supported by a UKRI Future Leaders Fellowship MR/V021117/1. Dr Frank Clemens would like to acknowledge Somashree Mondal for her assistance in creating the reference chapter. This work was supported by internal funding from Empa.

References

1 Cesarano, J. and Grieco, S. (1997). Robocasting: a new technique for the freeform fabrication of near-net-shape ceramics. *Mater. Technol.* 12 (3–4): 98–100.

2 Lewis, J.A., Smay, J.E., Stuecker, J., and Cesarano, J. (2006). Direct ink writing of three-dimensional ceramic structures. *J. Am. Ceram. Soc.* 89 (12): 3599–3609.

3 Smay, J.E. and Lewis, J.A. (2012). Solid free-form fabrication of 3-D ceramic structures. In: *Ceramics and Composites Processing Methods* (ed. N.P. Bansal and A.R. Boccaccini), 459–484. Hoboken, NJ: Wiley.

4 dos Santos, V.I., Chevalier, J., Fredel, M.C. et al. (2024). Ceramics and ceramic composites for biomedical engineering applications via Direct Ink Writing: Overall scenario, advances in the improvement of mechanical and biological properties and innovations. *Mater. Sci. Eng. R. Rep.* 161: 100841.

5 Ashwin, A.J. and Jafferson, J.M. (2021). State of the art direct ink writing (DIW) and experimental trial on DIW of HAp bio-ceramics. *Mater. Today Proc.* 46: 1298–1307.

6 Petit, C., Meille, S., Maire, E. et al. (2017). Fracture behavior of robocast HA/β-TCP scaffolds studied by X-ray tomography and finite element modeling. *J. Eur. Ceram. Soc.* 37 (4): 1735–1745.

7 Franco, J., Hunger, P., Launey, M.E. et al. (2010). Direct write assembly of calcium phosphate scaffolds using a water-based hydrogel. *Acta Biomater.* 6 (1): 218–228.

8 Jones, I.K., Seeley, Z.M., Cherepy, N.J. et al. (2018). Direct ink write fabrication of transparent ceramic gain media. *Opt. Mater.* 75: 19–25.

9 Li, Z., Li, J., Luo, H. et al. (2022). Direct ink writing of 3D piezoelectric ceramics with complex unsupported structures. *J. Eur. Ceram. Soc.* 42 (9): 3841–3847.

10 Sun, D., Lu, Y., and Karaki, T. (2023). Review of the applications of 3D printing technology in the field of piezoelectric ceramics. *Res. Chem. Mater.* 2 (2): 128–142.

11 Cramer, C.L., Ionescu, E., Graczyk-Zajac, M. et al. (2022). Additive manufacturing of ceramic materials for energy applications: road map and opportunities. *J. Eur. Ceram. Soc.* 42 (7): 3049–3088.

12 Álvarez, F., Cifuentes, A., Serrano, I. et al. (2022). Optimization of the sintering thermal treatment and the ceramic ink used in direct ink writing of α-Al_2O_3: characterization and catalytic application. *J. Eur. Ceram. Soc.* 42 (6): 2921–2930.

13 Bui, H.M., Großmann, P.F., Berger, A. et al. (2023). Comparison of direct ink writing and binder jetting for additive manufacturing of Pt/Al_2O_3 catalysts for the dehydrogenation of perhydro-dibenzyltoluene. *Chem. Eng. J.* 458: 141361.

14 Radogna, C., Serrano, I., Fargas, G. et al. (2024). Chemical etching optimization of 3D printed α-Al_2O_3 monoliths to enhance the catalytic applications. *J. Eur. Ceram. Soc.* 44 (12): 7189–7200.

15 Muth, J.T., Dixon, P.G., Woish, L. et al. (2017). Architected cellular ceramics with tailored stiffness via direct foam writing. *Proc. Natl Acad. Sci. USA* 114 (8): 1832–1837.

16 Feilden, E., Glymond, D., Saiz, E., and Vandeperre, L. (2019). High temperature strength of an ultra high temperature ceramic produced by additive manufacturing. *Ceram. Int.* 45 (15): 18210–18214.

17 Lakhdar, Y., Tuck, C., Terry, A. et al. (2021). Direct ink writing of boron carbide monoliths. *J. Eur. Ceram. Soc.* 41 (16): 76–92.

18 Sesso, M.L., Slater, S., Thornton, J., and Franks, G.V. (2021). Direct ink writing of hierarchical porous ultra-high temperature ceramics (ZrB_2•). *J. Am. Ceram. Soc.* 104 (10): 4977–4990.

19 Zhang, D., Jonhson, W., Herng, T.S. et al. (2022). High temperature co-firing of 3D-printed $AlZnO/Al_2O_3$ multi-material two-phase flow sensor. *J. Mater.* 8 (3): 710–718.

20 Franchin, G., Wahl, L., and Colombo, P. (2017). Direct ink writing of ceramic matrix composite structures. *J. Am. Ceram. Soc.* 100 (10): 4397–4401.

21 Li, M., Huang, S., Willems, E. et al. (2024). UV-curing assisted direct ink writing of dense, crack-free, and high-performance zirconia-based composites with aligned alumina platelets. *Adv. Mater.* 36 (5): 2306764.

22 Kemp, J.W., Diaz, A.A., Malek, E.C. et al. (2021). Direct ink writing of ZrB_2-SiC chopped fiber ceramic composites. *Addit. Manuf.* 44: 102049.

23 Studart, A.R. (2016). Additive manufacturing of biologically-inspired materials. *Chem. Soc. Rev.* 45 (2): 359–376.

24 Feilden, E., Ferraro, C., Zhang, Q. et al. (2017). 3D printing bioinspired ceramic composites. *Sci. Rep.* 7 (1): 13759.

25 Bargardi, F.L., Le Ferrand, H., Libanori, R., and Studart, A.R. (2016). Bio-inspired self-shaping ceramics. *Nat. Commun.* 7 (1): 13912.

26 Costakis, W.J., Rueschhoff, L.M., Diaz-Cano, A.I. et al. (2016). Additive manufacturing of boron carbide via continuous filament direct ink writing of aqueous ceramic suspensions. *J. Eur. Ceram. Soc.* 36 (14): 3249–3256.

27 Agrawal, R. and García-Tuñón, E. (2024). Interplay between yielding, 'recovery', and strength of yield stress fluids for direct ink writing: new insights from oscillatory rheology. *Soft Matter* 20 (37): 7429–7447.

28 Corker, A., Chi-Hin Ng, H., Poole, R.J., and García-Tuñón, E. (2019). 3D printing with 2D colloids: designing rheology protocols to predict 'printability' of soft-materials. *Soft Matter* 15 (6): 1444–1456.

29 Feilden, E., Blanca, E.G.-T., Giuliani, F. et al. (2016). Robocasting of structural ceramic parts with hydrogel inks. *J. Eur. Ceram. Soc.* 36 (10): 2525–2533.

30 Yang, L., Zeng, X., Ditta, A. et al. (2020). Preliminary 3D printing of large inclined-shaped alumina ceramic parts by direct ink writing. *J. Adv. Ceram.* 9 (3): 312–319.

31 Tubío, C.R., Nóvoa, J.A., Martín, J. et al. (2019). Broadband terahertz ZnO photonic crystals fabricated by 3D printing. *Ceram. Int.* 45 (5): 6223–6227.

32 George, M.C., Mohraz, A., Piech, M. et al. (2009). Direct laser writing of photoresponsive colloids for microscale patterning of 3d porous structures. *Adv. Mater.* 21 (1): 66–70.

33 Wei, L., Li, J., Zhang, S. et al. (2020). Fabrication of SiOC ceramic with cellular structure via UV-assisted direct ink writing. *Ceram. Int.* 46 (3): 3637–3643.

34 Román-Manso, B., Weeks, R.D., Truby, R.L., and Lewis, J.A. (2023). Embedded 3D printing of architected ceramics via microwave-activated polymerization. *Adv. Mater.* 35: 2209270.

35 Zhou, S., Tirichenko, I.S., Zhang, X. et al. (2022). Embedded 3D printing of multi-material composites. *Matter* 7: 668–684.

36 Huang, K., Elsayed, H., Franchin, G., and Colombo, P. (2021). Embedded direct ink writing of freeform ceramic components. *Appl. Mater. Today* 23: 101005.

37 Chen, Z., Li, Z., Li, J. et al. (2019). 3D printing of ceramics: a review. *J. Eur. Ceram. Soc.* 39 (4): 661–687.

38 del-Mazo-Barbara, L. and Ginebra, M.-P. (2021). Rheological characterisation of ceramic inks for 3D direct ink writing: a review. *J. Eur. Ceram. Soc.* 41 (16): 18–33.

39 Li, Z.L., Zhou, S., Saiz, E., and Malik, R. (2024). Ink formulation in direct ink writing of ceramics: a meta-analysis. *J. Eur. Ceram. Soc.* 44 (12): 6777–6796.

40 Nelson, A.Z. and Ewoldt, R.H. (2017). Design of yield-stress fluids: a rheology-to-structure inverse problem. *Soft Matter* 13 (41): 7578–7594.

41 Nelson, A.Z., Schweizer, K.S., Rauzan, B.M. et al. (2019). Designing and transforming yield-stress fluids. *Curr. Opinion Solid State Mater. Sci.* 23 (5): 100758.

42 Feilden, E. (2017). Additive manufacturing of ceramics and ceramic composites via robocasting. *Prog. Mater. Sci.* 116: 100736.

43 García-Tuñón, E., Feilden, E., Zheng, H. et al. (2017). Graphene oxide: an all-in-one processing additive for 3D printing. *ACS Appl. Mater. Interfaces* 9 (38): 32977–32989.

44 Chan, S.S.L., Sesso, M.L., and Franks, G.V. (2020). Direct ink writing of hierarchical porous alumina-stabilized emulsions: rheology and printability. *J. Am. Ceram. Soc.* 103 (10): 5554–5566.

45 Peng, E., Wei, X., Garbe, U. et al. (2018). Robocasting of dense yttria-stabilized zirconia structures. *J. Mater. Sci.* 53 (1): 247–273.

46 Larson, C.M., Choi, J.J., Gallardo, P.A. et al. (2016). Direct ink writing of silicon carbide for microwave optics. *Adv. Eng. Mater.* 18 (1): 39–45.

47 Lu, X., Lee, Y., Yang, S. et al. (2010). Solvent-based paste extrusion solid freeforming. *J. Eur. Ceram. Soc.* 30 (1): 1–10.

48 Deliormanlı, A.M. and Rahaman, M.N. (2012). Direct-write assembly of silicate and borate bioactive glass scaffolds for bone repair. *J. Eur. Ceram. Soc.* 32 (14): 3637–3646.

49 Eqtesadi, S., Motealleh, A., Miranda, P. et al. (2014). Robocasting of 45S5 bioactive glass scaffolds for bone tissue engineering. *J. Eur. Ceram. Soc.* 34 (1): 107–118.

50 Lewis, J.A. (2000). Colloidal processing of ceramics. *J. Am. Ceram. Soc.* 83 (10): 2341–2359.

51 Pierin, G., Grotta, C., Colombo, P., and Mattevi, C. (2016). Direct ink writing of micrometric SiOC ceramic structures using a preceramic polymer. *J. Eur. Ceram. Soc.* 36 (7): 1589–1594.

52 Chaudhary, R.P., Parameswaran, C., Idrees, M. et al. (2022). Additive manufacturing of polymer-derived ceramics: materials, technologies, properties and potential applications. *Prog. Mater. Sci.* 128: 100969.

53 Zhang, D., Peng, E., Borayek, R., and Ding, J. (2019). Controllable ceramic green-body configuration for complex ceramic architectures with fine features. *Adv. Funct. Mater.* 29 (12): 1807082.

54 Elizarova, I.S., Vandeperre, L., and Saiz, E. (2020). Conformable green bodies: plastic forming of robocasted advanced ceramics. *J. Eur. Ceram. Soc.* 40 (2): 552–557.

55 Gratson, G.M. and Lewis, J.A. (2005). Phase behavior and rheological properties of polyelectrolyte inks for direct-write assembly. *Langmuir* 21 (1): 457–464.

56 Wang, C., Zhang, Y., Zhang, B., and Yang, J. (2023). 3D printable ultra-highly porous and mechanically strong foam materials for multiple applications. *Adv. Funct. Mater.* 33 (7): 2205537.

57 Alexandridis, P. and Alan Hatton, T. (1995). Poly(ethylene oxide)-poly(propylene oxide)-poly(ethylene oxide) block copolymer surfactants in aqueous solutions and at interfaces: thermodynamics, structure, dynamics, and modeling. *Colloids Surf. A Physicochem. Eng. Asp.* 96 (1): 1–46.

58 Mortensen, K. (1996). Structural studies of aqueous solutions of PEO-PPO-PEO triblock copolymers, their micellar aggregates and mesophases; a small-angle neutron scattering study. *J. Phys. Condens. Matter* 8 (25A): A103–A124.

59 Hyun, K., Nam, J.G., Wilhellm, M. et al. (2006). Large amplitude oscillatory shear behavior of PEO-PPO-PEO triblock copolymer solutions. *Rheol. Acta* 45 (3): 239–249.

60 Newby, G.E., Hamley, I.W., King, S.M. et al. (2009). Structure, rheology and shear alignment of Pluronic block copolymer mixtures. *J. Colloid Interface Sci.* 329 (1): 54–61.

61 Rishav Agrawal, Patrick T. Spicer, Esther García-Tuñón (2025). Connecting bulk rheology, structural transitions and heterogeneous flow in Pluronic F127 micellar cubic liquid crystals using rheo-microscopy. *Journal of Colloid and Interface Science*, 699, 138226.

62 García-Tuñón, E., Agrawal, R., Ling, B., and Dennis, D.J.C. (2023). Fourier-transform rheology and printability maps of complex fluids for three-dimensional printing. *Phys. Fluids* 35 (1): 017113.

63 He, X., Xu, J., and Ji, W. (2022). Linking rheology, process parameters and formability for material extrusion and photo-polymerization combined process aiming at high-density ceramics. *J. Sol-Gel Sci. Technol.* 104 (2): 276–286.

64 Ewoldt, R.H. and McKinley, G.H. (2017). Mapping thixo-elasto-visco-plastic behavior. *Rheol. Acta* 56 (3): 195–210.

65 Donley, G.J., de Bruyn, J.R., McKinley, G.H., and Rogers, S.A. (2019). Time-resolved dynamics of the yielding transition in soft materials. *J. Non-Newtonian Fluid Mech.* 264: 117–134.

66 Donley, G.J., Singh, P.K., Shetty, A., and Rogers, S.A. (2020). Elucidating the G″ overshoot in soft materials with a yield transition via a time-resolved experimental strain decomposition. *Proc. Natl Acad. Sci. USA* 117 (36): 21945–21952.

67 Kamani, K.M. and Rogers, S.A. (2024). Brittle and ductile yielding in soft materials. *Proc. Natl Acad. Sci. USA* 121 (22): e2401409121.

68 Rogers, S.A. (2012). A sequence of physical processes determined and quantified in LAOS: an instantaneous local 2D/3D approach. *J. Rheol.* 56 (5): 1129–1151.

69 Nelson, A.Z., Bras, R.E., Liu, J., and Ewoldt, R.H. (2018). Extending yield-stress fluid paradigms. *J. Rheol.* 62 (1): 357–369.

70 Morrison, F.A. (2001). *Understanding Rheology*. New York: Oxford University Press.

71 Coussot, P. and Rogers, S.A. (2021). Oldroyd's model and the foundation of modern rheology of yield stress fluids. *J. Non-Newtonian Fluid Mech.* 295: 104604.

72 Oldroyd, J.G. (1947). A rational formulation of the equations of plastic flow for a Bingham solid. *Math. Proc. Camb. Philos. Soc.* 43 (1): 100–105.

73 Saramito, P. (2007). A new constitutive equation for elastoviscoplastic fluid flows. *J. Non-Newtonian Fluid Mech.* 145 (1): 1–14.

74 de Souza Mendes, P.R. and Thompson, R.L. (2013). A unified approach to model elasto-viscoplastic thixotropic yield-stress materials and apparent yield-stress fluids. *Rheol. Acta* 52 (7): 673–694.

75 Kamani, K., Donley, G.J., and Rogers, S.A. (2021). Unification of the rheological physics of yield stress fluids. *Phys. Rev. Lett.* 126 (21): 218002.

76 Mezger, T. (2006). *The Rheology Handbook: For Users of Rotational and Oscillation Rheometers*. Hannover: Vincentz Network.

77 Geffrault, A., Bessaies-Bey, H., Roussel, N., and Coussot, P. (2021). Extensional gravity-rheometry (EGR) for yield stress fluids. *J. Rheol.* 65 (5): 887–901.

78 Saengow, C., Sen, S., Yus, J., Lovrich, E.E., Hoika, A.G., Chang, K.M., Pfeil, A.A., Haug, N., Johnson, A.J.W., and Ewoldt, R.H. (2025) Stretching the printability metric in direct-ink writing with highly extensible yield-stress fluids. *Soft Condensed Matter*. arXiv:2501.12630. https://doi.org/10.48550/arXiv.2501.12630.

79 Herschel, W.H. and Bulkley, R. (1926). Konsistenzmessungen von Gummi-Benzollösungen. *Kolloid Zeitschrift* 39 (4): 291–300.

80 M'Barki, A., Bocquet, L., and Stevenson, A. (2017). Linking rheology and printability for dense and strong ceramics by direct ink writing. *Sci. Rep.* 7 (1): 6017.

81 Ewoldt, R.H., Johnston, M.T., and Caretta, L.M. (2015) Experimental challenges of shear rheology: how to avoid bad data, in *Complex Fluids in Biological Systems* (eds. Spagnolie, S.E.), Springer New York, pp. 207–241.

82 Saengow, C. and Giacomin, A.J. (2019). Review of nonlinear oscillatory shear flow notations and presentations: polymeric liquids. *Curr. Opin. Colloid Interface Sci.* 43: 26–38.

83 Ovarlez, G., Cohen-Addad, S., Krishan, K. et al. (2013). On the existence of a simple yield stress fluid behavior. *J. Non-Newtonian Fluid Mech.* 193: 68–79.

84 Divoux, T., Agoritsas, E., Aime, S. et al. (2024). Ductile-to-brittle transition and yielding in soft amorphous materials: perspectives and open questions. *Soft Matter* 20 (35): 6868–6888.

85 Hyun, K., Wilhelm, M., Klein, C.O. et al. (2011). A review of nonlinear oscillatory shear tests: analysis and application of large amplitude oscillatory shear (LAOS). *Prog. Polym. Sci.* 36 (12): 1697–1753.

86 Wei, P., Cipriani, C., Hsieh, C.-M. et al. (2023). Go with the flow: rheological requirements for direct ink write printability. *J. Appl. Phys.* 134 (10): 100701.

87 Ewoldt, R.H., Hosoi, A.E., and McKinley, G.H. (2008). New measures for characterizing nonlinear viscoelasticity in large amplitude oscillatory shear. *J. Rheol.* 52 (6): 1427–1458.

88 Ewoldt, R.H., Winter, P., Maxey, J., and McKinley, G.H. (2010). Large amplitude oscillatory shear of pseudoplastic and elastoviscoplastic materials. *Rheol. Acta* 49 (2): 191–212.

89 Ling, B., Agrawal, R., He, D. et al. (2024). 3D-printed porous organic cages for gas filtration: fabrication and flow simulations. *Adv. Funct. Mater.* 34 (46): 2405320.

90 Choi, J., Nettesheim, F., and Rogers, S.A. (2019). The unification of disparate rheological measures in oscillatory shearing. *Phys. Fluids* 31 (7): 073107.

91 Mours, M. and Winter, H.H. (1994). Time-resolved rheometry. *Rheol. Acta* 33 (5): 385–397.

92 Smay, J.E., Cesarano, J., and Lewis, J.A. (2002). Colloidal inks for directed assembly of 3-D periodic structures. *Langmuir* 18 (14): 5429–5437.

93 Liu, Y., Hildner, M., Roy, O. et al. (2023). On the selection of rheological tests for the prediction of 3D printability. *J. Rheol.* 67 (4): 791.

94 Sen, S. and Ewoldt, R.H. (2022). Thixotropic spectra and Ashby-style charts for thixotropy. *J. Rheol.* 66 (5): 1041–1053.

95 Afarani, H.T., Moser, N.H., Garboczi, E.J. et al. (2022). Print fidelity metrics for additive manufacturing of cement-based materials. *Addit. Manuf.* 55: 102784.

96 Ribeiro, A., Blokzijl, M.M., Levato, R. et al. (2017). Assessing bioink shape fidelity to aid material development in 3D bioprinting. *Biofabrication* 10 (1): 014102.

97 Ouyang, L., Yao, R., Zhao, Y., and Sun, W. (2016). Effect of bioink properties on printability and cell viability for 3D bioplotting of embryonic stem cells. *Biofabrication* 8 (3): 035020.

98 Coussot, P. (2014). Yield stress fluid flows: a review of experimental data. *J. Non-Newtonian Fluid Mech.* 211: 31–49.

99 Larson, R.G. (1999). *The Structure and Rheology of Complex Fluids.* New York: Oxford University Press.

100 Rabideau, B.D., Moucheront, P., Bertrand, F. et al. (2010). The extrusion of a model yield stress fluid imaged by MRI velocimetry. *J. Non-Newtonian Fluid Mech.* 165 (7): 394–408.

101 Poole, R.J. (2023). Inelastic and flow-type parameter models for non-Newtonian fluids. *J. Non-Newtonian Fluid Mech.* 320: 105106.

102 Divoux, T., Barentin, C., and Manneville, S. (2011). Stress overshoot in a simple yield stress fluid: an extensive study combining rheology and velocimetry. *Soft Matter* 7 (19): 9335–9349.

103 Divoux, T., Barentin, C., and Manneville, S. (2011). From stress-induced fluidization processes to Herschel-Bulkley behaviour in simple yield stress fluids. *Soft Matter* 7 (18): 8409–8418.

104 Geffrault, A., Bessaies-Bey, H., Roussel, N., and Coussot, P. (2023). Printing by yield stress fluid shaping. *Addit. Manuf.* 75: 103752.

105 Comminal, R., Serdeczny, M.P., Pedersen, D.B., and Spangenberg, J. (2019). Motion planning and numerical simulation of material deposition at corners in extrusion additive manufacturing. *Addit. Manuf.* 29: 100753.

106 Crump, S.S. (1991). Fused deposition modeling (FDM): putting rapid back into prototyping, 354–357.

107 Crump, S.S. (1992). Apparatus and method for creating three-dimensional objects. United States US5121329A, filed Oct. 30, 1989 and issued Jun. 9, 1992.

108 Crump, S.S., Comb, J.W., Jr, Priedeman, W.R. and Zinniel, R.L. (1996). Process of support removal for fused deposition modeling. United States US5503785A, filed Jun. 2, 1994 and issued Apr. 2, 1996.

109 Clemens, F.J. (2019). FDM Keramikdruck: Herstellung von keramischen Strukturen mit Hilfe von kommerziellen Filamentdruckern. *Keram. Z.* 71: 40–44.

110 Greul, M., Pintat, T., and Greulich, M. (1995). Rapid prototyping of functional metallic parts. *Comput. Ind.* 28 (1): 23–28.

111 Clemens, F. (2007) Thermoplastic extrusion for ceramic bodies. *Extrusion in Ceramics*, pp. 295–311.

112 Clemens, F., Schulz, J., Gorjan, L. et al. (2021). Debinding and sintering of dense ceramic structures made with fused deposition modeling. In: *Industrializing Additive Manufacturing*, 293–303. OUCI.

113 Gorjan, L., Reiff, L., Liersch, A., and Clemens, F. (2018). Ethylene vinyl acetate as a binder for additive manufacturing of tricalcium phosphate bio-ceramics. *Ceram. Int.* 15817–15823.

114 Clemens, F., Sarraf, F., Borzì, A. et al. (2023). Material extrusion additive manufacturing of advanced ceramics: towards the production of large components. *J. Eur. Ceram. Soc.* 43: 2752–2760.

115 Smirnov, A., Terekhina, S., Tarasova, T. et al. (2023). From the development of low-cost filament to 3D printing ceramic parts obtained by fused filament fabrication. *Int. J. Adv. Manuf. Technol.* 128 (1): 511–529.

116 Vukšić, M., Bek, M., Perše, L.S. et al. (2023). The role of paraffin wax on the properties and printability of ethylene vinyl acetate-based feedstocks for alumina fused filament fabrication. *Open Ceram.* 16: 100496.

117 He, Q., Jiang, J., Yang, X. et al. (2021). Additive manufacturing of dense zirconia ceramics by fused deposition modeling via screw extrusion. *J. Eur. Ceram. Soc.* 41 (1): 1033–1040.

118 Hung Hung, Y.M.X., Talou, M.H., and Camerucci, M.A. (2021). Porous structures printed by fused deposition of granules processed from a novel polyamide/alumina-based system. *Int. J. Appl. Ceram. Tech.* 18 (5): 1466–1477.

119 Hadian, A., Fricke, M., Liersch, A., and Clemens, F. (2022). Material extrusion additive manufacturing of zirconia parts using powder injection molding feedstock compositions. *Addit. Manuf.* 57: 102966.

120 Wick-Joliat, R. and Penner, D. (2023). Flexible interconnected ceramic parts 3D printed by two-component material extrusion with water-soluble support structures. *J. Eur. Ceram. Soc.* 43 (11): 4877–4884.

121 Krinitcyn, M., Kopytov, G., and Ryumin, E. (2024). Additive manufacturing of Ti_3AlC_2/TiC and Ti_3AlC_2/SiC ceramics using the fused granules fabrication technique. *J. Manuf. Mater. Process.* 8 (3): 123.

122 Bellini, A., Shor, L., and Guceri, S.I. (2005). New developments in fused deposition modeling of ceramics. *Rapid Prototyp. J.* 11 (4): 214–220.

123 Scheithauer, U., Schwarzer, E., Richter, H., and Moritz, T. (2015). Thermoplastic 3D printing - an additive manufacturing method for producing dense ceramics. *Int. J. Appl. Ceram. Tech.* 12 (1): 26–31.

124 Venkataraman, N., Rangarajan, S., Matthewson, M.J. et al. (2000). Feedstock material property – process relationships in fused deposition of ceramics (FDC). *Rapid Prototyp. J.* 6 (4): 244–253.

125 Lous, G.M., Cornejo, I.A., McNulty, T.F. et al. (2000). Fabrication of Piezoelectric Ceramic/Polymer Composite Transducers Using Fused Deposition of Ceramics. *J. Am. Ceram. Soc.* 83 (1): 124–128.

126 Giberti, H., Strano, M., and Annoni, M. (2016). An innovative machine for fused deposition modeling of metals and advanced ceramics. *MATEC Web Conf.* 43: 03003.

127 Boparai, K.S., Singh, R., and Singh, H. (2016). Development of rapid tooling using fused deposition modeling: a review. *Rapid Prototyp. J.* 22 (2): 281–299.

128 Gorjan, L., Tonello, R., Sebastian, T. et al. (2019). Fused deposition modeling of mullite structures from a preceramic polymer and γ-alumina. *J. Eur. Ceram. Soc.* 39: 2463–2471.

129 Nötzel, D., Eickhoff, R., and Hanemann, T. (2018). Fused filament fabrication of small ceramic components. *Materials* 11 (8): 1463.

130 Bach, M., Sebastian, T., Melnykowycz, M., Lusiola, T., Scharf, D., and Clemens, F. (2018). Additive manufacturing of piezoelectric 3-3 composite structures. *Industrializing Additive Manufacturing – Proceedings of Additive Manufacturing in Products and Applications – AMPA2017*, pp. 93–103.

131 Abel, J., Scheithauer, U., Janics, T. et al. (2019). Fused filament fabrication (FFF) of metal-ceramic components. *JoVE* 143: 57693.

132 Ramkumar, P.L. and Rijwani, T. (2022). Additive manufacturing of metals and ceramics using hybrid fused filament fabrication. *J. Braz. Soc. Mech. Sci. Eng.* 44 (10): 455.

133 Orlovská, M., Chlup, Z., Bača, Ľ. et al. (2020). Fracture and mechanical properties of lightweight alumina ceramics prepared by fused filament fabrication. *J. Eur. Ceram. Soc.* 40 (14): 4837–4843.

134 Sudan, K., Singh, P., Gökçe, A. et al. (2020). Processing of hydroxyapatite and its composites using ceramic fused filament fabrication (CF3). *Ceram. Int.* 46 (15): 23922–23931.

135 Vozárová, M., Neubauer, E., Bača, Ľ. et al. (2023). Preparation of fully dense boron carbide ceramics by fused filament fabrication (FFF). *J. Eur. Ceram. Soc.* 43 (5): 1751–1761.

136 Eltayeb, A., Graß, V., Lee, K. et al. (2023). Characterisation and thermochemical stability analysis of 3D printed porous ceria structures fabricated via composite extrusion Modelling. *Mater. Des.* 236: 112514.

137 Lieberwirth, C., Sarhan, M., and Seitz, H. (2018). Mechanical properties of stainless-steel structures fabricated by composite extrusion modelling. *Metals* 8 (2): 84.

138 Romero, E.S., Barocio, E., and Trice, R.W. (2024). Evaluating extrusion deposited additively manufactured fiber-reinforced thermoplastic polymers as carbon/carbon preforms. *Appl. Compos. Mater.* 31 (2): 399–419.

139 Kollenberg, W. and Nikolay, D. (2018). Additive Fertigung keramischer Bauteile. *Keram. Z.* 70 (1–2): 22–25.

140 Lusiola, T., Scharf, D., Graule, T., and Clemens, F. (2014). Low shear compounding process for thermoplastic fabrication of ferroelectric lead-free fibres. *J. Eur. Ceram. Soc.* 34 (10): 2265–2274.

141 Clemens, F.J., Hadian, A., and Fricke, M. (2022). Material extrusion-based additive manufacturing for ceramics using thermoplastic feedstocks. *CFI Ceram. Forum Int.* 96–100.

142 Quackenbush, C.L., French, K., and Neil, J.T. (1982). Fabrication of sinterable silicon nitride by injection molding. In: *Ceramic Engineering and Science Proceedings*, vol. 3, 20–34. Hoboken, NJ: Wiley.

143 Clemens, F., Sarraf, F., and Eckey, L. (2024). Extrusion based additive manufacturing of polymer derived ceramics. *Eighteenth European Inter-Regional Conference on Ceramics, CIEC-18*, pp. 1–5.

144 Ismael, M.R., Clemens, F., Bohac, W.M. et al. (2009). Effects of rheology on the interface of Pb(Zr, Ti)O_3 monofilament composites obtained by co-extrusion. *J. Eur. Ceram. Soc.* 29: 3015–3021.

145 Goodrich, J.E. and Porter, R.S. (1967). A rheological interpretation of torque-rheometer data. *Polym. Eng. Sci.* 7: 45–51.

146 Clemens, F.J., Wallquist, V., Buchser, W. et al. (2007). Silicon carbide fiber-shaped microtools by extrusion and sintering SiC with and without carbon powder sintering additive. *Ceram. Int.* 33: 491–496.

147 Cano, S., Gonzalez-Gutierrez, J., Sapkota, J. et al. (2019). Additive manufacturing of zirconia parts by fused filament fabrication and solvent debinding: selection of binder formulation. *Addit. Manuf.* 26: 117–128.

148 Nötzel, D., Eickhoff, R., Pfeifer, C., and Hanemann, T. (2021). Printing of zirconia parts via fused filament fabrication. *Materials* 14 (19): 5467.

149 Cogswell, F.N. (1972). Converging flow of polymer melts in extrusion dies. *Polym. Eng. Sci.* 12 (1): 64–73.

150 Bagley, E.B., Cabott, I.M., and West, D.C. (1958). Discontinuity in the flow curve of polyethylene. *J. Appl. Phys.* 29 (1): 109–110.

151 Sarraf, F., Hadian, A., Churakov, S.V., and Clemens, F. (2023). EVA-PVA binder system for polymer derived mullite made by material extrusion based additive manufacturing. *J. Eur. Ceram. Soc.* 43: 530–541.

152 Hadian, A., Zamani, C., Gorjan, L., and Clemens, F.J. (2019). Thermoplastic processing and debinding behavior of NbC-M2 high speed steel cemented carbide. *J. Mater. Process. Technol.* 263: 91–100.

153 Sarraf, F., Abbatinali, E., Gorjan, L. et al. (2021). Effect of MgO sintering additive on mullite structures manufactured by fused deposition modeling (FDM) technology. *J. Eur. Ceram. Soc.* 41: 6677–6686.

154 Gorjan, L., Galusca, C., Sami, M. et al. (2020). Effect of stearic acid on rheological properties and printability of ethylene vinyl acetate based feedstocks for fused filament fabrication of alumina. *Addit. Manuf.* 36: 101391.

155 Hanemann, T. and Heldele, R. (2011). Fatty acid surfactant structure-feedstock flow properties: correlation for high-pressure ceramic injection molding: fatty acid surfactant structure-feedstock flow properties. *Int. J. Appl. Ceram. Technol.* 8 (6): 1296–1304.

156 Rane, K., Farid, M.A., Hassan, W., and Strano, M. (2021). Effect of printing parameters on mechanical properties of extrusion-based additively manufactured ceramic parts. *Ceram. Int.* 47 (9): 12189–12198.

157 Cano, S., Lube, T., Huber, P. et al. (2020). Influence of the infill orientation on the properties of zirconia parts produced by fused filament fabrication. *Materials* 13 (14): 3158.

158 Smirnov, A., Nikitin, N., Peretyagin, P. et al. (2023). Experimental and statistical modeling for effect of nozzle diameter, filling pattern, and layer height

of FDM-printed ceramic–polymer green body on biaxial flexural strength of sintered alumina ceramic. *J. Composit. Sci.* 7 (9): 381.
159 Anitha, R., Arunachalam, S., and Radhakrishnan, P. (2001). Critical parameters influencing the quality of prototypes in fused deposition modelling. *J. Mater. Process. Technol.* 118 (1): 385–388.
160 Bellini, A., Güçeri, S., and Bertoldi, M. (2004). Liquefier dynamics in fused deposition. *J. Manuf. Sci. Eng.* 126 (2): 237–246.
161 Kaur, S., Riedel, R., and Ionescu, E. (2014). Pressureless fabrication of dense monolithic SiC ceramics from a polycarbosilane. *J. Eur. Ceram. Soc.* 34 (15): 3571–3578.
162 Lewis, J.A. (1997). Binder removal from ceramics. *Annu. Rev. Mater. Sci.* 27 (1): 147–173.
163 Rak, Z.S. (1999). New trends in powder injection moulding. *Powder Metall. Metal Ceram.* 38 (3): 126–132.
164 Heiber, J., Clemens, F., Graule, T., and Hülsenberg, D. (2005). Thermoplastic extrusion to highly-loaded thin green fibres containing Pb(Zr, Ti)O_3. *Adv. Eng. Mater.* 7: 404–408.
165 Salehi, M., Clemens, F., Graule, T., and Grobéty, B. (2012). Kinetic analysis of the polymer burnout in ceramic thermoplastic processing of the YSZ thin electrolyte structures using model free method. *Appl. Energy* 95: 147–155.
166 Salehi, M., Pfaff, E.M., Kaletsch, A. et al. (2015). Manufacturing of tubular dead-end membranes by continuous thermoplastic extrusion. *Int. J. Appl. Ceram. Technol.* 12: E13–E18.
167 Halloran, J.W., Hodge, J.D., Chandler, D. et al. (1992). Fabrication and properties of high-temperature superconducting wire by the green-fiber method. *J. Am. Ceram. Soc.* 75 (4): 903–907.
168 Calvert, P. and Cima, M. (1990). Theoretical models for binder burnout. *J. Am. Ceram. Soc.* 73 (3): 575–579.
169 Amaya, H.E. (1990). Solvent dewaxing: principles and application. *Adv. Powder Metall.* 3: 233–246.
170 Jardiel, T., Levenfeld, B., Jiménez, R., and Várez, A. (2009). Fabrication of 8-YSZ thin-wall tubes by powder extrusion moulding for SOFC electrolytes. *Ceram. Int.* 35 (6): 2329–2335.
171 Gonzalez-Gutierrez, J., Cano, S., Schuschnigg, S. et al. (2018). Additive manufacturing of metallic and ceramic components by the material extrusion of highly-filled polymers: a review and future perspectives. *Materials* 11 (5): 840.
172 Hadian, A., Koch, L., Koberg, P. et al. (2021). Material extrusion based additive manufacturing of large zirconia structures using filaments with ethylene vinyl acetate based binder composition. *Addit. Manuf.* 47: 102227.
173 Yang, X., Jia, C., Xie, Z. et al. (2013). Water-soluble binder system based on poly-methyl methacrylate and poly-ethylene glycol for injection molding of large-sized ceramic parts. *Int. J. Appl. Ceram. Technol.* 10 (2): 339–347.
174 Rolere, S., Soupremanien, U., Bohnke, M. et al. (2021). New insights on the porous network created during solvent debinding of powder injection-molded

(PIM) parts, and its influence on the thermal debinding efficiency. *J. Mater. Process. Technol.* 295: 117163.

175 German, R.M. (1987). Theory of thermal debinding. *Int. J. Powder Metall.* 23: 237–245.

176 LOGRASSO, B.K. (1990). Thermal debinding of injection molded powder compacts. *Powder Metall. Int.* 22: 17–22.

177 Thomas-Vielma, P., Cervera, A., Levenfeld, B., and Várez, A. (2008). Production of alumina parts by powder injection molding with a binder system based on high density polyethylene. *J. Eur. Ceram. Soc.* 28 (4): 763–771.

178 Zhao, M., Qiao, L., Zheng, J. et al. (2019). Investigation of the solvent debinding in the injection molding of ZrO_2 ceramics using LDEP, HDPE and wax binders. *Ceram. Int.* 45 (3): 3894–3901.

179 Gorjan, L., Lusiola, T., Scharf, D., and Clemens, F. (2017). Kinetics and equilibrium of eco-debinding of PZT ceramics shaped by thermoplastic extrusion. *J. Eur. Ceram. Soc.* 37: 5273–5280.

180 Scheithauer, U., Bergner, A., Schwarzer, E. et al. (2014). Studies on thermoplastic 3D printing of steel–zirconia composites. *J. Mater. Res.* 29 (17): 1931–1940.

181 Conzelmann, N.A., Gorjan, L., Sarraf, F. et al. (2020). Manufacturing complex Al_2O_3 ceramic structures using consumer-grade fused deposition modelling printers. *Rapid Prototyp. J.* 26: 1035–1048.

182 Ter Maat, J.H.H., Ebenhöch, J., and Sterzel, H.-J. (1992). Fast catalytic debinding of injection moulded parts. In: *4th International Symposium on Ceramic Materials and Components for Engines* (ed. R. Carlsson, T. Johansson, and L. Kahlman), 544–551. Dordrecht: Springer Netherlands.

183 Krug, S., Evans, J.R.G., and ter Maat, J.H.H. (2000). Residual stresses and cracking in large ceramic injection mouldings subjected to different solidification schedules. *J. Eur. Ceram. Soc.* 20 (14): 2535–2541.

184 Krug, S., Evans, J.R.G., and ter Maat, J.H.H. (2001). Transient effects during catalytic binder removal in ceramic injection moulding. *J. Eur. Ceram. Soc.* 21 (12): 2275–2283.

185 Stringari, G.B., Zupančič, B., Kubyshkina, G. et al. (2011). Time-dependent properties of bimodal POM – application in powder injection molding. *Powder Technol.* 208 (3): 590–595.

186 Attia, U.M. and Alcock, J.R. (2012). Fabrication of ceramic micro-scale hollow components by micro-powder injection moulding. *J. Eur. Ceram. Soc.* 32 (6): 1199–1204.

187 Gonzalez-Gutierrez (2014). Selection of appropriate polyoxymethylene-based binder for powder injection molding feedstock. *Zeitschrift Kunststofftechnik: J. Plast. Technol.*, 10 (5), 150–164.

188 Chartier, T., Ferrato, M., and Baumard, J.F. (1995). Supercritical debinding of injection molded ceramics. *J. Am. Ceram. Soc.* 78 (7): 1787–1792.

189 Chartier, T., Delhomme, E., and Baumard, J.-F. (1997). Mechanisms of binder removal involved in supercritical debinding of injection moulded ceramics. *J. Phys. III France* 7 (2): 291–302.

190 Chartier, T., Bordet, F., Delhomme, E., and François Baumard, J. (2002). Extraction of binders from green ceramic bodies by supercritical fluid: influence of the porosity. *J. Eur. Ceram. Soc.* 22 (9): 1403–1409.

191 Wu, Y., Si, W.J., and Miao, H.Z. (2008). Kinetics for supercritical CO_2 debinding of injection molded ZrO_2. *KEM* 368–372: 736–739.

192 Leriche, A., Cambier, F., and Hampshire, S. (2017). Sintering of ceramics. In: *Reference Module in Materials Science and Materials Engineering*. Elsevier.

193 Kemp, J.W., Hmeidat, N.S., and Compton, B.G. (2020). Boron nitride-reinforced polysilazane-derived ceramic composites via direct-ink writing. *J. Am. Ceram. Soc.* 103 (8): 4043–4050.

194 Zhou, S., Cai, Q., Tirichenko, I.S. et al. (2023). Additive manufacturing of Al_2O_3 with engineered interlayers and high toughness through multi-material co-extrusion. *Acta Mater.* 246: 118704.

195 Cai, Q., Meille, S., Chevalier, J. et al. (2023). 3D-printing of ceramic filaments with ductile metallic cores. *Mater. Des.* 225: 111463.

196 Fu, Z., Freihart, M., Schlordt, T. et al. (2017). Robocasting of carbon-alumina core-shell composites using co-extrusion. *Rapid Prototyp. J.* 23 (2): 423–433.

197 Xia, Y., Lu, Z., Cao, J. et al. (2019). Microstructure and mechanical property of Cf/SiC core/shell composite fabricated by direct ink writing. *Scr. Mater.* 165: 84–88.

198 Abel, J., Kunz, W., Michaelis, A. et al. (2022). Non-oxide CMC fabricated by fused filament fabrication (FFF). *Int. J. Appl. Ceram. Tech.* 19 (2): 1148–1155.

199 Cheng, Y., Zhang, Y., Zhang, F. et al. (2024). Using flexible hydrothermal carbon-coated continuous carbon fiber to fabricate Cf/ZrB2-SiC composites by fused deposition modeling and precursor infiltration pyrolysis. *J. Eur. Ceram. Soc.* 44 (2): 721–728.

200 Freudenberg, W., Wich, F., Langhof, N., and Schafföner, S. (2022). Additive manufacturing of carbon fiber reinforced ceramic matrix composites based on fused filament fabrication. *J. Eur. Ceram. Soc.* 42 (4): 1822–1828.

201 Cramer, C.L., Yoon, B., Lance, M.J. et al. (2022). Additive manufacturing of C/C-SiC ceramic matrix composites by automated fiber placement of continuous fiber tow in polymer with pyrolysis and reactive silicon melt infiltration. *J. Composit. Sci.* 6 (12): 359.

202 Mei, H., Yan, Y., Feng, L. et al. (2019). First printing of continuous fibers into ceramics. *J. Am. Ceram. Soc.* 102 (6): 3244–3255.

203 Yan, Y., Mei, H., Zhang, M. et al. (2022). Key role of interphase in continuous fiber 3D printed ceramic matrix composites. *Compos. A: Appl. Sci. Manuf.* 162: 107127.

204 Hadian, A., Duckek, J., Parrilli, A. et al. (2024). Additive manufacturing of fiber-reinforced zirconia-toughened alumina ceramic matrix composites by material extrusion-based technology. *Adv. Eng. Mater.* 26: 2302158.

205 Stiller, J.H.M., Nestler, D., Uhlmann, S. et al. (2024). Additive manufacturing of short fiber oxide ceramic matrix composite: process analysis and material properties. *Int. J. Appl. Ceram. Tech.* 21 (6): 3863–3875.

5

Powder-Bed Fusion (PBF)

Enrique Juste and Fabrice Petit

Research and Technological Support Department, Belgian Ceramic Research Center, 4, Avenue Gouverneur Cornez, 7000, Mons, Belgium

Powder-bed fusion (PBF), as defined by the ISO/ASTM 52900 standard, constitutes one of the seven additive manufacturing (AM) processes. This manufacturing process involves the layer-by-layer construction of 3D objects from a powder bed, using a high-energy beam to selectively melt the powder material following its prior deposition onto a build platform. While extensively researched and widely implemented for polymers and metals, PBF remains comparatively less prevalent for ceramics, despite its inherent advantages in shaping these materials.

Ceramics are fundamentally defined as inorganic, nonmetallic materials, with a diverse range of compositions, including oxides, nitrides, and carbides, as well as noncrystalline and partially crystalline glasses. Characterized by high hardness, abrasion resistance, elevated melting temperatures, and remarkable chemical stability, ceramics, however, exhibit inherent brittleness and are often associated with low electrical and thermal conductivities, which introduce significant challenges in their shaping and processing. Traditional ceramic industrial shaping methods, including molding, extrusion, injection molding, pressing, and machining, often involve more or less high costs as a function of the quantity of parts to produce and impose certain restrictions on the complexity of manufacturable geometries. PBF offers an alternative, enabling the production of advanced ceramic structures that would be challenging, if not impossible, to achieve through conventional techniques.

This chapter focuses on the selective PBF of ceramics, commonly referred to as selective laser sintering (SLS), selective laser melting (SLM), and electron beam melting (EBM). Following a concise historical overview of PBF and a discussion on appropriate terminological considerations, this chapter addresses the distinction between direct (single-step) and indirect (multi-step) PBF approaches for ceramics. Subsequent sections explore key process parameters, including the interaction of the energy beam with the ceramic powder bed, considerations about powder properties and spreading, and potential post-processing techniques. This discussion is supported by concrete examples drawn from various academic sources, including research articles, review papers, and doctoral theses.

The objective is to provide a comprehensive understanding of ceramic PBF, clarify the fundamental challenges associated with the process, and highlight the most recent advancements in the state of the art.

5.1 Short Historical Background of PBF

AM, also known as 3D printing, has revolutionized material fabrication by enabling the creation of complex geometries that were previously non-shapable with traditional subtractive or formative methods. Defined by ASTM/ISO 52900 [1], AM is the process of joining materials layer by layer, based on a 3D digital model, offering unparalleled flexibility in design. PBF is one of the current seven manufacturing processes described in the standard.

This paradigm shift was first envisioned in the early 1980s, with the advent of technologies targeting polymers and metals. In 1986, Carl R. Deckard patented what is now known as selective laser sintering (SLS), a cornerstone method in AM [2]. Originally, the process was developed to handle a wide array of materials, including polymers, metals, and ceramics. Deckard's innovation gained momentum when he, with his supervisor and later research partner, Prof. Joe Beaman, successfully demonstrated the first academic functional prototype of an SLS machine, nicknamed "Betsy." By the late 1980s, commercial interest in SLS expanded, securing investments and paving the way for its industrial adoption. SLS was originally developed by Nova Automation to further commercial development. In 1989, Nova Automation became DTM Corp. and was later acquired by 3D Systems in 2001.

While PBF process initial focus was primarily on polymers and metals, the field expanded to include ceramics due to their distinct properties. Regardless of the type of material being shaped from a powder bed using laser tools, the basic principle remains the same. PBF is an AM process that allows the construction of 3D objects layer by layer through the following steps:

1) **Digital slicing:** the 3D-CAD model of the desired object is sliced into a series of 2D cross sections defining the laser scanning paths for each layer.
2) **Powder-bed spreading:** a thin, uniform layer of powder is spread evenly onto the build platform.
3) **Beam fusion:** a high-powered energy beam (either laser or electrons), guided by a precisely controlled scanner, selectively binds the powder along the defined scan paths, creating the shape of the current layer.
4) **Platform lowering:** once a layer is completed, the build platform is lowered by a distance compatible with the layer thickness to prepare for the spreading of the next layer.
5) **Iterative building:** steps 2–4 are repeated iteratively, with each successive layer adding to the growing structure until the entire 3D object is fully shaped.
6) **Powder removal:** after the building is complete, any unbound powder is removed from the manufactured part. This excess powder can be collected

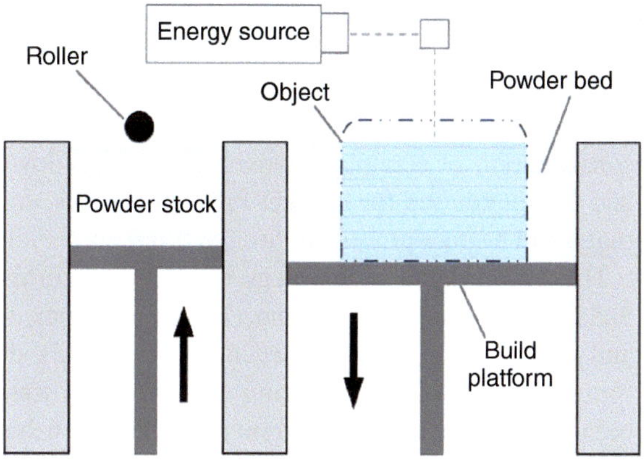

Figure 5.1 Schematic diagram of powder-bed fusion process. [3] / MDPI / CC BY 4.0.

and reused after treatments such as sieving for subsequent builds, minimizing material waste.

A schematic diagram of the PBF procedure is presented in Figure 5.1.

PBF offers several advantages in the context of ceramics processing, including:

- **Mold-free manufacturing:** the ability to fabricate near-net-shape ceramic parts directly from digital models without requiring molds, simplifying the production process, and reducing associated costs.
- **Complexity of design for free:** enhanced freedom in designing complex and customized components that would be challenging or impossible to achieve with traditional methods.
- **Minimal material waste:** AM reduces waste by depositing material only where needed, cutting costs, and supporting sustainable practices.

5.2 Terminology Focus

The original purpose of PBF was initially applied to polymers and metals, serving as a tool for rapid prototyping (RP). Over the years, its role has expanded beyond prototyping, establishing itself as a viable manufacturing method for complex and high-performance materials, including ceramics. However, the terminology originally developed for polymer and metal PBF does not always align with the unique characteristics of ceramic materials, leading to ambiguities and inconsistencies when describing ceramic PBF processes. PBF technology was first developed for low-melting-point materials, such as polymers, like ABS, PVC, PEEK, and PA, and various composites. The adaptation of this technology to metals, including aluminum, titanium alloys, steels, and copper, significantly broadened its industrial relevance. Ceramics, on the other hand, pose distinct challenges due to their high melting points, low thermal conductivity, and susceptibility to thermal shock. The direct application of metal-based PBF terminology to ceramics can be misleading, as ceramics behave fundamentally differently under high-energy beam irradiation. The introduction of binder-assisted PBF was a key development that facilitated the processing of refractory ceramic materials. Lakshminarayan et al. were among the pioneers in ceramic SLS, demonstrating the feasibility of fabricating complex alumina-based components using binders with lower melting points, such as ammonium phosphate ($NH_4H_2PO_4$) and boron oxide (B_2O_3), which enabled the consolidation of ceramic powders at relatively low temperatures. This approach laid the foundation for indirect PBF of ceramics, enabling the fabrication of parts that could be post-processed through binder removal and sintering [4, 5].

The conventional terminology used in PBF of metals and polymers, including SLS, SLM, laser beam melting (LBM), laser powder-bed fusion (L-PBF), EBM, and electron beam powder-bed fusion (EB-PBF), does not always precisely capture the mechanisms occurring in ceramic processing. This discrepancy arises because ceramics exhibit different consolidation behaviors under high-energy beam irradiation compared to metals and polymers. Kruth et al. [6] categorized PBF consolidation mechanisms into four primary types:

1) **Solid-state sintering:** Powder particles undergo diffusion-driven bonding at high temperatures without full melting. This method requires preheating the powder material and extended irradiation times (typically exceeding five seconds), making it a comparatively slow process.
2) **Chemically induced binding:** High-energy beam (generally laser) irradiation triggers a chemical reaction that facilitates interparticle bonding within the powder bed.
3) **Liquid-phase sintering and partial melting:** This occurs when one component of a multiphase powder-bed melts, acting as a binder for the remaining solid particles. Partial melting may also take place if a single-phase powder does not fully reach its melting point, leading to the formation of a porous structure of partially fused particles.
4) **Full melting:** The high-energy beam generates a complete melt pool within the processed volume, yielding a fully dense component.

In metal-based PBF, the distinction between SLS and SLM is straightforward:

- SLS refers to the diffusion-driven coalescence of particles below their melting point.
- SLM involves the full melting and subsequent resolidification of the material.

However, when dealing with ceramics, this classification becomes problematic. Many ceramic materials, particularly non-oxides, cannot be melted due to decomposition at high temperatures. Furthermore, the exact mechanisms governing ceramic PBF consolidation remain incompletely understood, and it is often unclear whether sintering, melting, or a combination of both occurs during such processes. This makes the terms "melting" and "sintering" unsuitable in many cases.

Given the inadequacies of conventional PBF terminology for ceramics, Grossin et al. (2021) [7] proposed the term "Powder-Bed Selective Laser Processing (PBSLP)" to provide a more flexible description of the process. Unlike SLS and SLM, which imply specific bonding mechanisms, PBSLP is defined based on processing criteria rather than consolidation mechanisms, making it applicable to a wide range of ceramic materials and process conditions. This term acknowledges that ceramic PBF may involve a combination of sintering, partial melting, and chemical reactions, without making overly restrictive assumptions about the dominant consolidation mechanism.

Throughout this chapter, PBSLP will be used as the preferred term when discussing ceramic powder-bed processing, particularly when the precise bonding mechanisms are unclear or involve multiple densification pathways.

In the field of ceramics, PBF can be classified into two main categories: direct and indirect methods. The direct process route produces near-net-shape components with varying degrees of density directly from the printer. In contrast, the indirect process route relies on a polymer binder and requires post-processing steps to achieve the final properties of ceramic parts. These methods will be detailed in the following sections.

5.3 Indirect – Powder-Bed-Selective Laser Processing (PBSLP)

Indirect PBSLP of ceramics involves the use of a sacrificial organic polymer binder, which is generally either mixed with the ceramic powder or applied as a coating on the individual ceramic particles. This binder material has a lower melting point than the ceramic phase. During laser processing, the polymer binder melts and acts as a temporary adhesive, consolidating the ceramic particles into a green part. This green part consists of a composite structure of ceramic particles embedded within the melted polymer. Subsequent post-processing steps are necessary. First, the polymer is removed during a debinding step. Then, a sintering step is performed to densify the ceramic part and achieve the desired mechanical properties [8, 9]. Figure 5.2 illustrates the flowchart of indirect PBSLP. As depicted in Table 5.1, many ceramics have already been processed using this approach. Different types of organic binders can be used as temporary binders, for example, waxes, thermoplastics, thermosets, and, sometimes, combinations such as a thermoset with semicrystalline PA-11 or wax with an amorphous thermoplastic.

The protocol illustrated in Figure 5.2 outlines the foundation for numerous studies employing this indirect approach to shape ceramics via indirect-PBSLP. Alumina has been extensively investigated by various research groups using different polymer binders, including stearic acid, polyamides (with contents ranging from 53 to 71 vol%), polystyrene (71 vol%), polypropylene (60 vol%), and carnauba wax mixed with low-density polyethylene [13]. The use of semicrystalline polymers such as polyamide and polypropylene is particularly relevant, as these materials are commonly employed in polymer-based SLS. During laser fusion, it is essential to preheat the powder bed. The preheating temperature is a function of the nature of the polymer. Semicrystalline thermoplastic polymers exhibit significant shrinkage during cooling as they transition from the molten to the solid state. This differential shrinkage can lead to considerable deformation in the fabricated parts. The crystallization temperature (T_c) is lower than the melting temperature (T_m), as illustrated

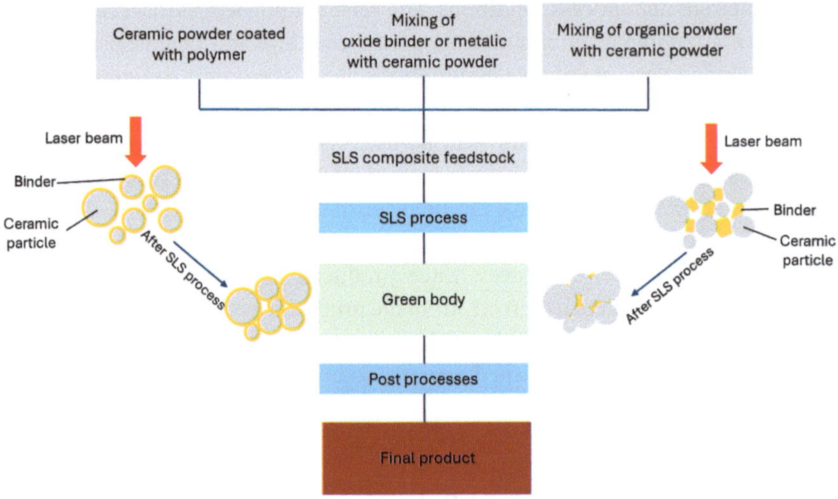

Figure 5.2 Flowchart of indirect PBSLP of ceramic parts. Adapted from Refs. [10–12].

Table 5.1 Example of organic polymer binders used in SLS process of ceramics.

	Organic binders		
	Wax	**Thermoset**	**Thermoplast**
Al_2O_3 [13]	Stearic		
Al_2O_3 [14]			Polymethyl metacrylate
Al_2O_3 [15]			Polyvinyl acetate
Al_2O_3 [16]			polystyrene
Al_2O_3–ZrO_2–TiC [17]		Unspecified	Polymethyl metacrylate
Apatite-mullite [18]			Unspecified acrylic binder
K_2O–Al_2O_3–SiO_2 [19]		Epoxy resin	
SiC [20]		Phenolic resin	
SiC [21]			Polymethyl metacrylate
ZrO_2 [8]			Polypropylene
3Y-TZP [22]		Epoxy resin	
ZrB_2 [23]	stearic		

in Figure 5.3. During the shaping process, the powder bed is preheated, and the temperature is maintained within the range between T_c and T_m. This temperature range is referred to as the "SLS window." The laser provides a small additional energy input to sinter the powder. These semicrystalline polymers exhibit a narrow "SLS window" (e.g. approximately 125–160 °C for polypropylene), which simplifies the shaping of polymer-ceramic powder mixtures under low laser power. In binder mixtures (e.g. carnauba wax + polypropylene), a low-viscosity binder is often selected.

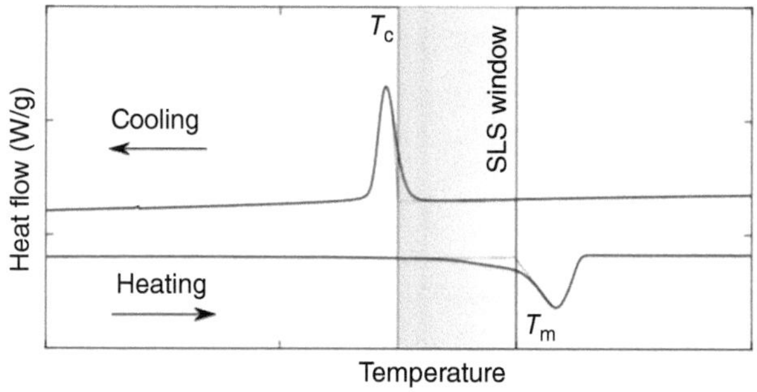

Figure 5.3 DSC curve of a semicrystalline polymer illustrating the SLS window (range between the crystallization – T_c and the melting temperature – T_m).

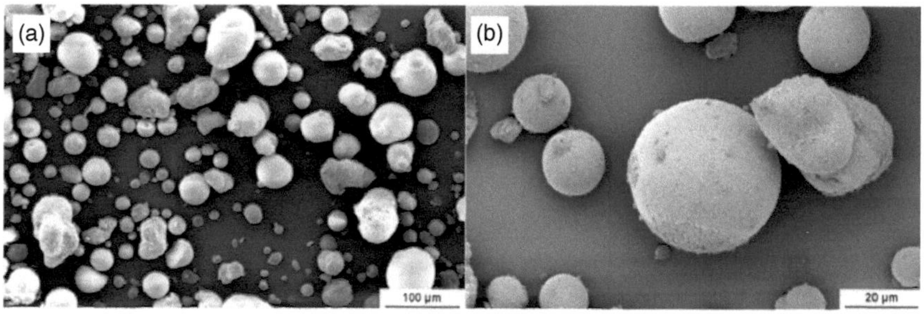

Figure 5.4 SEM micrograph of the PA-50 vol% Al_2O_3 composite microspheres at low (a) and high (b) magnification. Reference [8, 24] / KU Leuven, Groep Wetenschap & Technologie, Arenberg Doctoraatsschool.

Upon melting, such binders improve the packing arrangement of the ceramic particles in the powder bed, thereby increasing the relative density of the fabricated parts. For instance, feedstocks of micrometer-sized polymer/ceramic powder were prepared and used in SLS process by Shahzad [8] and Deckers et al. [9] through thermally induced phase separation (TIPS), as shown in Figure 5.4. Although the fabrication of 3D green parts was demonstrated, the green density of the parts was poor (36–54%), due to the low packing density of the spread powder bed. After sintering, the relative densities of Al_2O_3 or ZrO_2 ceramic parts (Figure 5.5) remained low (32–50%). In another study, Liu et al. [13] laser sintered alumina/stearic acid compounds with a CO_2 laser. Parts with relative densities up to 88% and a flexural strength of 255 ± 17 MPa were achieved. Post-processing techniques have been implemented to enhance the density of the final parts. This topic will be further discussed in Section 5.3.

In the indirect approach, binders are often polymers but can also be inorganic, such as metals or glasses with melting points lower than the ceramic phases. Organic binders are removed through post-treatment in a furnace, leaving only the ceramic phase. In contrast, inorganic binders cannot be eliminated. Instead, they become an integral part of the final structure, forming a composite material or reacting with another chemical phase to create a secondary material during a post-processing step.

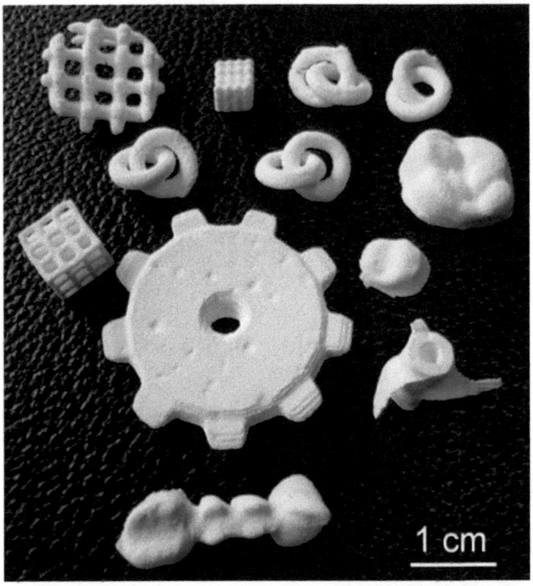

Figure 5.5 Porous alumina parts fabricated by the indirect SLS/debinding/ sintering process. Reference [8, 24] / KU Leuven, Groep Wetenschap & Technologie, Arenberg Doctoraatsschool.

For non-oxide ceramics that decompose before melting, the indirect method is particularly suitable. For instance, in the 1990s, Nelson et al. [15] demonstrated the feasibility of shaping silicon carbide (SiC) using SLS by coating SiC powder with PMMA and employing a CO_2 laser with a maximum power of 50W. Similarly, Hon et al. [25] processed a mixture containing 50 vol% polyamide and SiC using SLS. Meyers et al. [26] also shaped SiC but employed silicon (Si) as a binder (Figure 5.6). During laser treatment, the silicon melts and binds the SiC particles upon resolidification. A subsequent post-treatment step results in reaction bonded silicon carbide (RBSiC), achieving a flexural strength of up to 220 MPa. By the same route, Meyers et al. also shaped boron carbide by investigating reaction bonded boron carbide (RBBC) with a preliminary SLS step with silicon as binder. RBBC items exhibit a flexural strength

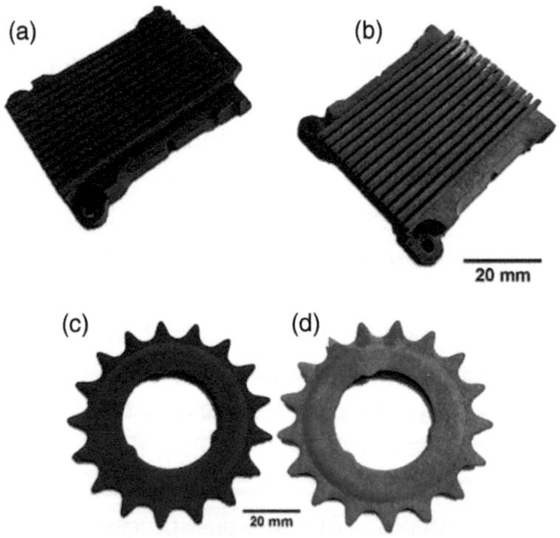

Figure 5.6 Porous SiC parts as laser-processed (a, c) and Si-SiC dense finished (b, d). [28] / with permission of Elsevier.

of up to 168 MPa and an elastic modulus of 300 GPa [27]. Whether for RBBC or RBSC, the parts manufactured have SiC or B_4C content of up to 84%.

5.4 Direct – Powder-Bed-Selective Laser Processing (PBSLP)

5.4.1 Electron Beam Melting (EBM)

Most academic studies on PBF have used a laser as the heat source. But recently, electron beams have also reported for ceramic PBF. This process, called EBM or electron beam-powder-bed fusion (EB-PBF), is a powder-bed technique that employs a focused electron beam to selectively melt powder particles layer by layer, forming solid components [28]. This approach requires electrically conductive materials. The electrons interact with the powder, and their kinetic energy is transferred into the powder bed, thus generating heat. Figure 5.7 illustrates the basic principle of the EB-PBF process with two different types of commercial machines. The electron beam power is governed by the cathode current (BC), while the energy of the beam is managed through a combination of BC and beam speed (BS). Additionally, the local energy deposition and the affected volume are a function of electron beam spot size and its focus offset (FO) [29]. The range of spot size is 200–1000 μm [30]. Smaller spot sizes enabling higher electron densities, EB-PBF is primarily used for 3D printing of metals and alloys due to their inherent electrical conductivity. However, considering the high energy density from the electron beam and the clean processing conditions offered by a vacuum or a controlled low-pressure helium atmosphere, EB-PBF is a viable option for processing more demanding materials such as ceramics, which are characterized by their high melting points. Certain ceramics exhibit sufficient electrical conductivity, making them potential candidates for EB-PBF processing. For example, titanium carbide (TiC) and tungsten carbide (WC) have electrical resistivities in the range of 178 and 63 μΩ.cm, respectively [31]. By comparison, Ti-6Al-4V alloy, a commonly processed material in EB-PBF, has a higher electrical resistivity of approximately 168 μΩ.cm [29].

EB-PBF offers several advantages over other PBF techniques. The electron beam has higher energy input and faster scanning speeds, thus allowing a more efficient processing of high-temperature and refractory materials. Additionally, the elevated powder-bed temperature during processing reduces thermal gradients, minimizing the risk of cracking and thermal stresses. However, the high energy and thermal requirements that make EB-PBF suitable for certain materials also introduce challenges when processing ceramics. Recent studies have mentioned EB-PBF to process advanced ceramics such as TiB_2 [32] and NbC [31]. Sjöström et al. explored EB-PBF of Al_2O_3 [29]. To enhance the inherently poor electrical conductivity of this metallic oxide, a 1 μm-thick nickel coating was applied to the powder particles. One of the most significant challenges in using EB-PBF for ceramics is the phenomenon of "smoking," where powder particles accumulate a negative charge from the electron beam, leading to Coulomb repulsion that destabilizes the powder bed

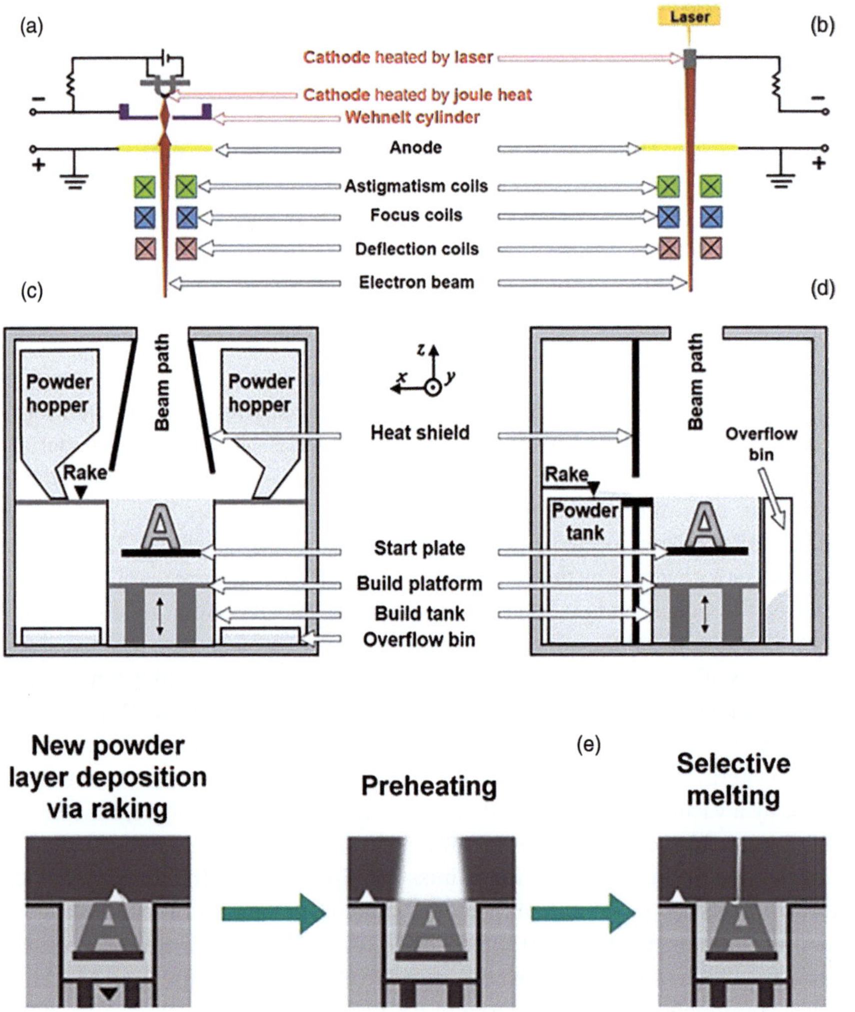

Figure 5.7 Schematic design of EB-PBF systems: triode versus diode. (a,c) Typical Arcam machine is equipped with a triode electron beam gun and powder hoppers as a powder feeder in the vacuum chamber. (b, d) Typical Freemelt ONE machine with a diode electron beam gun and two liftable powder tanks for build and powder feed in the vacuum chamber. (e) Schematic diagram of the EB-PBF process. [29] / Taylor & Francis / CC BY 4.0.

(Figure 5.8). This issue is particularly problematic for ceramics and other low electrical conductivity materials, which are unable to dissipate the accumulated charge efficiently. Smoking can result in powder displacement, defects in the final component, and even process interruptions.

To address the issue of smoking, several strategies have been explored. Preheating the powder bed with a defocused electron beam lightly sinters the particles, creating mechanical bridges that stabilize the powder bed and improve conductivity. Alternatively, applying a thin conductive coating to ceramic powders has been investigated as a method to enhance surface electrical conductivity, thereby reducing smoking sensitivity. In the case of Ni@Al_2O_3 particles, the Ni coatings are

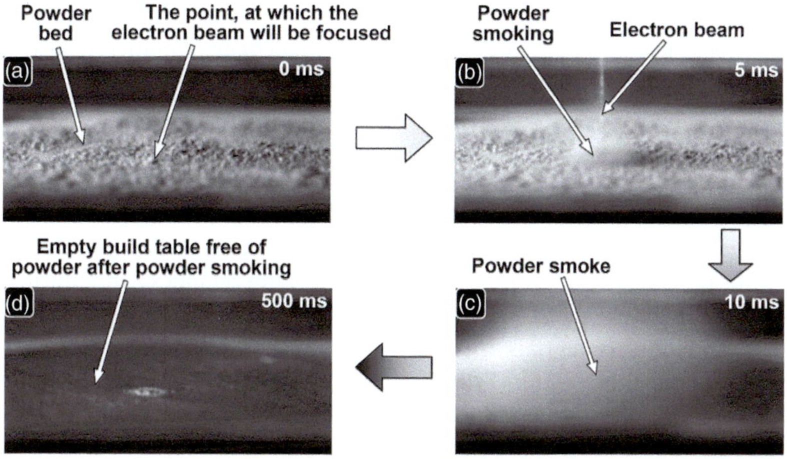

Figure 5.8 Example of powder layer smoking during EB-PBF processing. [29] / Taylor & Francis / CC BY 4.0.

designed to evaporate during processing and to minimally impact the chemistry and properties of the alumina [29]. Figure 5.9 shows an example of microstructure obtained with EB-PBF processing of Ni@Al_2O_3 powder.

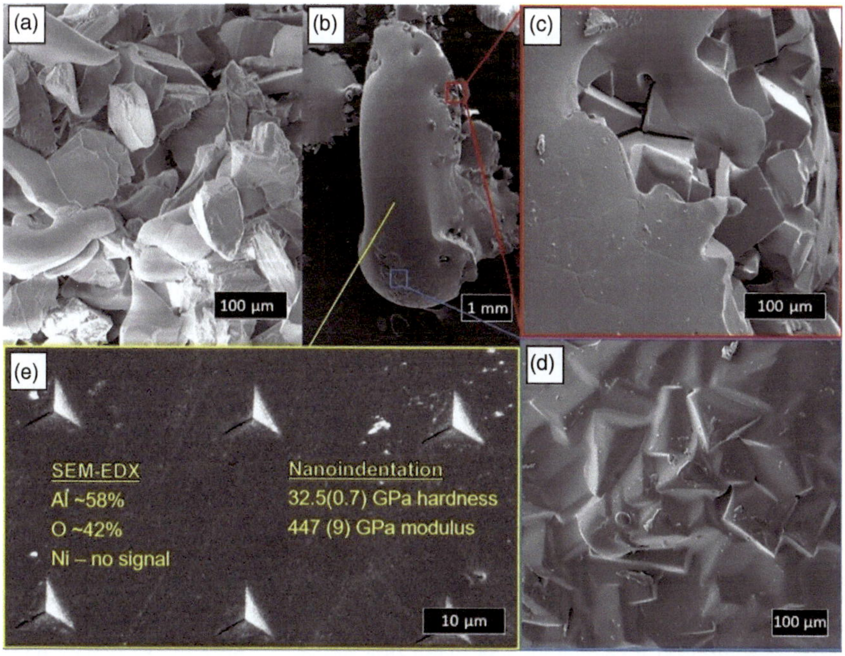

Figure 5.9 SEM micrographs of alumina samples; (a) sinter cake, (b) melted particle of consolidated powder, (c, d) close-up on isometric crystal formation on the melted particle, and (e) nanoindentations on a prepared cross section of a melted particle. [30] / Springer Nature / CC BY 4.0.

As with all processes, the determination of the suitable operating parameter window and the optimal parameters (BC, BS, and FO) is challenging. Another important parameter is the beam scanning strategy, which plays a role in controlling the thermal evolution of the layer or part being built. The thermal evolution of the material can be influenced by the preheating temperature (from 40 to 80% of the material's melting point), the applied energy density, and the selected scanning strategy. For ceramics, preheating is important to limit thermal gradients that may contribute to crack formation, as discussed in Section 5.4.2. Despite these challenges, EB-PBF holds significant potential for processing high-temperature ceramics and refractory materials. Studies have demonstrated the feasibility of using sintering aids to facilitate the consolidation of materials like cemented tungsten carbide (WC-13%Co [33]) or ultrahigh-temperature ceramic (ZrB_2–$ZrSi_2$ [34]). In these cases, the sintering aid melts under the electron beam, inducing liquid-phase sintering. Such approaches could pave the way for wider applications of EB-PBF in ceramic AM.

Research on EB-PBF for ceramics remains limited, with most studies focusing on overcoming fundamental barriers to processing. Innovations such as advanced powder treatments and tailored process strategies are critical for addressing the limitations of low-conductivity ceramics. While fully dense pure ceramics remain difficult to process, the development of composite materials and conductive additives offers potential solutions for expanding the applicability of EB-PBF. This process represents a promising advancement in AM, with the potential to unlock new possibilities for high-performance ceramic components.

5.4.2 Powder-Bed-Selective Laser Processing (PBSLP)

The use of a laser as an energy source is the most commonly applied method in PBF process. In the context of metals, PBSLP is more commonly referred to as SLM. Over recent years, SLM has emerged as the fastest-growing 3D-printing technology, primarily due to its capability to produce near-net-shape parts in a single-step process with final properties that often require no additional post-processing. In the field of ceramics, "direct-PBSLP" or "direct-SLS/SLM" involves the full sintering or melting of the powder bed to form a 3D solid part layer by layer. The term "direct" means the absence of polymers as binders, thereby eliminating the need for shaping and subsequent post-processing steps to complete ceramic 3D items. As such, direct PBSLP stands out among 3D-printing techniques for its ability to fabricate complex-shaped components directly from powder in a single step. However, processing ceramics with laser tools presents unique challenges compared to metals. These challenges arise from the intrinsically brittle nature of ceramics, their low resistance to thermal shock, and their high refractoriness.

In the literature, examples can be found of ceramics subjected to 3D shaping using the direct method of SLM or PBSLP. Many of these studies focus on alumina, alumina-zirconia, silica, and other oxide ceramics derived from these base ceramics. Bertrand et al. [35] investigated AM using this process to produce near-net-shape parts from various grades of yttria-stabilized zirconia with a fiber laser. Although the relative density remained low (56%), these parts demonstrated the feasibility of the 3D process. However, the surface roughness of the parts was high, and it was not possible to improve their relative density, even with post-processing densification treatments in conventional furnaces. Shishkovsky et al. [36] also fabricated porous alumina-zirconia parts using the same type of laser ($\lambda_{laser} = 1\ \mu m$), employing mixes with aluminum powder that oxidized under laser treatment to form alumina, which bonded with zirconia. In the early 2000s, Exner, Rugenfuss et al. [37, 38] developed and built a homemade custom system dedicated to Laser Micro Sintering (LMS), designed for shaping metals and ceramics. This machine was equipped with an Nd:YAG laser capable of sintering metallic and ceramic powders using stochastically distributed Q-switched pulses. The method significantly influenced how laser energy was distributed and absorbed by the powder, impacting both the manufacturing process quality and the properties of the produced parts. This technique minimized thermal accumulation effects, reducing the risks of overheating and excessive thermal gradients. Various ceramics, including Si-SiC, SiC, and SiO_2, were successfully shaped into small parts with high resolution. In the case of SiO_2, the consolidation mechanism could not be clearly identified due to its low absorption in the near-infrared (NIR) range. A reactive sintering mechanism was hypothesized [39]. The authors did not report any specific properties or relative density values for the fabricated parts.

Several research groups have chosen to develop custom equipment for PBSLP of ceramics due to the limitations of commercial systems and the inherent properties of ceramics that lead to defects during the process. For example, to mitigate thermal shocks caused by repeated rapid heating and cooling cycles during laser scanning, Hagedorn et al. [38, 39] proposed using a CO_2 laser to preheat the powder bed to high temperatures (up to 1700 °C) and a Nd:YAG laser to provide the additional energy required for melting (Figure 5.10a). In their work, Hagedorn et al. [40, 41] demonstrated the feasibility of melting the eutectic composition of 58.5 wt%

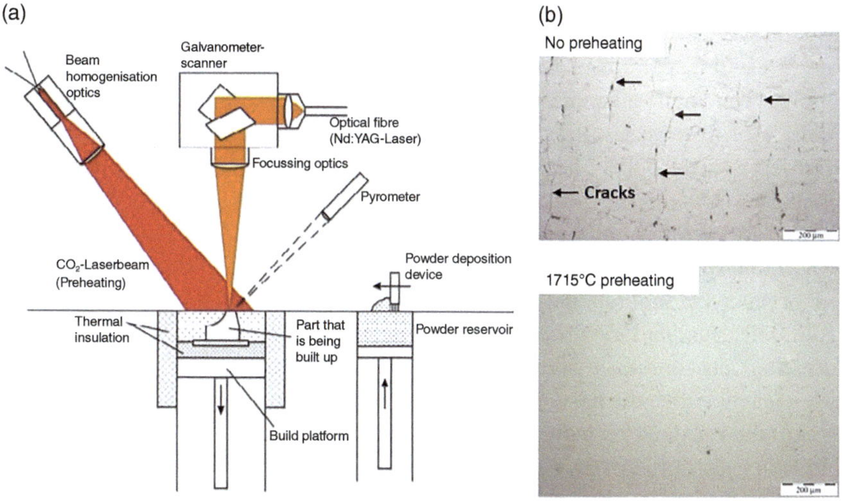

Figure 5.10 Schematic illustration of the experimental setup of SLM with ceramics, and the resulting microstructure without and with preheating. [44] / with permission of Emerald Publishing Limited.

Al_2O_3 and 41.5 wt% ZrO_2 via PBF. By preheating the powder bed just below its melting temperature with the CO_2 laser, they were able to reduce the formation of cracks typically caused by thermal shocks during the laser scanning cycles on conventional commercial systems. The 3D objects fabricated from this eutectic mixture were nearly fully dense (Figure 5.10b), with microstructures featuring fine grains and zirconia dendrites. While small cracks were observed, they did not exceed 50 μm, which corresponds to the layer thickness used in this study. These cracks were attributed to the thermal shock between the powder feeder and the preheated area created by the CO_2 laser. To address this issue, the authors introduced a buffer zone preheating step, heating the powder up to 400 °C before its introduction into the second preheated area via the CO_2 laser. This adjustment eliminated microstructural cracking. The resulting 3D objects exhibited flexural strengths of up to 500 MPa. However, the fabricated parts displayed significant surface roughness due to the extensive thermally affected zone created by the laser preheating. Adjacent grains in the powder bed were strongly fused to the irradiated region, contributing to the observed surface irregularities. Despite these challenges, the dual-laser system provided a promising pathway for improving the quality and mechanical performance of ceramic parts produced via PBSLP.

Florio et al. [42] also developed custom equipment for academic purposes to obtain a better understanding of the phenomena occurring during the laser irradiation of ceramics. The key advantage of their laboratory setup is its flexibility, as the type of laser used can be easily changed, and the equipment can be instrumented for further analysis [43]. In their study, they mainly used a green short-pulsed nanosecond (ns) laser (λ_{laser} = 532 nm) to process alumina granules doped with a small quantity of iron oxide (hematite) as an absorption promoter. Using this setup, they successfully fabricated 3D objects with relative densities of up to 90% and achieved a bi-axial flexural strength of 25 MPa, as measured by the ball-on-three-balls test.

Other authors have opted to use commercial equipment and adapt ceramic powders to meet the hardware requirements. This approach was employed by Juste et al. [44], who used an SLM system designed for metal-LPBF to process alumina. The main adaptations involved modifying powder properties such as particle size distribution, flowability, and, most importantly, optical absorbance at the laser wavelength. They utilized a spherical powder from a graphite-based colloidal suspension incorporated into an alumina slurry prior to spray drying. The resulting gray powder improved the optical coupling efficiency with a Nd:YAG laser (λ_{laser} = 1.064 μm). Following laser treatment, near-net shape 3D alumina objects were produced with a relative density over 97%. In this study, the equipment used was initially qualified for metals and operated under a neutral atmosphere. Laser processing was conducted at a low partial pressure of oxygen (<5000 ppm). Upon exiting the machine, the parts appeared gray. After post-treatment in an air furnace, the alumina regained its white color. However, it was not possible to increase the relative density of the parts, even with heat treatment at 1650 °C to remove cracks and porosity (Figure 5.11). This method was later adopted by several other research groups with variations in the powder preparation routes. For instance, Verga et al. [45] fabricated alumina toughened zirconia (ATZ) 3D components via PBSLP

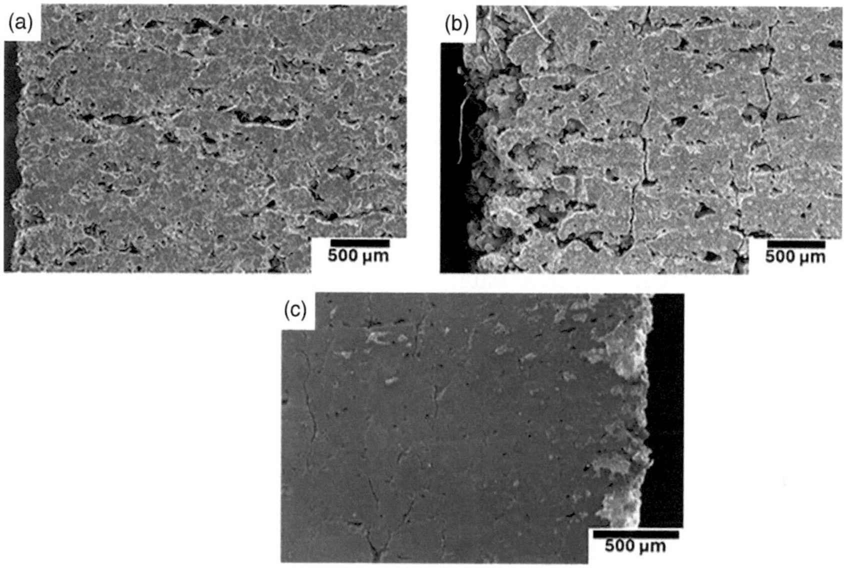

Figure 5.11 Example of microstructures resulting from PBSLP of alumina powder (a) as printed and (b) with post-thermal treatment at 1650 °C/2 h. (a and b) [45] / with permission of Springer Nature. (c) As printed after parameters optimization courtesy of BCRC.

using spray-dried powder containing a pyrolyzable binder. After granulation, the powder was calcinated under a reducing atmosphere, resulting in the thermal decomposition of organic matter and a uniform carbon distribution within the granulate structure. Following laser treatment, near-net shape parts were obtained with relative densities over 90% and a four-point bending strength of 31 ± 11 MPa. Similarly, Ferrage et al. [46, 47] applied this methodology by mixing graphite powders with yttria fully stabilized zirconia (ZrO_2-$8Y_2O_3$, YSZ). It was found that PBSLP enabled the reproducible fabrication of simple YSZ components with good geometrical accuracy, mechanical strength, and a relative density of 96.5%. Shaker [48] also used a similar approach with hydroxyapatite (HA) powder to develop bone substitutes with custom macroporosity for skull implants. HA parts were manufactured with channels down to 750 μm in diameter. In the PBSLP process, using carbon as an absorption enhancer is interesting due to its biocompatibility and chemical inertness with HA.

The application of PBSLP to glass and silica materials has been explored, demonstrating both the potential and challenges of processing amorphous structures using laser-based AM. Authors such as Wang et al. [49] and Tang et al. [50] studied silica powder shaping with a CO_2 laser ($\lambda_{laser} = 10.6$ μm). The silica powders employed were not pure and contained traces of metals such as Zr, Ti, Al, Ba, Fe, Mg, or alumina. These secondary phases facilitated the shaping process either by melting the metallic species or by forming a low-temperature eutectic phase between alumina and silica. The mechanical performance of the fabricated parts remained low but was sufficiently promising to consider their use for metal casting molds. Silica was also shaped using an Nd:YAG laser and using a carbon additive to process controlled 3D porous structures [51]. Among these studies, Datsiou et al. [52, 53] investigated the direct laser processing of soda lime silica glass powder without the use of absorption promoters for a laser wavelength $\lambda_{laser} = 1064$ nm. Following laser irradiation, the resulting structures remained fully amorphous, with an overall porosity of approximately 12%, primarily consisting of closed pores. The fabricated components exhibited a flexural strength of up to 6.9 MPa. Microstructural analysis revealed the presence of unmelted particles and cracks within the material, despite maintaining the build platform at a temperature of 250 °C throughout the fabrication process. Although the successful production of complex 3D geometries was achieved, the resulting objects lacked transparency. This opacity was attributed to several factors, including incomplete fusion of powder particles, residual porosity, and the adhesion of satellite powder particles on the surface due to the heat-affected zone induced by laser irradiation. These microstructural defects pose limitations for applications requiring optical clarity. Nevertheless, the findings from this study highlight the feasibility of processing glass through PBF, particularly for applications where transparency is not a primary requirement. Given the inherent chemical stability of soda lime silica glass, such fabricated structures could be well-suited for industries requiring high chemical resistance, such as chemical processing, pharmaceuticals, and catalysis.

Monton et al. successfully fabricated SiC components using a commercial PBF system and commercially available silicon carbide powder. Their approach relied on a combined numerical and experimental analysis to determine the optimal operating parameters that deliver the required thermal energy for effective SiC sintering. Through this method, they achieved the production of geometrically complex SiC structures with relative densities of up to 82% (Figure 5.12). However, despite the success in forming SiC parts, the fabricated components contained up to 5 wt% residual silicon and 1 wt% carbon. These secondary phases were attributed to the decomposition of SiC during laser irradiation, a phenomenon exacerbated by localized overheating, which is inherently linked to the operational limitations of the commercial laser system used. In their subsequent studies, Monton et al. sought to further refine their process by functionalizing the SiC powder surface to enhance its sinterability. The approach involved grafting monomer-derived SiC ceramics onto the powder particles. This surface modification significantly reduced the energy input required for SiC processing, thereby lowering the peak temperatures reached during laser irradiation. As a result, the extent of SiC decomposition into elemental Si and C was reduced by a factor of two compared to untreated powders [55–58].

For hard materials such as cemented tungsten carbide (WC-Co), SLM has also gained significant attention as a viable fabrication method. However, the specific conditions of SLM, including high localized laser power and vacuum chamber operation, introduce several challenges [59]. Among the most critical issues are decarburization, the formation of unwanted η-phase, and the occurrence of cracking. These defects stem from the high solidification rate of the melt pool and the mismatch in thermal expansion coefficients between WC and its metallic binder, typically Co, Ni, or Fe. The resulting thermal gradients induce significant mechanical stresses, leading to crack formation during cooling. While reducing laser intensity can alleviate

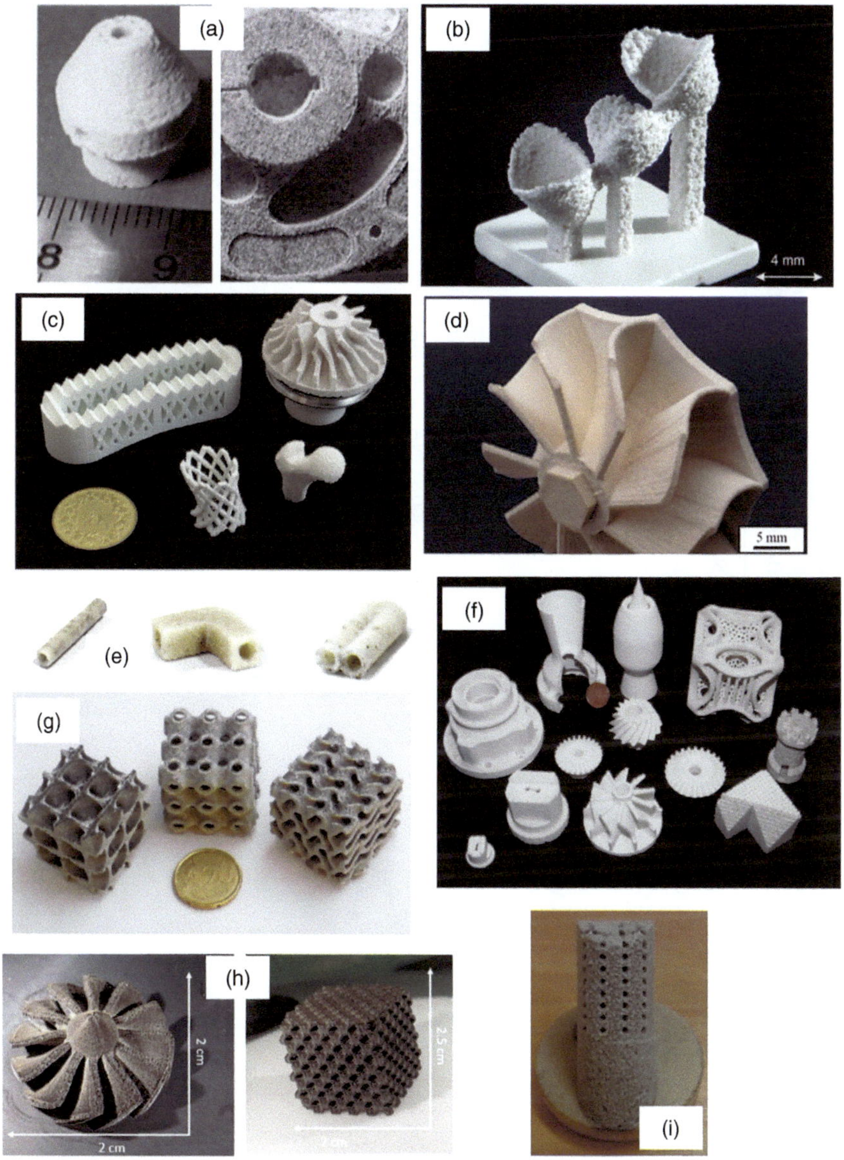

Figure 5.12 Ceramic components manufacturing by PBSLP. (a) Zirconia. [36] / with permission of Elsevier. (b) 58.5 wt% Al_2O_3 and 41.5 wt% ZrO_2. [41] / Elsevier / CC BY-NC-ND 3.0. (c) alumina toughened zirconia [46] / with permission of Elsevier. (d) 80 wt.% zirconia/20 wt.% alumina. [44] / with permission of Emerald Publishing Limited. (e) Soda lime silica glass. [53] / Elsevier / CC BY 4.0. (f, g) Alumina. [45] / with permission of Springer Nature, [66] / Springer Nature/ CC BY 4.0. (h) silicon carbide. [56] / Springer Nature/ CC BY 4.0. (i) Hydroxyapatite. [48] / Centre for Direct Scientific Communication / CC BY 4.0.

Table 5.2 Examples of the process conditions and final properties of ceramic parts by PBSLP.

Materials	Laser wavelength (μm)	Laser power (W)	Scanning speed (mm/s)	Laser spot size (μm)	Hatch space (μm)	Layer thickness (μm)	Energy density (J/mm^3)	Relative density (%)	Mechanical properties	Comments
ZrO_2-Y_2O_3 [35]	NA	50	1250–2000	NA	20–40	30	NA	56	NA	—
ZrO_2-8Y_2O_3 [46]	1065	30–100	10–1500	NA	30–80	80–180	NA	96.5	NA	—
Al_2O_3-ZrO_2 in eutectic ratio [40]	1064	60	200	200	50	50	30	~100	$\sigma_f = 500$ MPa	CO_2 Laser preheating induced crack-free sample
Alumina toughened zirconia [45]	1064	34	86	NA	175	40	NA	>90	$\sigma_f = 31 \pm 11$ MPa (4 point bending strength)	Laser irradiation executed under nitrogen atmosphere
Al_2O_3 [65]	1070	10–18.5	20–100	170	125	100	100–300	NA	NA	Al_2O_3 mixed with Fe_2O_3 or MnO_2
Al_2O_3 [44]	1064	125–150		69		100	NA	97.5	NA	Al_2O_3 mixed with graphite
Al_2O_3 [54]	1070	50–210	100–400	NA	30–50	100	90–100	94.5	$\sigma_C = 140.8 \pm 11.6$ MPa Compressive strength	Laser irradiation executed under argon atmosphere Pores and cracks
soda lime silica glass [52]	NA	20–100	5–60	20	100–450	70–120	60–140	NA	6.2–6.9 MPa	—
Silica sand [50]	10.6	120	120	NA	200	150		NA	13 MPa	—
SiSiC [66]	1.064	12–15	50–200	50	77	30		NA	$\sigma_f = 162$ MPa	Reaction-bonded silicon carbide
SiC [55]	1060	30–50	50–500	70	8.75 – 87.5	20–40	180–600	81	NA	—
WC-Co [67]	NA	60–140	20–100	NA	130–230	30–50	185–1667	>95	NA	—

cracking, it may also lead to residual porosity, affecting the mechanical performance of the final component [60]. Eliminating these defects remains a primary focus of ongoing research [61]. Studies have shown that optimizing laser parameters (power, scanning speed, hatch space), binder composition, and grain morphology can yield relative densities exceeding 96% [62]. More recently, advancements in SLM technology have introduced high-temperature powder-bed preheating, with temperatures ranging from 650 to 950 °C. Literature reports indicate that this approach significantly reduces crack formation, improving the structural integrity of WC-Co components. However, challenges remain in eliminating undesirable phases, even when utilizing advanced equipment [63, 64]. Further research is needed to refine processing conditions and enhance the quality of PBSLP-fabricated cemented tungsten carbide components for industrial applications.

These advancements demonstrate the potential of viable processes to manufacture complex 3D ceramic parts, paving the way for enhanced material integrity and performance in laser-based PBF AM (Table 5.2).

5.5 Process Parameters PBF

Regardless of whether the direct or indirect approach is used, PBF processes are highly complex and influenced by numerous factors, including the properties of the powder or feedstock, the characteristics of the equipment, and the manufacturing strategies employed. This complexity arises from the underlying phenomena involved in the consolidation mechanisms driven by the interaction between the material and the energy beam (photons or electrons). In PBF processes, the interaction between the energy beam and the powder creates highly dynamic and intricate conditions at the microscopic level. These conditions significantly influence the microstructure, mechanical properties, and geometric accuracy of the manufactured parts. Process parameters represent a set of critical variables that must be thoroughly understood and precisely controlled to achieve optimal shaping. Since laser-based processes are the most widely used, the following section will focus on the parameters specific to these techniques.

5.5.1 Powder-Bed Features

Powder-bed packing density is a critical factor influencing the properties of the 3D-built part. A high and homogeneous packing density is essential for achieving parts with high density after laser processing, along with uniform microstructure and consistent properties. For indirect PBF, the packing density ensures the stability of green parts during fabrication, minimizing the risk of deformation or collapse under process conditions. Therefore, the optimization of the final properties of the components is linked to the combination of the powder deposition technique and the characteristics of the particles as illustrated in Figure 5.13 [68, 69].

These parameters can be broadly grouped into three main categories, each of which interacts with the others.

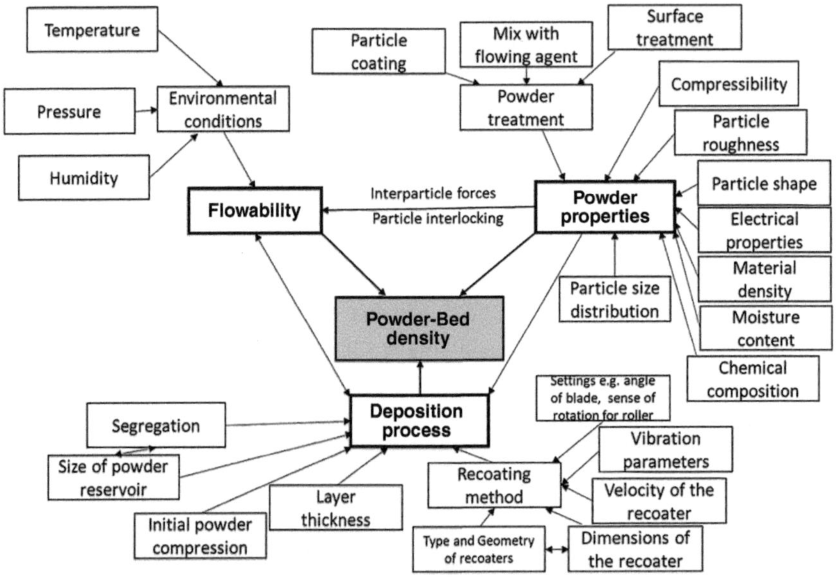

Figure 5.13 Parameters influencing powder-bed density. [69] / with permission of Elsevier.

5.5.1.1 Powder Properties

The shape, surface roughness, particle size, and particle size distribution are critical factors for the effective operation of powder beds. Fine particles are more reactive; however, they present significant challenges. For particles below 10 μm, van der Waals forces dominate, leading to agglomeration that hinders flowability and powder spreading. In contrast, larger particles improve flowability but may result in lower relative density when used alone. Furthermore, particle size must not exceed the set layer thickness, as larger particles can disrupt the layering process. This relationship highlights the need for an optimal particle size distribution tailored to the chosen layer thickness. A bimodal particle size distribution, where finer particles fill the voids between coarser ones, promotes denser and more homogeneous packing. Achieving maximum powder packing density requires an optimal size ratio between coarse and fine particles. The literature provides various models for determining the ideal proportions of powder mixtures to achieve the highest possible granular packing densities for AM processes based on powder beds. Typically, bimodal size distributions with a particle size ratio of approximately (1:10) and a fine particle proportion of around 30 wt% are frequently cited. This combination leads to powder beds with packing densities ranging from 60% to 85% [70, 71]. The surface quality of printed parts is also heavily influenced by particle size distribution and flowability. Powders with high flowability facilitate uniform and smooth layer deposition, which is critical for producing parts with superior surface finishes [72].

- **Flowability:**

Powder flowability is a critical parameter for achieving dense and homogeneous powder beds. Flowability is influenced by both the intrinsic properties of the powder and operational conditions (or room processing conditions), such as humidity and temperature, which can affect electrostatic and van der Waals forces [68]. The shape and size of particle distribution play a key role in determining flowability. Spherical particles provide better flowability and compactness than irregularly shaped particles, increasing friction, which reduce powder flowability and its ability to spread uniformly across the bed [73]. Similarly, particle surface roughness significantly impacts flowability; lower surface roughness reduces interparticle friction, thereby enhancing flow behavior. A narrow particle size distribution also improves powder flowability. Consequently, ceramic powders used in PBSLP are often prepared through spray drying, a process that enables precise control over both particle size and shape distribution. The powder feeding system within the build area varies depending on the machine design. The two most common systems are the hopper-based feeding system and the elevator-based feeder. Powder flowability has a significant impact on the dosing capability of these systems. Specifically, powders with poor flowability are more challenging to dose using a hopper than an elevator-based feeder, which is better suited for handling such powders. Flowability is also important during the de-powdering step, which involves removing unirradiated powder surrounding and within the cavities of the fabricated part. Poor flowability can hinder this process, particularly in complex geometries.

Powder flowability and rheology can be evaluated using various techniques, including Hall flow meters, tapped density measurements (to calculate the Hausner ratio), angle of repose, powder rheometers, and rotating drum flow analyzers [74, 75]. These methods provide quantitative insights into the powder's flow characteristics and its suitability for use in AM powder-based processes. Nevertheless, no single method is sufficient for the complete characterization of a powder. A comprehensive understanding of powder flow behavior under various operating conditions requires the use of multiple techniques, tailored to the specific requirements and operational parameters depending on the spreading system available on the used PBF machine [7, 68].

- **Spreading process:**

The design of powder spreading devices plays an important role in determining the homogeneity and packing density of the powder bed, directly influencing the quality of parts produced in PBF processes. The choice of spreading mechanism depends on the specific machine design, with the most common systems being wipers and rollers.

A wiper, a simple thin blade, is a commonly used powder spreading device. The wiper drags and spreads the powder from the powder feeding zone across the build plate and previous spread layer. Blade geometry is a critical factor in determining the quality of the spread layer, with flat, round, or sharp edges tested to optimize surface smoothness [76–79]. The speed of the wiper also plays a significant role, as higher speeds can negatively impact powder-bed density by reducing the compaction of the layer.

Rollers are widely used in PBF machines due to their ability to spread powder uniformly while applying minimal force to the particles. Key parameters such as the linear and rotational speeds of the roller, its rotation direction, and its diameter significantly influence the homogeneity and packing density of the powder bed. Optimization of these parameters ensures consistent layer deposition, which is critical for achieving high relative density and defect-free printed parts.

Controlled vibration can also be integrated into spreading systems. The incorporation of transverse and longitudinal vibrations into powder spreading systems, such as wipers and rollers, is an alternative to improving the quality of the powder bed. These oscillatory motions are particularly beneficial for spreading fine or agglomerated powders, where achieving a uniform layer remains challenging. The addition of vibrations enhances the spreading process by facilitating the dispersion of agglomerated particles and promoting a more uniform distribution of powder across the build plate. Furthermore, the oscillatory motion induces particle rearrangement within the powder bed, resulting in a higher packing density [77, 80].

Beyond the conventional roller and wiper designs, innovative spreading systems have been developed to enable advanced functionalities. For example:

- **Ring recoater:** as described by Exner et al. [38], [81], this system allows for multi-material deposition, offering new possibilities for fabricating functionally graded materials or components with tailored properties.
- **Selective powder deposition (SPD) recoater:** this advanced design employs drums that selectively deposit fine powder voxels in a line-at-once manner, enabling the creation of layers composed of multiple powder materials. Such systems facilitate the fabrication of multi-material parts with distinct zones of different materials [82].

Achieving optimal parameters typically requires a combination of analytical approaches and iterative trial-and-error testing directly on the equipment, as analytical methods alone remain insufficient.

5.5.2 Hardware Parameters

Numerous parameters related to the laser system significantly impact the relative density and the quality of the fabricated part. These parameters must be adapted according to the nature of the materials used and their features. Fine-tuning of these parameters is required to ensure the performance of the part.

Machine selection for laser PBF requires careful consideration of fixed laser parameters like wavelength and beam shape. Although dynamic beam shape control is an emerging technology with the potential to enhance productivity [83], it is not yet a standard feature. The selection of the laser wavelength is a critical consideration, as the user must ensure sufficient optical coupling between the laser wavelength and the material being processed. This is particularly important for ceramics, as many oxide ceramics exhibit low absorption in the NIR range ($\lambda_{laser} \approx 1\ \mu m$) while absorbing significantly more in the far-infrared ($\lambda_{laser} \approx 10\ \mu m$), as shown in Table 5.3. However, most PBF machines utilize Nd:YAG or Yb-doped fiber lasers ($\lambda_{laser} \approx 1.06\ \mu m$), which are widely adopted due to their effectiveness in processing metals with high precision, enabled by beam spot sizes on the order of

Table 5.3 Absorptance (*A*) of materials at two laser wavelengths ($\lambda = 1.06$ and $10.6\,\mu m$).

	A	
Material	$\lambda = 1.06\,\mu m$	$\lambda = 10.6\,\mu m$
Cu	0.59	0.26
Fe	0.64	0.45
Ti	0.77	0.59
Sn	0.66	0.23
Al_2O_3	0.03	0.96
8YSZ	0.02	—
SiO_2	0.04	0.96
ZnO	0.02	0.94
SnO	0.05	0.95
BaO	0.04	0.92
CuO	0.11	0.76
SiC	0.78	0.66
TiC	0.82	0.46
WC	0.82	0.48

According to Refs. [84, 85].

tens of microns. In contrast, CO_2 lasers, operating at longer wavelengths, are less commonly used in such systems mainly due to their large laser spot size and the limited availability of open-architecture machines.

To overcome this limitation, several researchers have explored the addition of absorption promoters to enhance the optical coupling of ceramics at wavelengths around $\lambda_{laser} \sim 1\,\mu m$. For instance, Juste et al. [44] identified graphite as an effective absorption enhancer, as it significantly improves optical coupling in the NIR range without requiring large additions (typically $0.05–1_{vol}\%$). Moreover, under these processing conditions, graphite does not react with alumina. Similar approaches have been reported by other researchers: Ferrage et al. for Y_2O_3-stabilized ZrO_2 [46], Chang et al. for silica [51], and Verga et al. for zirconia [45]. Moniz et al. also investigated the use of carbon and β-SiC in combination with alumina [86], while Florio, Pfeiffer et al. tested alumina and zirconia-toughened alumina (ZTA) with various additives such as MnO_2 and Fe_2O_3 [42, 43, 87]. Relative densities up to 97% were reached in these different works.

Once the laser wavelength has been selected and sufficient optical coupling has been ensured, the key processing parameters are described below and illustrated in Figure 5.14.

- **Laser power:** This parameter refers to the energy applied to the powder bed via the laser source per second. In PBF machines, several types of lasers are used, such as continuous wave, pulsed, or frequency-modulated ones. Insufficient laser

Table 5.4 Relative densities achieved with alumina as a function of E_s and overlapping.

Sample	Relative density (%)	Energy density per surface unit – E_S (J/mm^2)	Overlapping
S3	0	5.3	−477
S6	0	6.6	−477
S9	0	7.9	−477
S2	57.2	5.1	−44
S5	67.3	6.3	−44
S8	76.9	7.6	−44
S1	92.9	13.3	28
S4	97.5	16.6	28
S7	97.3	19.9	28

Adapted from Ref. [44].

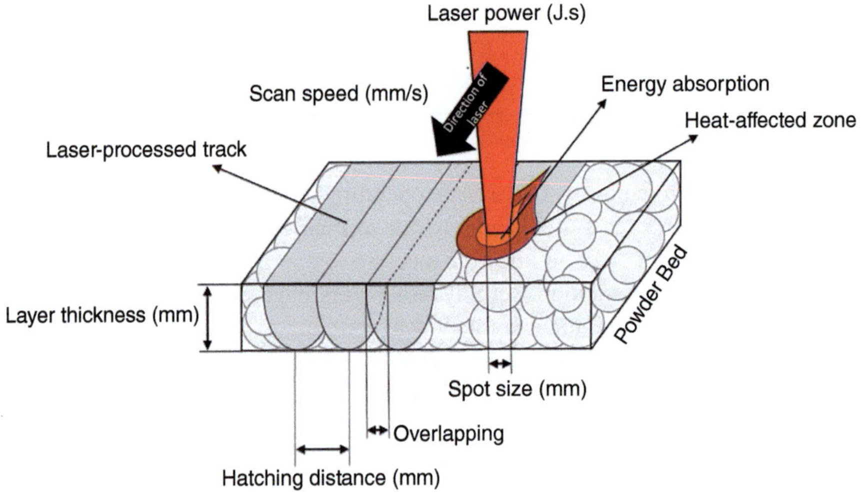

Figure 5.14 Scheme of variable parameters involved during the PBSLP process and from which the VED can be calculated. Reference [7] / Elsevier / CC BY 4.0.

power leads to incomplete melting or sintering of the powder bed, whereas excessive laser power generates large melt pools and can cause material evaporation.

- **Scanning speed:** It controls the duration of laser exposure in a given area onto the powder bed. If it is too low, it may cause overheating and material evaporation, while excessively high speeds lead to insufficient melting and weak bonding.

Achieving a balance between scanning speed and laser power is essential to ensure proper energy input, preventing the formation of defects such as porosity due to either excessive or insufficient melting.

- **Hatch space or hatching distance:** It is the distance between the centers of successive parallel laser beam tracks. This parameter is important for balancing part quality and process efficiency. The suitable hatch space depends on the laser spot size. If the hatching distance is too small, excessive overlap between laser tracks can lead to local overheating and defects. Conversely, if the hatching distance is too large, insufficient overlap between tracks may result in unreacted powder, weak bonding, and porosity between tracks. Optimizing the hatching distance is essential to ensure the fabrication of high-quality parts while maintaining an efficient build rate.
- **Layer thickness:** This parameter refers to the distance between consecutive layers and is a key factor in balancing build rate, resolution, and inter-layer bonding. Increasing the layer thickness reduces the number of layers required, thereby accelerating the build process and minimizing recoater wear. However, excessive layer thickness can compromise part resolution and lead to insufficient bonding between layers, potentially causing structural weaknesses or part failure. Conversely, thinner layers enhance resolution and improve inter-layer adhesion but result in longer production times. Optimizing layer thickness involves finding the highest possible value that maintains adequate bonding while ensuring part integrity and resolution. Layer thickness is also dependent on powder particle size distribution and the characteristics of the recoating system.

Laser power, scanning speed, and hatch space are fundamental parameters that collectively govern the energy transfer to the powder bed. Adjustments of these parameters can change the quality of the final manufactured items. Studies have shown that a combination of high laser power and low scanning speed can mitigate balling effect (which is the formation of spheroidal beads during laser irradiation) due to surface tension and lack of wetting of the previous layer [88, 89].

A common research practice to represent the laser heating effect is to consider the laser energy density *VED* per volume unit, which can be calculated through Eq. (5.1).

$$\text{VED} = \frac{P}{\text{v.h.t}}\ (\text{J.mm}^{-3}) \tag{5.1}$$

where VED is the volumetric energy density in J/mm^3, P is the laser power (J/s), v is the scanning speed (mm/s), h is the hatch space (mm), and t is the powder layer thickness (mm).

Sometimes alternative expressions of energy density are used as metrics such as E_l: linear energy Eq. (5.2) or E_s: energy density per surface unit Eq. (5.3). The use of the laser beam diameter can sometimes be found instead of hatch space in Eq. (5.1).

$$E_l = \frac{P}{\text{v}}\ (\text{J.mm}) \tag{5.2}$$

$$E_S = \frac{P}{v.h}\ (\text{J.mm}^{-2}) \tag{5.3}$$

Laser energy density is widely used as a reference metric in laser-PBF processes; however, it has inherent limitations. While it provides a general framework for parameter selection, it does not fully account for the complex physical interactions that occur during laser irradiation. Heat transfer mechanisms, material responses, and process stability must also be considered to achieve a more precise understanding of energy input and its effects. The actual energy transferred to the powder bed is governed by multiple variables beyond laser energy density. Factors such as laser beam diameter, focalization onto the powder-bed surface, scanning strategy, powder-bed compaction, and melt pool dynamics significantly influence heat transfer and material behavior. These parameters affect the extent of fusion, thermal gradients, and defect formation, making laser energy density an insufficient predictor of process outcomes. Additionally, it does not account for effects such as spattering, powder displacement, or localized overheating, all of which are particularly critical in ceramic processing due to their high melting temperatures and low thermal shock resistance. These effects lead to variations in densification and microstructural uniformity that cannot be solely dictated by energy density calculations.

While increasing laser energy density generally improves material fusion, excessive values can introduce defects due to material evaporation or excessive shrinkage. Empirical studies demonstrate that while laser power, scanning speed, and hatch spacing collectively influence densification, their effects are not proportionally represented in energy density calculations. Research findings suggest that laser power has the most significant impact on final part quality, followed by hatch spacing and scanning speed.

Another important consideration is the variation in energy absorption caused by the penetration depth of the laser beam into the powder bed. Multiple reflections of the laser beam can occur due to the optical property of the powder and powder-bed characteristics (packing density, particle shape), inducing different material responses and the size of the sintering/melting zone for the same energy density [90, 91].

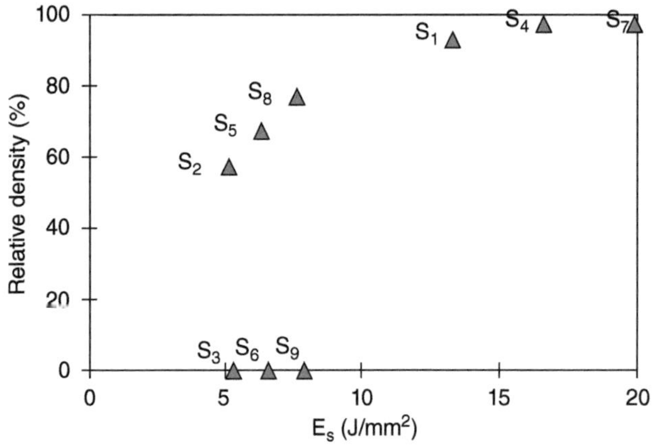

Figure 5.15 Change of parts relative density as a function of laser energy density per surface unit (E_S) and overlap (see value in Table 5.4). Adapted from Ref. [44].

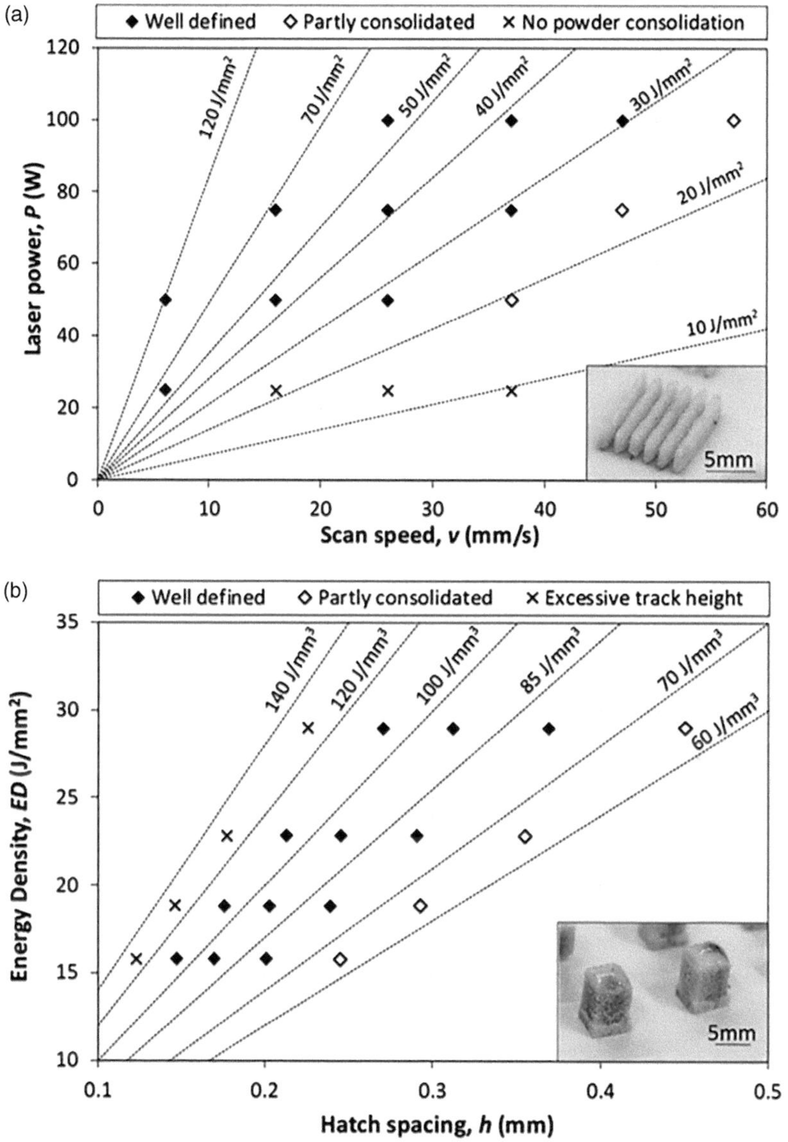

Figure 5.16 Example of process maps for the building of soda lime glass items: (a) thin walls and (b) cubes. [54] / John Wiley & Sons / CC BY 4.0.

Identical energy density values can result in significantly different densification behaviors depending on laser overlap, spot size, or beam focus conditions. Indeed, Juste et al. [44] have demonstrated that even for very close energy densities, the achieved relative densities of built parts can differ significantly due to an unsuitable overlap ratio of the laser beam track (Figure 5.15 and 5.16).

Given these limitations, an effective optimized scanning strategy for ceramic laser PBF must integrate multiple interdependent factors beyond laser energy density. A robust process model should account for additional parameters such as laser path strategy, beam spot size, powder-bed properties, and material-specific characteristics, including thermal conductivity and reflectivity. Relying solely on energy density as a predictive or optimization metric may lead to misleading conclusions, ultimately affecting part quality, repeatability, and process efficiency. As proposed by several authors [52, 53], the use of process maps can be a useful tool to determine a PBSLP processing window as a function of parameter settings.

Scanning strategy refers to the arrangement of laser scan vectors within a single layer and across multiple successive layers. This parameter significantly influences the thermal gradients induced by laser beam movement on the surface of the object being fabricated. Proper selection of the scanning strategy is critical, as it must be adapted not only to the geometry of the part but also to the specific section undergoing laser exposure at any given time. Moreover, the strategy must align with the appropriate energy density for the material and its deposition conditions to prevent localized overheating, which may cause material evaporation, or defects such as thermal stress-inducing cracks.

The impact of scanning strategy on material properties has been investigated by Abdelmoula et al., who examined its influence on the characteristics of alumina parts fabricated under identical laser parameters [92] as well as on silicon carbide [93]. Figure 5.17 illustrates different scanning strategies and the resulting printed objects. Despite maintaining constant laser power, scanning speed, and hatch spacing, significant variations in part properties were observed. The relative densities of the produced samples ranged from 67% to 88%, while the surface roughness (R_a) varied from 115 to 225 μm depending on the chosen scanning strategy. Improper selection of the scanning strategy can also lead to unintended surface texturing if it does not match the energy density requirements. Variations in temperature distribution within the layer may cause excessive melting in some regions, while others remain insufficiently processed, leading to localized evaporation or severe distortions in the printed object. In extreme cases, the resulting parts may exhibit substantial degradation and no longer resemble the original 3D model.

To address these challenges, numerical modeling serves as a valuable tool for optimizing scanning strategies. Abdelmoula et al. [92] proposed a new generation of scanning strategies based on space-filling mathematical curves to achieve better thermal management during laser exposure. This approach aims to enhance process stability, minimize thermal defects, and improve the overall quality of ceramic parts produced through PBF (Figure 5.18).

One aspect that is often overlooked in literature is the selection of the build support in PBF. As in other AM technologies, ensuring failure-free deposition of the first layer remains a critical challenge. The choice of the base plate plays a significant role in the success of the build, influencing both the adhesion of the initial layers and the ease of part removal after fabrication. For a component to be successfully constructed, the first layers must properly fuse and adhere to the build platform. Therefore, the substrate must possess adequate absorbance and melting characteristics under the process conditions. Its composition should be similar, if

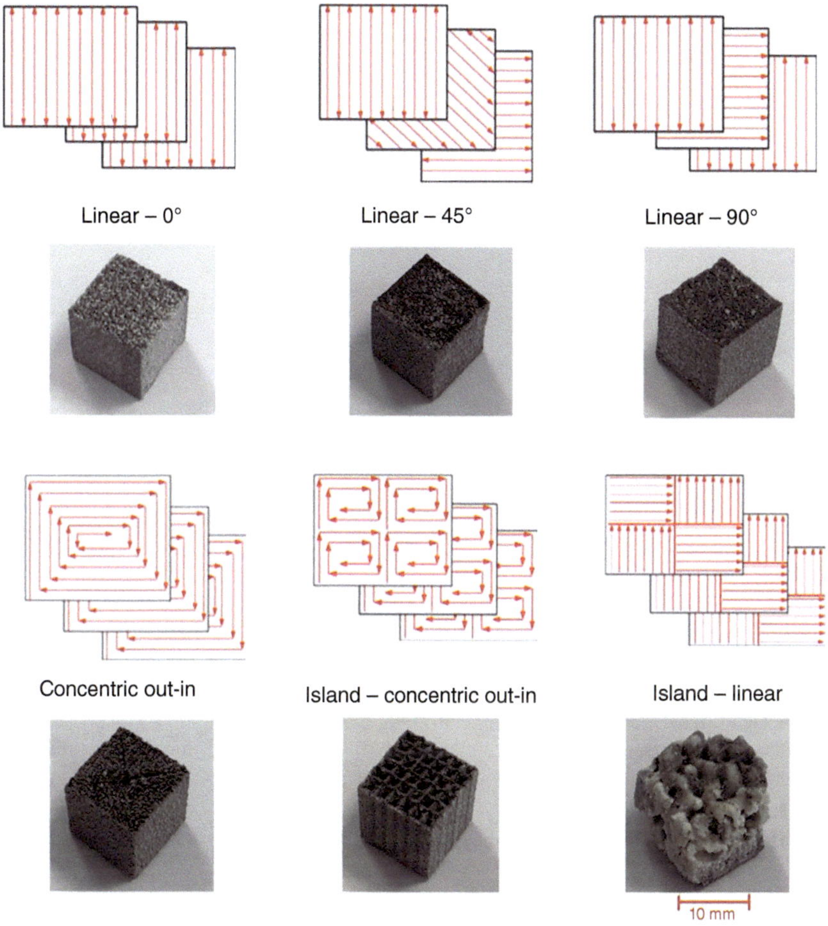

Figure 5.17 Investigation of scan strategy effect on alumina sample printed with different scanning strategies using 95 W laser power, 200 mm/s scanning speed, 50 μm hatching space. [93] / MDPI / CC BY 4.0.

not identical, to that of the material being processed to prevent contamination and minimize residual stresses that could lead to cracking. In laser PBF, matching the substrate and feedstock composition is a common practice to ensure strong adhesion and compatibility with thermal expansion characteristics. However, the choice of build support is not only dictated by adhesion requirements but also by the necessity of part removal during post-processing. While a strong bond between the part and the base plate ensures stability during manufacturing, excessive adhesion can complicate separation. Strategies such as controlled interfacial bonding, the use of sacrificial layers, or optimizing process parameters to achieve a balance between adhesion and removability are essential to facilitate efficient part extraction while maintaining structural integrity [45, 48, 52].

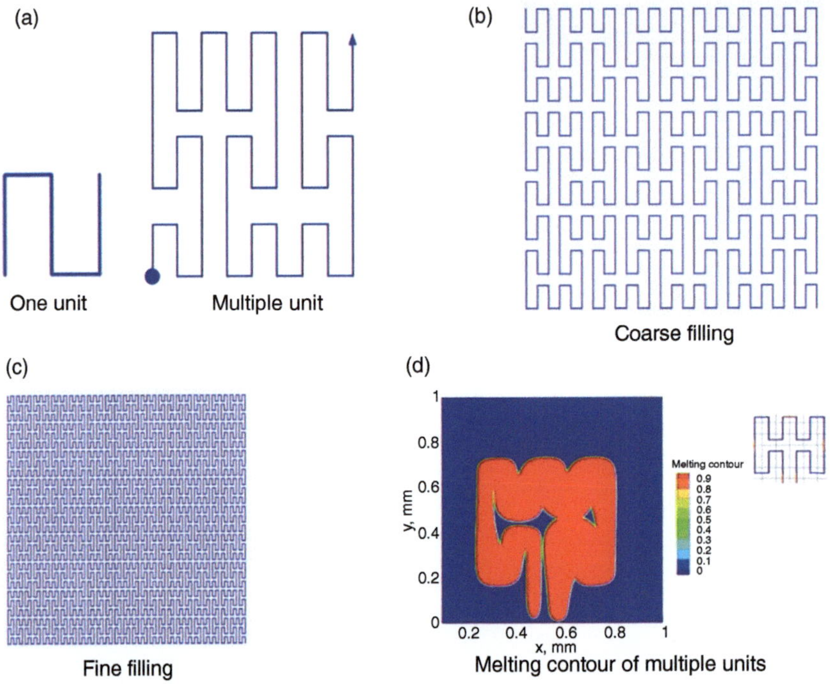

Figure 5.18 Proposition of new scanning strategy based on space-filling curve (a) one and multiple units, (b) coarse filling, (c) fine filling, and (d) melting contour of multiple units. [93] / MDPI / CC BY 4.0.

5.6 Heat Treatment and Post-Processing

In general, ceramics fabricated through the direct PBF approach do not require post-processing. The produced parts are near-net shape, and their properties remain unchanged after printing, as further modification is not feasible. Unlike metals or polymer-based AM, ceramic components processed via direct PBF exhibit limited potential for post-treatment densification. Studies conducted by Bertrand et al. [35] and Juste et al. [44] have demonstrated that increasing the relative density of yttria-stabilized zirconia (YSZ) and alumina parts, respectively, through conventional sintering post-treatment is not possible. When subjected to furnace treatment at conventional sintering temperatures, the relative density of these ceramics remains unchanged. However, one notable effect observed during post-treatment is the restoration of their original coloration and the formation of cracks. Parts initially appear dark upon removal from the PBF process, but revert to a white color after thermal exposure in an oven. This color transformation has been attributed to the presence of oxygen vacancies introduced during laser irradiation, depending on operating conditions such as atmosphere and laser parameters. About crack formation, this phenomenon may be due to relaxation of thermal stresses due to laser-induced heating and cooling cycle. This difficulty in densifying ceramics through post-treatment suggests that full melting occurs during laser exposure.

Consequently, no further microstructural rearrangement is possible at temperatures lower than those reached during the PBF process itself. This observation highlights the fundamental difference between ceramics and other material classes in AM, where sintering or secondary thermal treatments can typically enhance density and mechanical properties. In the case of oxide ceramics, the equilibrium state is already reached during the laser process, leaving post-processing with minimal impact beyond superficial modifications such as color correction. These insights emphasize the need for careful parameter selection during the initial processing phase, as the microstructure and final properties of ceramic parts are effectively locked in at the moment of solidification.

In the case of indirect PBSLP, at least two thermal treatment steps are required (debinding and sintering) after laser processing to finalize the part. The first thermal treatment is necessary to remove the transient organic binder used during fabrication. However, the indirect approach introduces two major challenges: residual porosity in the final parts and significant shrinkage during sintering. The shrinkage is not always isotropic, leading to distortions in the final shape.

Several post-processing techniques (Figure 5.19) can be employed to enhance relative density and mitigate shrinkage. These techniques may be applied individually or sequentially to optimize the mechanical performance of the fabricated parts. The main post-treatment approaches include:

- **Infiltration:** a ceramic-loaded suspension, composed of the same material as the initial feedstock, is used to fill the remaining porosity. This process can be performed with or without additional pressure to enhance infiltration efficiency.
- **Isostatic pressing:** various types of isostatic pressing have been reported in the literature to enhance densification, including cold isostatic pressing (CIP), quasi isostatic pressing (QIP), warm isostatic pressing (WIP), and hot isostatic pressing (HIP).

Shahzad et al. [8] and Deckers et al. [9] investigated the sequential and combined application of these post-treatments to improve the performance of complex

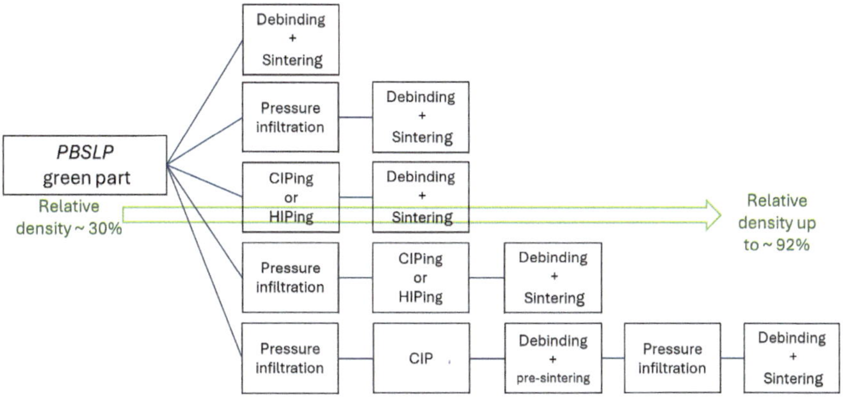

Figure 5.19 Different post-processing routes could be applied in indirect PBSLP. Adapted from Refs. [8, 94].

ceramic components fabricated via the indirect SLS method. For instance, a relative density of approximately 93% was achieved for an alumina/polyamide feedstock shaped by SLS, followed by QIP at 20 MPa and 165 °C for 15 minutes. Another example, a study involved an alumina/polypropylene feedstock processed via SLS and subsequently subjected to pressure infiltration in the green state at 13 MPa. This was then followed by WIP at 64 MPa and 135 °C. This combination of post-treatments increased the relative density to 88%, compared to only 39% without post-processing. The resulting parts (processed through pressure infiltration and WIP) exhibited a volumetric shrinkage of 62%. After sintering, the average flexural strength reached 148 ± 22 MPa. A similar approach was applied to a yttria-stabilized zirconia (3YSZ)/polypropylene feedstock, yielding a final density of up to 92% after sintering.

Despite the use of various post-processing strategies to densify ceramic feedstocks bound with organic binders in the indirect PBSLP method, achieving fully dense parts remains unattainable. However, it has been demonstrated that crack-free microstructures can still be obtained through optimized post-processing techniques, which significantly enhance the material's performance and structural integrity (Figure 5.20).

When employing inorganic binders, such as metals, in SLM for ceramics, two primary post-processing scenarios emerge. The first involves a reactive post-treatment aimed at converting the binder into a ceramic phase, whereas the second scenario, which does not involve post-treatment, results in a metal-ceramic composite. The selection of either approach significantly affects the final material properties, including mechanical strength and thermal stability.

Streek et al. [96], as well as Ahlhelm et al. [97] demonstrated the fabrication of silicon-infiltrated silicon carbide (Si-SiC) components via LMS. Their process

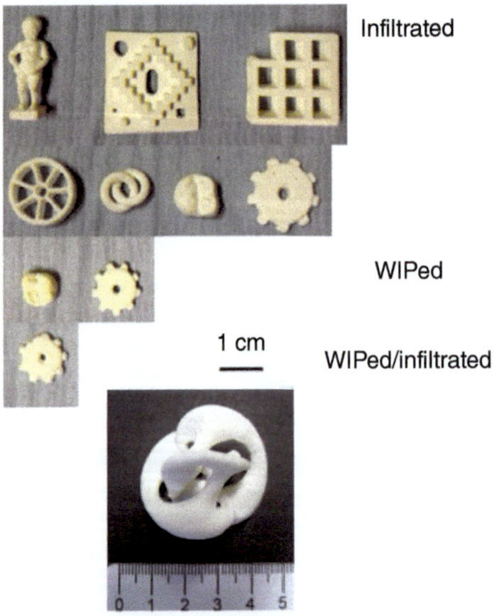

Figure 5.20 Examples of a part obtained by combining SLS and post-treatment. [96] / with permission of Elsevier and [97] / Göller Verlag / CC BY 4.0.

Figure 5.21 SiSiC turbine rotor. [99] / Göller Verlag / CC BY 4.0.

utilized a powder mixture consisting of silicon carbide and 41% silicon (Figure 5.21). This approach resulted in a composite structure in which the residual silicon phase remained embedded within the final part, contributing to its mechanical stability and functional characteristics. A different approach was explored by Stierlen et al. [98] and Evans et al. [99, 100], who developed silicon carbide components using an indirect fabrication route. In their method, preforms were initially produced via SLS from a SiC feedstock that had been coated with a polymeric binder. These preforms subsequently underwent pyrolysis, a process that decomposed the organic phase, leaving behind a porous carbon skeleton containing microcracks. The subsequent step involved liquid silicon infiltration (LSI), in which molten silicon penetrated the preform's porosities and fissures. This step was followed by siliconization, where a reaction occurred between the residual carbon and silicon, leading to the formation of a β-SiC phase. This method ensured enhanced densification while preserving structural integrity.

Meyers et al. pursued a similar reactive approach, processing a powder mixture of SiC and Si via SLM. The porous preforms obtained through SLM were subsequently infiltrated with a phenolic resin, which, upon pyrolysis, generated a substantial amount of carbon within the remaining porosity. Subsequent LSI treatment facilitated additional siliconization, resulting in parts with up to 84% SiC content (α-SiC + β-SiC). However, achieving such a high yield of SiC required multiple cycles of infiltration and pyrolysis. Despite these optimizations, residual silicon remained in the final structure, which negatively impacted some mechanical properties. Depending on the number of infiltration/pyrolysis cycles performed, the fabricated components exhibited a flexural strength of up to 220 ± 7 MPa and Young's modulus of 285 ± 5 GPa. The same research group extended their methodology to fabricate RBBC components using a feedstock comprising boron carbide

(B_4C), silicon carbide, and silicon. The SLM process resulted in a structurally stable preform containing these constituents. Following resin infiltration, pyrolysis, and subsequent LSI, a reaction between boron carbide and silicon led to the formation of additional β-SiC, as well as a ternary $B_xSi_yC_z$ phase. The final RBBC composite demonstrated notable mechanical performance, with a flexural strength of 168 ± 15 MPa and Young's modulus of 300 ± 2 GPa [26].

A major challenge when using inorganic binders lies in optimizing their content to ensure structural integrity while maintaining high mechanical performance. Reducing residual binder content is critical for improving ceramic conversion efficiency. Striking a balance between binder content, processing conditions, and final mechanical properties remains an important area for further investigation in PBF-based ceramic processing.

5.7 Conclusion

This chapter provides an overview of PBF methods reported in the literature for shaping ceramic components.

Two primary routes are distinguished in PBF for ceramics: the indirect and direct methods. The indirect approach, which relies on sacrificial binders, offers greater flexibility for shaping both oxide and non-oxide ceramics but requires multiple processing steps. In contrast, the direct method enables faster production with higher resolution, though it poses greater challenges related to material consolidation.

In the direct method, two approaches currently exist, using high-energy beams from either an electron or a laser source. While electron beam technology presents significant potential, its applicability is limited by the low electrical conductivity of many ceramics. However, recent advancements in powder treatment have addressed this constraint. PBF using lasers as an energy source has been more extensively studied, benefiting from developments in metal and polymer processing. However, literature does not always clearly distinguish between melting and sintering mechanisms in ceramic powders, leading some researchers to propose the term "Powder-Bed-Selective Laser Processing" (PBSLP) as a more appropriate classification. Despite recent progress, PBSLP remains largely at the research and development stage due to the inherent properties of ceramics. However, certain applications have been explored, including alumina or silica-based molds for metal casting that withstand thermal shocks up to 600 °C (in the case of alumina), as well as complex structures in silica and soda lime glass. These examples highlight the potential of PBSLP for applications requiring high chemical and thermal stability. A key finding from recent studies is that while PBSLP enables the fabrication of dense ceramic components, their mechanical properties and microstructural characteristics still fall short of those achieved through conventional methods. Direct fabrication is particularly affected by the formation of columnar grains, anisotropic properties, and residual porosity. Additionally, the high thermal gradients inherent to the process exacerbate the risk of cracking, making it difficult to produce defect-free parts. To address these challenges, several strategies have been explored. Process parameter optimization and the addition of absorption enhancers have led to improved densification efficiency. Moreover, controlling the temperature during the process allows for near-full density to be achieved, but often results in parts with lower surface quality. The indirect approach, combining PBSLP with post-processing techniques such as infiltration and isostatic pressing (CIP, WIP, HIP), has demonstrated significant potential for enhancing the final mechanical properties, achieving relative densities of up to 92%. Post-processing techniques will also play an important role in bridging the performance gap between ceramics produced via PBSLP and those fabricated through conventional methods. The integration of tailored thermal treatments, specific infiltration strategies, and composite formulations can improve mechanical strength and functional properties, broadening the scope of viable applications. Looking ahead, the advancement of ceramic PBSLP will depend on several key research areas. Enhancing equipment capabilities will be critical, including improved thermal control, the adoption of laser wavelengths better suited to ceramics, and real-time process monitoring to detect and correct defects. Numerical modeling will also become an essential tool to reduce the number of experimental iterations required for process optimization.

Ultimately, the industrialization of ceramic PBSLP will depend on its ability to deliver economically viable solutions with reproducible performance. Despite existing challenges, the growing interest in this technology, coupled with advances in materials science, laser processing, and AM, suggests promising prospects for applications in high-performance sectors such as aerospace, biomedical engineering, and advanced industrial manufacturing.

References

1 ISO (2021). ISO/ASTM 52900:2021(en) Additive manufacturing — General principles — Fundamentals and vocabulary.

2 Deckard, C.R. (1989). Method and apparatus for producing part by selective sintering. 4,863,538, Sep. 05.

3 Bai, L., Gong, C., Chen, X. et al. (2019). Additive manufacturing of customized metallic orthopedic implants: materials, structures, and surface modifications (1 September 2019). *MDPI AG. http://* https://doi.org/10.3390/met9091004.

4 Lakshminarayan, U. and Marcus, H.L. (1991). Microstructural and mechanical properties of Al2O3/P2O5 and Al2O3/B2O3 composties fabricated by selective laser sintering. *International Solid Freeform Fabrication Symposium*, Austin Texas (USA) August 12-14, 1991.

5 Lakshminarayan, U., Ogrydiziak, S., and Marcus, H.L. Selective laser sintering of ceramic materials Austin, *Austin Texas (USA), 6–8 August 1990*.

6 Kruth, J.P., Mercelis, P., Van Vaerenbergh, J. et al. (2005). Binding mechanisms in selective laser sintering and selective laser melting. *Rapid Prototyp. J.* 11 (1): 26–36. https://doi.org/10.1108/13552540510573365.

7 Grossin, D., Montón, A., Navarrete-Segado, P. et al. (2021). A review of additive manufacturing of ceramics by powder bed selective laser processing (sintering/

melting): calcium phosphate, silicon carbide, zirconia, alumina, and their composites. *Open. Ceram.*, 5 October 2020, 100073, https://doi.org/10.1016/j.oceram.2021.100073

8 Shahzad, K. (2013). Powder-based indirect selective laser sintering of ceramics. Doctoral thesis, KU Leuven.

9 Deckers, J. (2013). Selective laser sintering and melting as additive manufacturing methods to produce alumina parts '3D printing' of ceramics. Doctoral thesis, KU Leuven.

10 Chen, Z., Li, Z., Li, J. et al. (2019). 3D printing of ceramics: a review. *J. Eur. Ceram. Soc.* 39 (4): 661–687. https://doi.org/10.1016/j.jeurceramsoc.2018.11.013.

11 Chen, A.-N., Wu, J.-M., Shi, Y.-S. et al. (2017). High-performance ceramic parts with complex shape prepared by selective laser sintering: a review. *Adv. Appl. Ceram.* 117: 1–18.

12 Deckers, J., Shahzad, K., Vleugels, J., and Kruth, J.-P. (2012). Isostatic pressing assisted indirect selective laser sintering of alumina components. *Rapid Prototyp. J.* 18 (5): 409–419. https://doi.org/10.1108/13552541211250409.

13 Liu, Z.H., Nolte, J.J., Packard, J.I. et al. (2007). Selective laser sintering of high-density alumina ceramic parts. *Proc. 35th Int. MATADOR 2007 Conf.(Taipei, Taiwan)* 3: 351–354. https://doi.org/10.1007/978-1-84628-988-0_79.

14 Subramanian, K., Vail, N., Barlow, J., and Marcus, H. (1995). Selective laser sintering of alumina with polymer binders. *Rapid Prototyp. J.* 1 (2): 24–35. https://doi.org/10.1108/13552549510086844.

15 Tang, H.-H., Chiu, M.-L., and Yen, H.-C. (2011). Slurry-based selective laser sintering of polymer-coated ceramic powders to fabricate high strength alumina parts. *J. Eur. Ceram. Soc.* 31 (8): 1383–1388.

16 Cardon, L., Deckers, J., Verberckmoes, A. et al. (2013). Polystyrene-coated alumina powder via dispersion polymerization for indirect selective laser sintering applications. *J. Appl. Polym. Sci.* 128 (3): 2121–2128. https://doi.org/10.1002/app.38388.

17 Bai, P.-K., Cheng, J., Liu, B. et al. (2005). Selective Laser Sintering of polymer-coated $Al_2O_3/ZrO_2/TiC$ ceramic powder. *Trans. Nonferrous Metals Soc. China* 15 (2): 261–265.

18 Goodridge, R.D., Wood, D.J., Ohtsuki, C., and Dalgarno, K.W. (2007). Biological evaluation of an apatite-mullite glass-ceramic produced via selective laser sintering. *Acta Biomater.* 3: 221–231.

19 Liu, J., Zhang, B., Yan, C., and Shi, Y. (2010). The effect of processing parameters on characteristics of selective laser sintering dental glass-ceramic powder. *Rapid Prototyp. J.* 16 (2): 13–145. https://doi.org/10.1108/13552541011025861.

20 Stevinson, B.Y., Bourel, D.L., and Beaman Jr., J.J. (2006). Support free infiltration of selective laser sintered (SLS) silicon carbide preforms. *Proceedings of SFF Conference, Austin, Texas, USA, 14–16 August 2006.*

21 Christian Nelson, J., Vail, N.K., Barlow, J.W. et al. (1995). Materials and interfaces selective laser sintering of polymer-coated silicon carbide powders. *Ind. Eng. Chem. Res.* 34: 1641–1651.

22 Chen, F., Wu, J.-M., Wu, H.-Q. et al. (2018). Microstructure and mechanical properties of 3Y-TZP dental ceramics fabricated by selective laser sintering combined with cold isostatic pressing. *Int. J. Lightweight Mater. Manuf.* 1 (4): 239–245.

23 Leu, M.C., Adamek, E.B., Huang, T. et al. (2008). Freeform fabrication of zirconium diboride parts using selective laser sintering. *19th Annual International Solid Freeform Fabrication Symposium, SFF 2008*, 4-6 August 2008 , pp. 194–205.

24 K. Shahzad, J. Deckers, S. Boury, B. Neirinck, J. P. Kruth, and J. Vleugels, “Preparation and indirect selective laser sintering of alumina/PA microspheres,” *Ceram Int*, vol. 38, no. 2, pp. 1241–1247, Mar. 2012, doi: https://doi.org/10.1016/j.ceramint.2011.08.055.

25 Gill, T. and Hon, B. (2003). Selective laser sintering of SiC/polyamide matrix composites. *CIRP Ann. Manuf. Technol.* 52 (1): 173–176. https://doi.org/10.1016/S0007-8506(07)60558-7.

26 Meyers, S., De Leersnijder, L., Vleugels, J., and Kruth, J.P. (2018). Direct laser sintering of reaction bonded silicon carbide with low residual silicon content. *J. Eur. Ceram. Soc.* 38 (11): 3709–3717. https://doi.org/10.1016/j.jeurceramsoc.2018.04.055.

27 Meyers, S., Vinãs, M.T., Kruth, J., and Vleugels, J. (2022). Laser powder bed fusion as a net-shaping method for reaction bonded SiC and B4C. *Virtual Phys. Prototyp.* 1–10. https://doi.org/10.1080/17452759.2022.2077222.

28 Fu, Z. and Körner, C. (2022). *Actual State-of-the-Art of Electron Beam Powder Bed Fusion*. Taylor and Francis Ltd. https://doi.org/10.1080/26889277.2022.2040342.

29 Sjöström, W., Botero, C., and Jimenez-Piqueo, E. (2024). Melting ceramic Al_2O_3 powder by electron beam powder bed fusion. *Prog. Addit. Manuf.* 9 (5): 1523–1535. https://doi.org/10.1007/s40964-024-00601-4.

30 Suard, M. (2006). *Characterization and Optimization of Lattice Structures Made by Electron Beam Melting*. Univ Grenoble.

31 Wennersten, K., Xu, J., Armakavicius, N. et al. (2024). Feasibility of melting NbC using electron beam powder bed fusion. *Adv. Eng. Mater.* 26 (6): https://doi.org/10.1002/adem.202301388.

32 Xu, J., Wennersten, K., Fu, Z. et al. (2024). In-melt electron analysis to accelerate process exploration of ceramics: electron beam melting of TiB_2. *Materialia (Oxf.)* 38: 102243. https://doi.org/10.1016/j.mtla.2024.102243.

33 Konyashin, I., Hinners, H., Ries, B. et al. (2019). Additive manufacturing of WC-13%Co by selective electron beam melting: achievements and challenges. *Int. J. Refract. Met. Hard Mater.* 84: 105028. https://doi.org/10.1016/j.ijrmhm.2019.105028.

34 Keerti, V., Pasagada, V., Yang, N., and Xu, C. (2021). Electron beam sintering (EBS) process for ultra-high temperature ceramics (UHTCs) and the comparison with traditional UHTC sintering and metal electron beam melting (EBM) processes. *Ceram. Int.* 48 (7): 10174–10186. https://doi.org/10.1016/j.ceramint.2021.12.229.

35 Bertrand, P., Bayle, F., Combe, C. et al. (2007). Ceramic components manufacturing by selective laser sintering. *Appl. Surf. Sci.* 254 (4): 989–992. https://doi.org/10.1016/j.apsusc.2007.08.085.

36 Shishkovsky, I., Yadroitsev, I., Bertrand, P., and Smurov, I. (2007). Alumina-zirconium ceramics synthesis by selective laser sintering/melting. *Appl. Surf. Sci.* 254 (4): 966–970. https://doi.org/10.1016/j.apsusc.2007.09.001.

37 P. Regenfuss, A. Streek L. Hartwing, S. Klötzer, "Advancements in laser micro sintering," in *the Third International WLT*, Munich, Germany, June 13th–16th, 2005.

38 Exner, H., Regenfuss, P., Streek, A. et al. (2006). Laser micro sintering of ceramic materials. *cfi/Ber. DKG* 83 (13): 45–52.

39 Streek, A., Regenfuß, P., Süß, T. et al. (2008). Laser micro sintering of SiO_2. *Proc. SPIE* 6985: 69850Q–69850Q–8. https://doi.org/10.1117/12.787121, St. Petersburg, Russian Federation.

40 Hagedorn, Y.C., Wilkes, J., Meiners, W. et al. (2010). Net shaped high performance oxide ceramic parts by selective laser melting. *Phys. Procedia* 5: 587–594. https://doi.org/10.1016/j.phpro.2010.08.086.

41 Wilkes, J., Hagedorn, Y.-C., Meiners, W., and Wissenbach, K. (2013). Additive manufacturing of alumina zirconia ceramic compounds by selective laser melting. *Rapid Prototyp. J.* 19: https://doi.org/10.1108/13552541311292736.

42 Florio, K., Pfeiffer, S., Makowska, M. et al. (2019). An innovative selective laser melting process for hematite-doped aluminum oxide. *Adv. Eng. Mater.* 21 (6): https://doi.org/10.1002/adem.201801352.

43 Florio, K. (2021). Powder bed fusion of ceramic oxides: process development and process understanding. *ETH Zurich* https://doi.org/10.3929/ethz-b-000544358.

44 Juste, E., Petit, F., Lardot, V., and Cambier, F. (2014). Shaping of ceramic parts by selective laser melting of powder bed. *J. Mater. Res.* 29 (17): 2086–2094. https://doi.org/10.1557/jmr.2014.127.

45 Verga, F., Borlaf, M., Conti, L. et al. (2020). Laser-based powder bed fusion of alumina toughened zirconia. *Addit. Manuf.* 31 (June 2019): 100959. https://doi.org/10.1016/j.addma.2019.100959.

46 Ferrage, L., Bertrand, G., and Lenormand, P. (2018). Dense yttria-stabilized zirconia obtained by direct selective laser sintering. *Addit. Manuf.* 21 (February): 472–478. https://doi.org/10.1016/j.addma.2018.02.005.

47 Ferrage, L. (2018). *Elaboration d'un assemblage céramique-métal par fusion / frittage sélectif (ve) d'un lit de poudre à l'aide d ' un laser Nd:YAG*. University of Toulouse.

48 Shaker, E. (2017). *Fabrication directe d'implants osseux biocéramiques sur mesure par le procédé de fusion laser sélective sur lit de poudre : élaboration de la poudre d'hydroxyapatite, fabrication de pièces architecturées et étude de l'interaction rayonnement / matière*. Université de recherche Paris Sciences et Lettres-MINES ParisTech.

49 Wang, X.H., Fuh, J.Y.H., Wong, Y.S., and Tang, Y.X. (2003). Laser sintering of silica sand – mechanism and application to sand casting mould. *Int. J. Adv. Manuf. Technol.* 21: 1015–1020. https://doi.org/10.1007/s00170-002-1424-x.

50 Tang, Y., Fuh, J.Y.H., Loh, H.T. et al. (2003). Direct laser sintering of a silica sand. *Mater. Des.* 24: 623–629. https://doi.org/10.1016/S0261-3069(03)00126-2.

51 Chang, S., Li, L., Lu, L., and Fuh, J.Y.H. (2017). Selective laser sintering of porous silica enabled by carbon additive. *Materials* 10 (11).

52 Datsiou, K.C., Spirrett, F., Ashcroft, I. et al. (2021). Laser powder bed fusion of soda lime silica glass: optimisation of processing parameters and evaluation of part properties. *Addit. Manuf.* 39 (March): https://doi.org/10.1016/j.addma.2021.101880.

53 Datsiou, K.C., Saleh, E., Spirrett, F. et al. (2019). Additive manufacturing of glass with laser powder bed fusion. *J. Am. Ceram. Soc.* 102 (8): 4410–4414. https://doi.org/10.1111/jace.16440.

54 Abdelmoula, M., Küçüktürk, G., Juste, E., and Petit, F. (2024). Enhancing powder bed fusion of alumina ceramic material: a comprehensive study from powder tailoring to mechanical performance evaluation. *Int. J. Adv. Manuf. Technol.* https://doi.org/10.1007/s00170-024-13158-x.

55 Montón Zarazaga, A., Abdelmoula, M., Kucukturk, G. et al. (2022). Process parameters investigation for direct powder bed selective laser processing of silicon carbide parts. *Prog. Addit. Manuf.* 7: 1307–1322. https://doi.org/10.1007/S40964-022.

56 Montón, A., Abdelmoula, M., Küçüktürk, G. et al. (2021). Experimental and numerical study for direct powder bed selective laser processing (sintering/melting) of silicon carbide ceramic. *Mater. Res. Express* 8 (4): 45603. https://doi.org/10.1088/2053-1591/abf6fc.

57 Abdelmoula, M., Küçüktürk, G., Grossin, D. et al. (2023). Direct selective laser sintering of silicon carbide: realizing the full potential through process parameter optimization. *Ceram. Int.* 49 (20): 32426–32439. https://doi.org/10.1016/j.ceramint.2023.07.189.

58 Montón Zarazaga, A. (2021). *Fabrication additive de céramiques à base de carbure de silicium par traitement laser sélectif sur lit de poudre: de la poudre fonctionnalisée à la pièce finale.* Institut National Polytechnique de Toulouse.

59 Yang, Y., Zhang, C., Wang, D. et al. (2020). Additive manufacturing of WC-Co hardmetals: a review. *Int. J. Adv. Manuf. Technol.* 108: 1653–1673. https://doi.org/10.1007/s00170-020-05389-5.

60 Grigoriev, S., Tarasova, T., Gusarov, A. et al. (2019). Possibilities of manufacturing products from cermet compositions using nanoscale powders by additive manufacturing methods. *Materials* 12 (20): https://doi.org/10.3390/ma12203425.

61 Khmyrov, R.S., Shevchukov, A.P., Gusarov, A.V., and Tarasova, T.V. (2017). Phase composition and microstructure of WC-Co alloys obtained by selective laser melting. *Mech. Ind.* 18 (7): https://doi.org/10.1051/meca/2017059.

62 Mostafaei, A., Elliott, A.M., Barnes, J.E. et al. (2020). Binder jet 3D printing-process parameters, materials, properties, and challenges. *Prog. Mater. Sci.* 100684. https://doi.org/10.1016/j.pmatsci.2020.100684.

63 Fries, S., Genilke, S., Wilms, M.B. et al. (2020). Laser-based additive manufacturing of Wc–Co with high-temperature powder bed preheating. *Steel Res. Int.* 91 (3): 1900511. https://doi.org/10.1002/srin.201900511.

64 Tomas J., Schubert T., Bernthaler T., Merkel M., Schneider, G., and Sellmer, D. (2018). Laser sintering of tungsten carbide cutter. *Proc. of the 3rd Intl. Conf. on Progress in Additive Manufacturing (Pro-AM* 2018*)*, pp. 297–302, may 14-17, 2018, Nanyang, Singapore.

65 Florio, K., Puccio, D., Vigano, G., and Pfeiffer, S. (2021). Process characterization and analysis of ceramic powder bed fusion. *Int. J. Adv. Manuf. Technol.* 117: 2105–2116. https://doi.org/10.1007/s00170-021-07625-y/Published.

66 S. Meyers, J.-P. Kruth, and J. Vleugels, "Direct Selective Laser Sintering of Reaction Bonded Silicon Carbide,"In proceding Solid Freeform Fabrication Symposium, Austin, Texas (USA), August 10-12, 2015.

67 Uhlmann, E., Bergmann, A., and Gridin, W. (2015). Investigation on additive manufacturing of tungsten carbide-cobalt by selective laser melting. *Proc. CIRP* 35: 8–15. https://doi.org/10.1016/j.procir.2015.08.060.

68 Diener, S., Zocca, A., and Günster, J. (2021). Literature review: methods for achieving high powder bed densities in ceramic powder bed based additive manufacturing. *Open. Ceram.* 8 (August): 100191. https://doi.org/10.1016/j.oceram.2021.100191.

69 Spierings, A.B., Voegtlin, M., Bauer, T., and Wegener, K. (2016). Powder flowability characterisation methodology for powder-bed-based metal additive manufacturing. *Prog. Addit. Manuf.* 1 (1–2): 9–20. https://doi.org/10.1007/s40964-015-0001-4.

70 W. Du, X. Ren, Y. Chen, C. Ma, M. Radovic, and Z. Pei, "Model guided mixing of ceramic powders with graded particle sizes in binder jetting additive manufacturing," in *ASME 2018 13th International Manufacturing Science and Engineering Conference at College Station, College Station Texas (USA)*, 2018, pp. 1–9.

71 N. P. Karapatis, G. Egger, P. E. Gygax, and R. Glardon, "Optimization of powder layer density in selective laser sintering," in *Int. Solid Freeform Fabr. Symp. at Austin Texas (USA)*, August 9-11 1999. 10.26153/tsw/746.

72 Meier, C., Weissbach, R., Weinberg, J. et al. (2019). Critical influences of particle size and adhesion on the powder layer uniformity in metal additive manufacturing. *J. Mater. Process. Technol.* 266: 484–501. https://doi.org/10.1016/j.jmatprotec.2018.10.037.

73 ASM International Handbook Committee, P. W. Lee (Ed.), *Powder Metal Technologies and Applications*, vol. 7. Materials Park, OH: ASM International, 1998. ISBN: 0-87170-387-4.

74 Freeman, R. (2007). Measuring the flow properties of consolidated, conditioned and aerated powders – a comparative study using a powder rheometer and a rotational shear cell. *Powder Technol.* 174 (1–2): 25–33. https://doi.org/10.1016/j.powtec.2006.10.016.

75 ISO 4490 (2025). Poudres métalliques — Détermination du temps d'écoulement au moyen d'un entonnoir calibré (appareil de Hall) (https://www.iso.org/obp/ui#iso:std:iso:4490:ed-5:v1:en; accessed 27 January 2025.

76 Beitz, S., Uerlich, R., Bokelmann, T. et al. (2019). Influence of powder deposition on powder bed and specimen properties. *Materials* 12 (2): https://doi.org/10.3390/ma12020297.

77 Miao, G., Du, W., Pei, Z., and Ma, C. (2022). *A Literature Review on Powder Spreading in Additive Manufacturing*. Elsevier B.V. http://doi.org/10.1016/j.addma.2022.103029.

78 Wu, Q., Qiao, C., Wang, J. et al. (2022). Adaptability investigations on bottom modified blade in powder spreading process of additive manufacturing. *Addit. Manuf.* 49: https://doi.org/10.1016/j.addma.2021.102477.

79 Capozzi, L.C., Sivo, A., and Bassini, E. (2022). *Powder spreading and spreadability in the additive manufacturing of metallic materials: a critical review*. Elsevier Ltd. https://doi.org/10.1016/j.jmatprotec.2022.117706.

80 Roy, N.K. and Cullinan, M.A.; μ-SLS of Metals: Design of the Powder Spreader, Powder Bed Actuators and Optics for the System," *in Proceedings of the 26th Annual International Solid Freeform Fabrication Symposium (SFF 2015)*, Austin, TX, USA, Aug. 10–12, 2015, pp. 134–155.

81 Petsch, T.; Regenfuß, P.; Ebert, R.; Hartwig, L.; Klötzer, S.; Brabant, T.; Exner, H. Industrial laser micro sintering. In Proceedings of the 23rd International Congress on Applications of Laser and Electro-Optics, Congress Proceedings, San Francisco, California (USA), 4–7 October 2004.

82 Powder Recoater Aerosint (2025). Dual-metal L-PBF wtih Aerosint powder recoater. https://aerosint.com (accessed January 2025).

83 Martinsen, R., Kingsbury, A., Lowder, D.T., and Kliner, D.D. (2024). Dynamic beam shaping: unlocking productivity for cost-effective. *Laser Beam Powder Bed Fus.* 10: 173–183.

84 Tolochko, N.K., Laoui, T., Khlopkov, Y.V. et al. (2000). Absorptance of powder materials suitable for laser sintering. *Rapid Prototyp. J.* 6 (3): 155–160.

85 Hellrung, D., Yeh, L.-Y., Depriereux, F. et al. High-accuracy micromachining of ceramics by frequency-tripled Nd:YAG lasers. *Proc. SPIE 3618, Laser Applications in Microelectronic and Optoelectronic Manufacturing IV*. San Jose, CA (USA), July 15th 1999.

86 Moniz, L., Colin, C., Bartout, J.D. et al. (2018). Laser beam melting of alumina: effect of absorber additions. *JOM* 70 (3): 328–335. https://doi.org/10.1007/s11837-017-2716-9.

87 Pfeiffer, S., Makowska, M., Florio, K. et al. (2020). Selective laser melting of thermal pre-treated METAL oxide doped aluminum oxide granules. *Open. Ceram.* 100007. https://doi.org/10.1016/j.oceram.2020.100007.

88 Li, R., Liu, J., Shi, Y. et al. (2012). Balling behavior of stainless steel and nickel powder during selective laser melting process. *Int. J. Adv. Manuf. Technol.* 59 (9): 1025–1035. https://doi.org/10.1007/s00170-011-3566-1.

89 Tolochko, N.K., Mozzharov, S.E., Yadroitsev, I. et al. (2004). Balling processes during selective laser treatment of powders. *Rapid Prototyp. J.* 10 (2): 78–87. https://doi.org/10.1108/13552540410526953.

90 Wang, X.C. and Kruth, J.-P.Energy absorption and penetration in selective laser sintering: a ray tracing model. *MMT-2000: International Conference on Mathematical Modeling and Computer* (November 2000), pp. 673–682, Ariel, Israe, November 13-15 2000.

91 Boley, C.D., Mitchell, S.C., Rubenchik, A.M., and Wu, S.S.Q. (2016). Metal powder absorptivity: modeling and experiment. *Appl. Opt.* 55 (23): 6496. https://doi.org/10.1364/ao.55.006496.

92 Abdelmoula, M., Küçüktürk, G., Juste, E., and Petit, F. (2022). Powder bed selective laser processing of alumina: scanning strategies investigation. *Appl. Sci.* 12 (2): https://doi.org/10.3390/app12020764.

93 Abdelmoula, M., Zarazaga, A.M., Küçüktürk, G. et al. (2022). Scanning strategy investigation for direct powder bed selective laser processing of silicon carbide ceramic. *Appl. Sci. (Switz.)* 12 (2).

94 K. Shahzad, J. Deckers, Z. Zhang, J. P. Kruth, and J. Vleugels, "Additive manufacturing of zirconia parts by indirect selective laser sintering," *J Eur Ceram Soc*, vol. 34, no. 1, pp. 81–89, Jan. 2014, doi: https://doi.org/10.1016/j.jeurceramsoc.2013.07.023.

95 J. K. J.-P. Deckers Jan Vleugels, "Additive manufacturing of ceramics: A review," *Journal of Ceramic Science and Technology*, vol. 5, no. 4, pp. 245–260, 2014, [Online]. Available: https://lirias.kuleuven.be/handle/123456789/459498.

96 Streek, A., Regenfus, P., Ullmann, F. et al. (2006). Processing of silicon carbide by laser micro sintering. *17th Solid Freeform Fabrication Symposium, SFF Austin, Texa (USA), August14-16, 2006*, pp. 349–358.

97 Ahlhelm, M., Richter, H.-J., and Haderk, K. (2013). Selective Laser Sintering as an Additive Manufacturing Method for Manufacturing Ceramic Components. *Journal of Ceramic Science and Technology* 4 (1): 33–40. https://doi.org/10.4416/JCST2012-00024.

98 Stierlen, P. and Eyerer, P. (1999). Si/SiC-Ceramic prototypes via LS2 1 – process (liquid silicon infiltration of laser sintered C-SiC parts). 1999 *International Solid Freeform Fabrication Symposium,* Austin, Texas (USA), *August* 9-11, 1999.

99 Evans, R.S., Bourell, D.L., Beaman, J.J., and Campbell, M.I. (2003). Reaction bonded silicon carbide: SFF, process refinement and applications," in *2003 International Solid Freeform Fabrication Symposium,* Austin, Texas (USA), August 4-6, 2003.

100 Bourell, D.L., Evans, R.S., and Barrows, S.L. (2005). High temperature infiltration of non-metallic particles produced by selective laser sintering. *Mater. Sci. Forum* 475 (9): 2861–2866.

6

Directed Energy Deposition (DED) of Ceramics

Amit Bandyopadhyay, Nathaniel W. Zuckschwerdt, and Susmita Bose

W. M. Keck Biomedical Materials Research Lab, School of Mechanical and Materials Engineering, Washington State University, 405 NE Spokane Street 99164-2920, Pullman, WA, USA

6.1 Introduction

Directed energy deposition (DED) is an additive manufacturing (AM) process characterized by the simultaneous deposition and fusion of materials using a focused energy source. Unlike powder bed fusion (PBF) processes, where the material is selectively melted in a layer of powder deposited prior, DED delivers the material directly to a melt pool created by the energy source [1]. This technique enables the creation of three-dimensional (3D) structures by depositing each layer. A focused energy source in DED applications, typically a laser or electron beam, generates a melt pool on a substrate surface while a feedstock, usually in powder form, is precisely delivered into this molten region through nozzles. The material rapidly solidifies as the deposition head moves according to a predetermined path, creating a layer. These layers are successively deposited to build complex 3D ceramic components [2, 3].

The solidification dynamics in ceramic DED involve rapid cooling rates exceeding 10^3–10^5 °C/s. This rapid solidification can produce nonequilibrium microstructures, metastable phases, and fine grain sizes that distinguish DED-processed ceramics from those manufactured using conventional techniques. The resulting microstructure significantly influences the final components' mechanical, thermal, and functional properties [4–6]. DED offers several distinct advantages for ceramic manufacturing, including creating compositionally graded structures, repairing existing components, and directly fabricating fully dense parts without post-processing sintering steps. These capabilities make DED particularly valuable for high-performance ceramic applications in the aerospace, energy, biomedical, and defense sectors [7, 8].

The evolution of DED technology for ceramic materials has built upon foundations established primarily for metal processing while overcoming unique challenges inherent to ceramic properties. The development of ceramic DED systems has progressed through several stages of technological innovation and material expansion. The initial DED systems utilized CO_2 lasers due to their high-power capabilities and strong absorption by oxide ceramics. As technology advanced, Nd:YAG lasers were introduced, offering better beam quality and fiber delivery potential despite reduced absorption by many ceramics. Most recently, high-power fiber lasers have become predominant in ceramic DED, providing superior beam quality, higher efficiency, and enhanced reliability [9]. This evolution in laser sources has been critical for processing ceramics with characteristically high melting points and low thermal conductivity. Concurrently, powder delivery systems evolved from simple, gravity-fed, mechanical wheel designs to more sophisticated approaches. Developing vibration-assisted feeders [10], pressurized carrier gas systems, and eventually coaxial nozzle designs [11, 12] significantly improved deposition precision and material efficiency for ceramic powders, which typically exhibit poor flowability. Multi-jet powder delivery systems later enabled simultaneous deposition of multiple materials, facilitating functionally graded ceramics and ceramic-metal hybrid structures [13–16]. Process monitoring and control capabilities have also advanced substantially, incorporating thermal imaging, melt pool monitoring, and closed-loop control systems that adjust laser parameters in real time [17]. These improvements have been essential for achieving consistent properties in DED-processed ceramics. Substrate preheating systems have also been developed to mitigate thermal stress-induced cracking in ceramic components [18, 19].

From a materials perspective, the development of ceramic DED has seen a progressive expansion in processable materials. Early efforts focused primarily on alumina (Al_2O_3) due to its extensive industrial applications. Researchers gradually developed process parameters that could produce dense alumina components despite initial challenges, resulting in porous structures with extensive cracking [1, 20]. Zirconia (ZrO_2) became the second primary ceramic material successfully processed by DED. Its lower thermal conductivity initially posed significant challenges for crack-free fabrication, but optimized processing parameters enabled successful processing. ZrO_2 quickly became essential for dental and biomedical applications due to its biocompatibility and mechanical properties [21]. Later, researchers expanded ceramic DED capabilities to include more complex oxide ceramics such as mullite ($3Al_2O_3{\cdot}2SiO_2$) and spinel ($MgAl_2O_4$) [22, 23]. These multicomponent oxides demonstrated the potential for processing ceramic systems with more complex phase behaviors and thermal properties.

A significant milestone was achieved with the successful DED processing of non-oxide ceramics, beginning with silicon carbide (SiC). These materials require strict atmospheric control during processing to prevent degradation. Developing specialized environmentally controlled chambers and inert gas delivery systems made this expansion possible. Recent advancements in DED technology have successfully demonstrated the fabrication of ceramic composites and functionally graded materials [14, 24, 25]. This progress is particularly notable in metal–ceramic functionally graded materials, which eliminate sharp interfaces and minimize residual stresses. These innovative systems capitalize on DED's inherent design flexibility to produce components with spatially varied compositions

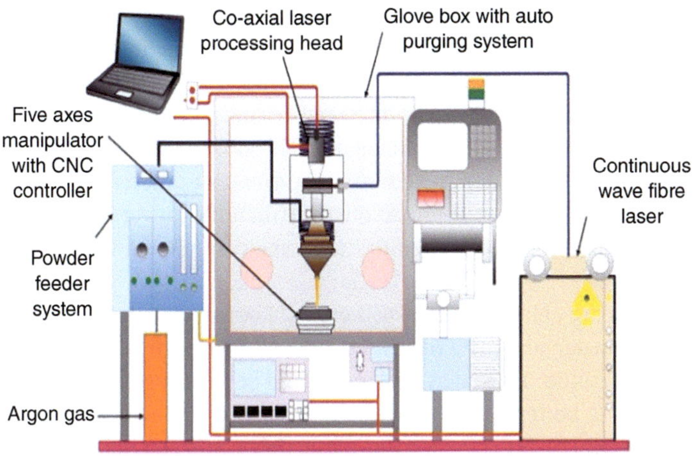

Figure 6.1 Schematic of a complete DED system. [27] / with permission of SAGE Publications.

that conventional manufacturing methods cannot achieve. Research indicates these developments have underscored DED's distinctive capability for creating compositionally varied structures with customized properties that meet specific performance requirements [26].

DED ceramic fabrication technology has evolved to incorporate ultra-high temperature ceramics (UHTCs), including zirconium diboride (ZrB_2), hafnium carbide (HfC), titanium carbide (TiC), titanium nitride (TiN), zirconium nitride (ZrN), and their composite formulations [27–32]. These exceptional materials, characterized by melting points above 3000 °C, demanded fundamental innovations in equipment systems and processing protocols. The successful implementation of DED for these challenging materials has expanded possibilities for components operating in extreme environments, particularly within aerospace and defense sectors, where thermal resistance is essential. Throughout this material evolution, each new ceramic system has required specific adaptations to DED technology and processing parameters, driving continuous innovation in the field. The progression from simple oxide ceramics to complex multi-material systems demonstrates DED's challenges and unique capabilities for advanced ceramic manufacturing. Figure 6.1 demonstrates the schematic of a laser-DED system.

6.2 Advantages and Limitations of DED

DED processes have many advantages over those of conventional manufacturing. As with most AM processes, the shapes of the structures can be as complicated as desired, with some limitations. For instance, internal cavities can be easily manufactured with the layer-by-layer deposition of the structure. However, for more complicated structures, a 5-axis or robotic arm can aid in overhang structures that might print too large an overhang angle to print on the prior layer. DED processes also allow for multi-material manufacturing through the use of multiple feeders. This is an excellent advantage as it allows for in situ manufacturing of two or more materials, leading to structures with site-specific properties. DED of ceramics also brings advantages, from the ability to repair damaged ceramics to the rapid prototyping of complex geometry. Another advantage of DED with ceramics is that the process can be done with a single step, with minimal or no post-processing. These lead to simpler processes that require optimization of the manufacturing parameters to achieve this single-step process compared to traditional ceramic manufacturing processes.

Repair and restoration are areas in which DED excels. The ability to take the feed material into a powder feeder, and print on a substrate of the same or similar material makes DED more versatile. Thus, that substrate could be a damaged high-value ceramic part that needs to be repaired by either replacing the damaged portion of the part or, if the part is broken into two or more parts, they could be rejoined using DED to weld the broken parts back together. This ability to build on a pre-existing structure also adds the capability of creating more complicated parts based on previously manufactured components. This allows for more complex structures where a conventionally manufactured ceramic part can be taken and modified to have several materials, whether through a coating of the structure or building a new part onto the structure consisting of a new material or the same material as the bulk structure.

Direct production of near-net shapes is another capability that DED has over other ceramic processing methods. This is due to how the material is processed during the manufacturing with the DED process. The printed ceramic is dense, with the material entirely melted and then resolidified during printing. Meanwhile, with other processes, the part is formed in a green part and then sintered to densify it. Densification of the green ceramic part causes a lot of shrinking and possible changes in the part's geometry, with the possibility of cracking and other defects. Thus, the ability to create a part that is close to the final desired dimensions would make it easier to manufacture with minimal post-processing.

The multi-material capabilities of the DED process allow more advanced integration of parts with site-specific properties. Being able to print two materials during the same process greatly reduces the part's processing time, making it unnecessary to take multiple steps to achieve a multi-material part. These multi-material parts manufactured with the DED process don't need to be sintered after printing, allowing materials that might have drastically different densification cycles to be joined together in a full melt rather than depending on solid-state diffusion. DED is also a process that allows for large amounts of rapid prototyping and iterations within a short period. DED can print one part at a time or as many as can fit on the substrate. With each group of parts, wide ranges of parameters can be tested till a smaller range is determined, wherein the next group could start fine-tuning all individual parameters till the best parameters are found for that material or part geometry.

Now, with all of these advantages of DED, there are also limitations of the DED process for ceramics. Between the thermal stress management and the range of materials that can be used, quite a few limitations have yet to be overcome.

Thermal stress management is one of the most significant limitations of the DED process due to the localized high-energy input. With the DED process, the localized high-energy input, combined with the low thermal conductivity of ceramics, leads to steep thermal gradients. These thermal gradients cause issues with the ceramic's cracking. This is a significant issue, particularly with poor thermal shock resistance ceramics. The surface finish quality is challenging due to the type of raw material being powder. During the printing process, excess powder is delivered to the melt pool, and some of that powder ends up adhering to the exterior of the melt pool. This powder that ends up on the melt pool does not cause many issues when printing the infill of the structure. However, this does cause a rough exterior when scanning along the exterior perimeter. There is also a slight variation in the thickness and height of the melt pool, leading to lower dimensional precision than that of conventional processes for ceramics. Despite the advantage of the ability to do rapid prototyping, DED does not have the same throughput as traditional manufacturing methods for ceramics. DED also has a higher cost per unit for large production volumes, as there is little to do in reducing processing costs with DED, and only material costs can be adjusted. This is limited to simple parts, as more complicated parts become easier to manufacture with DED than traditional ceramic manufacturing. The DED of ceramics is limited to specific materials, as the process has a particular criterion that needs to be met to be successful. The chosen ceramic can't have an extremely high melting point, a narrow melting range, or problematic phase transformation during rapid thermal cycling, as these particular issues have yet to be overcome in DED processing of ceramics.

6.3 Powder Characteristics

DED processes have emerged as versatile AM technologies capable of producing near-net-shape components with complex geometries. Unlike PBF methods that require highly spherical powders with excellent flowability, DED processes can accommodate a broader range of powder characteristics. DED processes typically utilize powder size distributions in the range of 44 to 75μm, although this can be adjusted based on specific processing requirements and equipment configurations. Some research has utilized powders as small as 15μm up to 150μm [14, 15]. This flexibility stems from the fundamental difference in powder delivery mechanisms, where DED uses gas-assisted powder feeding directly into the melt pool rather than requiring uniform powder bed formation.

The powders' morphology in DED processes can significantly influence processing outcomes, though with different constraints than other AM methods. While spherical powders produced via gas atomization remain preferred for optimal flow through powder feed systems, DED can successfully utilize nonspherical or irregular powders, provided consistent feed rates can be maintained. DED systems generally tolerate more significant deviations from ideal sphericity than powder bed technologies. This tolerance can potentially enable lower-cost powder production methods, including water atomization for specific applications, though careful system calibration is necessary. The chemical composition and surface oxidation of powders play crucial roles in DED processing outcomes and final part properties. Proper powder handling protocols, including inert gas storage and controlled exposure environments, are essential to preserve powder quality throughout the build process.

The internal porosity of feedstock powders represents a significant concern for DED processes, mainly when targeting high-performance applications with demanding mechanical requirements. This is particularly relevant for DED processes utilizing larger powder particle sizes, as internal porosity is more prevalent in coarser powders (diameter >70μm) typically used in DED systems. Gas atomization parameters significantly influence internal porosity formation, with the "bag breakup and collapse" mechanism identified as a primary source of entrapped gas in larger powders, potentially requiring process optimization or powder screening protocols to minimize these defects. In addition, pre-processing treatments are typically recommended to remove surface moisture from hygroscopic ceramic powders, particularly alumina, zirconia and many glass-ceramic compositions.

6.4 DED Processing-Related Challenges

Though DED is a great tool in a manufacturing environment, quite a few significant issues limit the process for large-scale operations. The process mainly utilizes powder as the material feedstock, with the size of these particles being on the micron scale; there are handling issues. Handling the powder is hazardous and requires a range of safety precautions to be in place to ensure that accidents don't happen. These powders require personal protective equipment (PPE) such as gloves, fire-retardant clothing, full-face respirators, and antistatic devices. Some powders cannot be handled in a normal atmosphere due to issues with the volatility of the small particle sizes. Dispersion of the powder into the air is also a limiting factor as a build-up on the surroundings where the powder is being mixed or moved from container to container could cause contamination of equipment or other powders, as well as electrically conductive powders, causing issues with electronics through shorting of the electronic components when the powder ingresses into these systems.

Most DED systems are focused on the manufacturing of metal components. Thus, the printing of ceramics needs some modifications to the equipment itself. For instance, when printing metals, the system would be set up to cool the substrate to remove heat from the system, as the temperature of the part needs to be controlled such that there is sufficient cooling to solidify the material in the desired location. However, ceramics have cooling rate issues as they are rather temperamental about high thermal gradients. These thermal gradients can cause cracking in ceramics with poor thermal shock resistance. Thus, cooling the substrate would not be optimal; heating the substrate would greatly benefit from using normal DED systems. This heating of the build plate would heat the printed structure, decreasing the thermal gradient between the melt pool and the layers below.

DED systems are running in inert environments where oxygen levels are kept at a minimum to reduce the chance of oxidation on and in a part. This limits the size of the parts that can be manufactured, as the system would need to be in an enclosed space that can still be accessed by the operator for maintenance and changing the substrate for a new part to be printed. The size of the part is also limited by the size of the passthrough, the antechamber, as without the use of the antechamber, the entire building chamber would need to be depressurized, and the inert environment would be lost. Thus, there are limitations on how large the build chamber can be so that it can be accessed and economical to keep in an inert environment, as well as how large the footprint of the machine is, such that it does not take up much space in the workplace.

6.5 Processing via DED

The DED process can manufacture two main overarching types of structures. Monolithic structures consist of a single bulk structure, where dopants and other additives can make a single consistent microstructure. Composite structures manufactured using DED usually consist of a metal matrix with a ceramic reinforcement; however, there can also be composites consisting of a ceramic matrix with higher strength ceramics that are used and don't form a solution as the dopants would with the monolithic style.

6.6 Monolithic Ceramics via DED

With material considerations, the first type of ceramics produced is oxide ceramics, particularly alumina. This process was expanded rapidly from the simpler process of printing oxide ceramics to coatings of non-oxide ceramics and then to large structures. Initial work was done with alumina, showing the ability of DED to fabricate a fully dense ceramic. With its high energy input and high thermal gradients, the DED process caused the microstructure to form columnar grains in the build direction in Figure 6.2. This is due to where the cooling of the melt pool is happening. The structure was cooled through the substrate, causing the grain formation in the initial layer; these grains then acted as a nucleation site for the successive layers [1]. The density found in this work for the as-fabricated alumina was 94%, but increased to 98% with a heat treatment of 1600 °C at five hours [1]. This showed that nearly dense fabrication of ceramics was possible with DED. The mechanical properties of these structures also showed great success. The compressive strength in the build direction of the as-fabricated alumina was 123 ± 9 MPa, while that of the heat-treated sample was 159 ± 8 MPa [1]. The compressive strength in the direction normal to the build direction in the as-fabricated and heat-treated samples was 229 ± 11 MPa and 276 ± 5 MPa, respectively [1]. These values did end up in the low range of what would be expected from alumina for compression, but that does lead to room for improvement.

(a) (b)

Figure 6.2 Alumina is manufactured by DED, (a) manufacture parts, (b) microstructure shows columnar grain growth. [1] / with permission of John Wiley & Sons.

A non-oxide ceramic processed with DED was lead zirconate titanate (PZT). This proved to have some difficulty during the printing process due to the volatilization of the Pb with its lower melting point. The samples showed a rougher exterior than the alumina samples in Figure 6.2. The density of these samples was found to be lower than that of the alumina, with a theoretical density of ~90% [34].

One major area in which DED has shown great abilities is the area of ceramic coatings on metal components. These coatings can aid in several applications, from wear resistance to high-temperature resistance. Wear-resistant coatings are important for increasing the lifespan of a part that may have the desired properties in terms of mechanical strength and fatigue life but have issues in environments where it could be exposed to extreme amounts of wear, decreasing the usable lifespan of the part. The same applies to the high-temperature coatings, where the metal itself would be great for the use case, except for the environmental conditions in which it would be used. High-temperature coatings protect the metal from oxidation at elevated temperatures, where the base metal would begin to oxidize and lose its mechanical properties. Work with titanium has been studied extensively with high-temperature and wear-resistant coatings due to its high strength-to-weight ratio and use in many industries. In the biomedical field, titanium, particularly Ti6Al4V (Ti64), is widely used, as it is biocompatible and fatigue resistant. However, it lacks bioactivity and wear resistance, which are crucial in applying joint replacement components [35, 36].

To increase the bioactivity of the titanium, coatings of hydroxyapatite (HA) have been used and done with the use of plasma spray deposition, but certain work has been done looking at a thin coating of hydroxyapatite with DED before the plasma spraying is done to improve the bond strength of the coating. The initial coating was done with a premixed powder of 3 wt.% HA and a balance of Ti64 [37]. This was deposited onto a Ti64 substrate, giving a compositional gradient layer of HA and Ti64. Then, a full HA coating was deposited onto the intermediate layer using plasma spraying. This method for coating by applying a HA coating to titanium increased the bond strength from approximately 25 MPa with just the plasma-sprayed HA to 40 MPa with the gradient layer [37]. To aid in wear resistance during higher load-bearing applications, titanium has been coated with DED with several different ceramics, including zirconia and several others that will be covered. These ceramic coatings aid in increasing the lifespan of components that need to endure extended deployment in inhospitable environments.

Zr-coatings were printed first onto a commercially pure titanium (CPTi) substrate in an environment with an argon atmosphere, keeping the O_2 content under 10 ppm [21]. The Zr coating was then remelted under a controlled oxygen environment using a laser to oxidize the Zr and form ZrO_2 on the surface of the Zr. This coating decreased the wear rate from 0.021 on the CPTi plate to 0.001 on the oxidized Zr coating processed at 100 W at a wear distance of 240 m [21]. Another coating attempted to create a composite coating with Ti64 and boron nitride (BN) to reinforce the matrix's TiB/TiB2 and TiN intermetallic phases. The feedstock consisted of premixed powders at certain weight percentages of BN powder and Ti64 powder. At 15 wt.% of BN in the Ti64 powder, the surface hardness increased from 147±22 on the Ti substrate to 877±75 $HV_{1.0}$ [21]. This was effective due to the DED process causing in situ alloying of the mixed powders during the printing process in the melt pool.

6.7 Metal-Matrix Composites via DED

As discussed briefly in the last coating, DED is excellent for the in situ alloying of mixed powders, which adds great ability to create metal-matrix composites (MMCs). These MMCs can improve the bulk properties of many materials in terms of yield strength, hardness, and wear resistance [15]. DED also allows these MMCs to be put in specific locations on the structure during a single process where the part is being manufactured. Depending on the properties of the added ceramic, it could affect the processing parameters such that the input power would need to be increased to a point that it exceeds the capability of the energy source due to the required energy to melt the metal and adhere the ceramic in the matrix. It is also limited by the amount of ceramic that can be put into the matrix without cracks forming in the structure because of the increase of hard, brittle intermetallics between the metal and the ceramic components.

Manufacturing MMCs with DED is done by taking the base metal powder, adding the ceramic reinforcement powder, and mixing the two powders at the desired percentage of the ceramic reinforcement. The mixed powder is then placed into the powder feeder and fed to the deposition head to be deposited. The energy source would then melt the metal powder and some of the ceramic, causing alloying between the ceramic's components and the metal, forming intermetallics. Around the exterior of the ceramic particles that remain in the metal matrix, there is a reaction region that bonds the ceramic to the metal such that it is not just a mechanical attachment, but also a chemical attachment of the ceramic particle to the metal matrix.

This process can be used to expand the high-temperature capabilities of metals. Work was done on CpTi with boron nitride to improve its high-temperature oxidation properties. Two weight percentages were used, 3 wt.% BN and 6 wt.% BN [13].

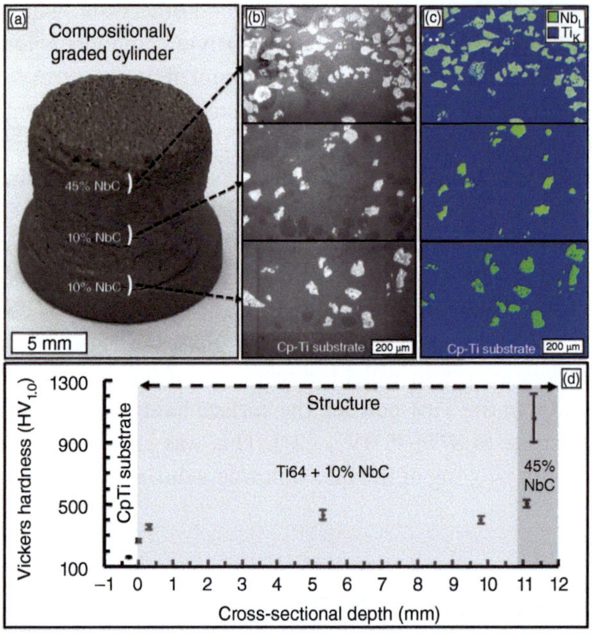

Figure 6.3 A metal-matrix composite consisting of Ti64 and NbC at varying percentages of NbC as the cylinder height increases, and a microhardness plot. (a) As printed cylinder. (b) SEM imaging of different sections. (c) Overlaid EDS maps. (d) Plot of microhardness of the sample. [24] / with permission of John Wiley & Sons.

The oxidation rate was examined using thermogravimetric analysis (TGA), where a small sample is placed into the TGA and held at a high temperature while the weight of the sample is continuously measured, and a flow of air goes through the testing equipment. Thus, as the sample oxidizes, the weight of the sample will increase as the exterior oxidizes. This treatment showed a ~55% decrease in the average instantaneous parabolic constant from the control sample to the 6 wt.% BN [13].

Several different carbides have been used as reinforcements in Ti64. One of which is niobium carbide (NbC). This was done to enhance multiple properties of the Ti64, such as wear resistance, hardness, and high-temperature oxidation of the sample. Two different weight loadings of the NbC were done, the first was 5 wt.% and the second was 10 wt.%. A cylinder was also manufactured with a graded composition from 0 to 45 wt.% NbC, Figure 6.3. The microhardness showed no significant difference in the hardness of the samples, but nanoindentation of a wear-tested sample showed a significant difference between the Ti64 substrate and the 5 wt.% NbC of an increase of 28% [24]. The wear rate was reduced by 24.7% under a 3N normal loading [24]. This shows a reduction in the wear rate and an increase in the hardness of the matrix. As seen in Figure 6.3, as the percentage of NbC increases, the amount of particles remaining in the matrix increases, and particles can easily be identified in the sections' energy dispersive X-ray spectroscopy (EDS) mapping [24]. Thus, graded structures were fabricated using DED, leading to more system capabilities for making parts with site-specific properties.

Finally, radial structures can be manufactured with a solid metal core and MMC on the shell; these can also be swapped around, and other structures can be tested rather than only a core and shell. Using a system with multiple powder feeders, one can be loaded with Ti64 powder, then the second is loaded with Ti64 and 5 wt.% boron carbide. These two powders were printed on their sections of the part on each layer by switching between the two feeders at the end of the core section as displayed in Figure 6.4 [14]. Two monolithic samples were also printed, the control being Ti64 and the first treatment being the composite of Ti64 + 5 wt.% B_4C. Uniaxial compression testing and Vickers microhardness of these samples were done. The compression data showed that the radial structure maintained approximately the same strength as the monolithic composite, a ~48% increase over the control, despite the large amount of cracks and porosity in the radial composite [14]. Hardness of the matrix also increased in the composite by ~58% in the shell of the radial composite compared to that of the monolithic Ti64 cylinders.

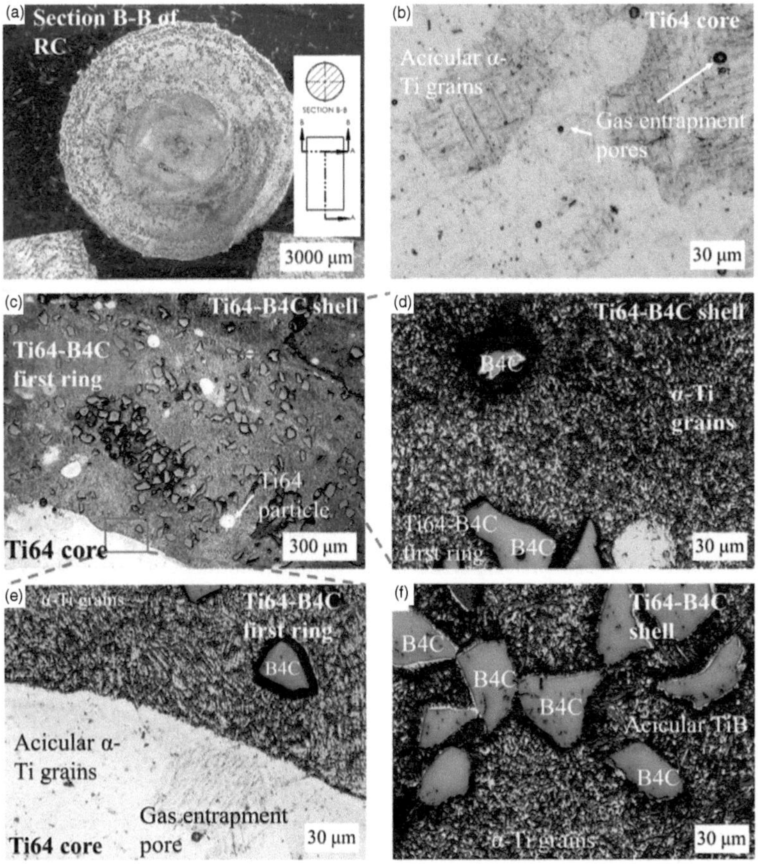

Figure 6.4 A horizontal cross section of a Ti64 and Ti64 + B_4C radial composite that is printed as a cylinder. (a) Low magnification showing the entire cross section and the different core and shell. (b–f) High magnification shows the individual sections of the radial composite's microstructure. [14] / AccScience Publishing / CC BY 4.0.

In Figure 6.4, the in situ alloying of the ceramic components into the Ti64 matrix can be identified with a great microstructure change. The normal needle-like α'-Ti grain structure seen in AM Ti64 parts can be seen in the core section of the sample, while that of the shell displays a cellular growth of the α-Ti [14]. Adhesion of the B_4C particles into the matrix can also be seen in the reaction zone around the B_4C particles with TiB/TiB_2 growing into the matrix from the particles [14].

6.8 Current Challenges and Future Directions

This chapter briefly highlights research on laser-based DED of melt-cast monolithic ceramic structures or metal-matrix composites with ceramic reinforcements. Although monolithic ceramic structures can be fabricated via laser-DED due to the excellent laser absorption and poor thermal conductivities of many ceramic materials, only small parts can be made using this approach due to poor thermal shock resistance. Unlike metals, many ceramics degrade or dissociate at high temperatures instead of melting. Those ceramics are unsuitable for laser-DED processing. When alumina ceramics were processed in an inert Ar environment, oxygen-deficient alumina was the final product. Further high-temperature heat treatments were necessary to enrich the alumina ceramics with oxygen to achieve good mechanical properties [1]. The DED process does not offer support structures like PBF-based AM. When ceramics are processed in laser-DED, feature definition is also poor. Complex shapes are difficult to process using this approach. However, if laser-PBF can be used, ceramics with fine features can be manufactured as long as they are small.

The advantages of laser-DED for ceramics are in manufacturing multi-material structures. Laser-DED can process multiple ceramic compositions at different layers [38]. As DED can use multiple hoppers to deliver different powder compositions, layered ceramic structures with varying compositions have been processed via DED with alumina and zirconia ceramics. More interestingly, the same concept can be used to manufacture not just ceramic-to-ceramic compositional gradations but metal-to-ceramic compositional gradations [39, 40]. Laser-DED was used to process 100% Ti6Al4V to 100% Al_2O_3 structures. A similar concept was also shown in 100% stainless steel 304 to 100% Vanadium carbide structures [41]. In such structures, compositional variations can be achieved and tailored site-specifically to manufacture structures with unique functionalities. An example of such structures is diamond-reinforced metal–ceramic composites for cutting tool applications [42]. Diamond dust in the metal matrix made the composite suitable for metal cutting operations.

The future of DED of ceramics needs specific applications. There are enough publications to show that it is possible and that ceramics with various compositions can be manufactured using DED. However, which compositions are useful to manufacture via DED and why is a question that needs to be answered first. Another aspect is what post-processing steps will be needed to make these structures operational. Laser-processed ceramic coatings with one to three layers of deposition that can add site-specific properties may show promise for this technology in the coming days. Multi-material metal–ceramic composites also have a significant potential. During space exploration, in situ resource utilization (ISRU) has to be critical, and the DED of Lunar and Martian dust will be vital to manufacture the first generation of parts to help colonize for human missions [2, 43, 44]. Laser processing of Lunar and Martian regoliths has been demonstrated on a small and large scale, and it is expected to be a viable technology used for future space missions. Finally, the DED of ceramics is exciting, full of challenges, and holds significant potential.

6.9 Summary

This chapter discusses laser-DED of advanced ceramic structures. Laser-DED has been used to process melt-cast ceramic parts from powder feedstock. Although monolithic ceramic structures can be fabricated via laser-DED due to their excellent laser absorption and poor thermal conductivities of many ceramic materials, only small parts can be made using this approach due to poor thermal shock resistance of ceramics. Unlike metals, many ceramics degrade or dissociate at high temperatures instead of melting. Those ceramics are unsuitable for laser-DED processing. The advantages of laser-DED for ceramics are in direct manufacturing of multi-material structures. Laser-DED can process multiple ceramic compositions at different layers. More interestingly, the same concept can be used to manufacture not just ceramic-to-ceramic compositional gradations but also metal-to-ceramic compositional gradations. The future of DED of ceramics needs specific applications. There are enough publications to show that it is possible and that ceramics with various compositions can be manufactured using DED. However, which compositions are useful to manufacture via DED and why are questions that need to be answered first. This chapter discussed historical developments of direct ceramic processing via laser-DED, properties, applications, challenges, and future directions.

Acknowledgments

We would like to acknowledge the support of Dr. Yanning Zhang while writing this book chapter. His comments were valuable in drafting this document.

References

1 Balla, V.K., Bose, S., and Bandyopadhyay, A. (2008). Processing of bulk alumina ceramics using laser engineered net shaping. *Int. J. Appl. Ceram. Technol.* 5 (3): 234–242. https://doi.org/10.1111/j.1744-7402.2008.02202.x.

2 Bose, S., Akdogan, E.K., Balla, V.K. et al. (2024). 3D printing of ceramics: advantages, challenges, applications, and perspectives. *J. Am. Ceram. Soc.* 107 (12): 7879–7920. https://doi.org/10.1111/jace.20043.

3 Das, M., Balla, V.K., Basu, D. et al. (2010). Laser processing of SiC-particle-reinforced coating on titanium. *Scr. Mater.* 63 (4): 438–441. https://doi.org/10.1016/J.SCRIPTAMAT.2010.04.044.

4 Sahasrabudhe, H., Soderlind, J., and Bandyopadhyay, A. (2016). Laser processing of in situ TiN/Ti composite coating on titanium. *J. Mech. Behav. Biomed. Mater.* 53: 239–249. https://doi.org/10.1016/J.JMBBM.2015.08.013.

5 Yang, C., Wang, T., Jiao, F. et al. (2024). Fabrication of carbon fiber reinforced SiC composites based on laser directed energy deposition. *Ceram. Int.* 50 (23): 50713–50725. https://doi.org/10.1016/J.CERAMINT.2024.09.416.

6 Roy, M., Bandyopadhyay, A., and Bose, S. (2008). Laser surface modification of electrophoretically deposited hydroxyapatite coating on titanium. *J. Am. Ceram. Soc.* 91 (11): 3517–3521. https://doi.org/10.1111/j.1551-2916.2008.02698.x.

7 Dass, A. and Moridi, A. (2019). State of the art in directed energy deposition: from additive manufacturing to materials design. *Coatings* 9 (7) MDPI AG: https://doi.org/10.3390/COATINGS9070418.

8 Svetlizky, D., das, M., Zheng, B. et al. (2021). Directed energy deposition (DED) additive manufacturing: physical characteristics, defects, challenges and applications. *Mater. Today* 49: 271–295. https://doi.org/10.1016/J.MATTOD.2021.03.020.

9 Varshney, A.K., Mainuddin, G.S., and Nayak, J. (2024). High power lasers for directed energy applications: developments and challenges. *Infrared Phys. Technol.* 136: 105064. https://doi.org/10.1016/J.INFRARED.2023.105064.

10 Wu, H., Pritchet, D., Wolff, S. et al. (2020). A vibration-assisted powder delivery system for additive manufacturing – an experimental investigation. *Addit. Manuf.* 34: 101170. https://doi.org/10.1016/J.ADDMA.2020.101170.

11 Singh, A., Kapil, S., and Das, M. (2020). A comprehensive review of the methods and mechanisms for powder feedstock handling in directed energy deposition. *Addit. Manuf.* 35: 101388. https://doi.org/10.1016/J.ADDMA.2020.101388.

12 Guan, X. and Zhao, Y.F. (2020). Numerical modeling of coaxial powder stream in laser-powder-based directed energy deposition process. *Addit. Manuf.* 34: 101226. https://doi.org/10.1016/J.ADDMA.2020.101226.

13 Avila, J.D. and Bandyopadhyay, A. (2019). Influence of boron nitride on reinforcement to improve high temperature oxidation resistance of titanium. *J. Mater. Res.* 34 (7): 1279–1289. https://doi.org/10.1557/jmr.2019.11.

14 Zuckschwerdt, N.W. and Bandyopadhyay, A. (2024). Multi-material structures of Ti6Al4V and Ti6Al4V-B4C through directed energy deposition-based additive manufacturing. *Mater. Sci. Addit. Manuf.* 3 (3): 3571. https://doi.org/10.36922/msam.3571.

15 Traxel, K.D. and Bandyopadhyay, A. (2020, https://doi.org/10.1016/j.addma.2019.101004). Influence of in situ ceramic reinforcement towards tailoring titanium matrix composites using laser-based additive manufacturing. *Addit. Manuf.* 31: 101004.

16 Bandyopadhyay, P.P., Balla, V.K., Bose, S., and Bandyopadhyay, A. (2007). Compositionally graded aluminum oxide coatings on stainless steel using laser processing. *J. Am. Ceram. Soc.* 90 (7): 1989–1991. https://doi.org/10.1111/j.1551-2916.2007.01651.x.

17 Haley, J., Karandikar, J., Herberger, C. et al. (2024). Review of in situ process monitoring for metal hybrid directed energy deposition. *J. Manuf. Process.* 109: 128–139. https://doi.org/10.1016/J.JMAPRO.2023.12.004.

18 He, S., Park, S., Shim, D.S. et al. (2023). Effect of substrate preheating on the microstructure and bending behavior of WC-Inconel 718 composite coating synthesized via laser directed energy deposition. *Int. J. Refract. Met. Hard Mater.* 115: 106299. https://doi.org/10.1016/J.IJRMHM.2023.106299.

19 Cai, Y., Xiong, J., Chen, H., and Zhang, G. (2023). A review of in-situ monitoring and process control system in metal-based laser additive manufacturing. *J. Manuf. Syst.* 70: 309–326. https://doi.org/10.1016/J.JMSY.2023.07.018.

20 Pfeiffer, S., Florio, K., Puccio, D. et al. (2021). Direct laser additive manufacturing of high performance oxide ceramics: a state-of-the-art review. *J. Eur. Ceram. Soc.* 41 (13): 6087–6114. https://doi.org/10.1016/J.JEURCERAMSOC.2021.05.035.

21 Balla, V.K., Xue, W., Bose, S., and Bandyopadhyay, A. (2009). Laser-assisted Zr/ZrO2 coating on Ti for load-bearing implants. *Acta Biomater.* 5 (7): 2800–2809. https://doi.org/10.1016/J.ACTBIO.2009.03.032.

22 Pappas, J.M. and Dong, X. (2020). Porosity characterization of additively manufactured transparent $MgAl_2O_4$ spinel by laser direct deposition. *Ceram. Int.* 46 (5): 6745–6755. https://doi.org/10.1016/J.CERAMINT.2019.11.164.

23 Wu, D., Zhao, D., Niu, F. et al. (2020). In situ synthesis of melt-grown mullite ceramics using directed laser deposition. *J. Mater. Sci.* 55 (27): 12761–12775. https://doi.org/10.1007/s10853-020-04938-3.

24 Avila, J.D. and Bandyopadhyay, A. (2022). Niobium carbide reinforced-Ti6Al4V composites via directed energy deposition. *Int. J. Appl. Ceram. Technol.* 19 (2): 1061–1073. https://doi.org/10.1111/ijac.13907.

25 Heer, B., Zhang, Y., and Bandyopadhyay, A. (2021). Processing and thermal diffusivity measurements of compositionally graded Al-12Si to Al_2O_3 structures. *Mater. Lett.* 290: 129466. https://doi.org/10.1016/J.MATLET.2021.129466.

26 Ratnala, D.C., Andersson, J., and Joshi, S. (2023). Development of functionally graded metal-ceramic systems by directed energy deposition: a review. *Mater. Sci. Forum* 1107: 105–110. https://doi.org/10.4028/p-4ekAtd.

27 X. Lv, Z. Zhan, and H. Cao, "Microstructure evolution and mechanical properties of needle-like ZrB2 reinforced Cu composites manufactured by laser direct energy deposition", *Micromachines (Basel)*, vol. 13, no. 2, p. 212, Jan. 2022, doi: 10.3390/mi13020212.

28 W. Jeong *et al.*, "Ultra-high temperature ceramic (HfC) reinforcement of laser powder-directed energy deposited inconel 718: Microstructural evolution and tensile properties at room and high temperatures", *Compos B Eng*, vol. 297, p. 112281, May 2025, doi: 10.1016/j.compositesb.2025.112281.

29 S. Mihai, F. Baciu, R. Radu, D. Chioibasu, and A. C. Popescu, "In Situ fabrication of TiC/Ti–Matrix composites by laser directed energy deposition", *Materials*, vol. 17, no. 17, p. 4284, Aug. 2024, doi: 10.3390/ma17174284.

30 X. Huang, K. Yang, X. Zou, C. T. Kwok, and L. Pan, "The role of TiN inoculant on microstructure and properties in wire arc directed energy deposition of super duplex stainless steels", *Mater Charact*, vol. 214, p. 114090, Aug. 2024, doi: 10.1016/j.matchar.2024.114090.

31 Z. Tai *et al.*, "In-situ synthesis of high-performance TiN/TC4 sandwich structures via nitrogen-controlled laser directed energy deposition", *Mater Sci Eng: A*, vol. 896, p. 146231, Mar. 2024, doi: 10.1016/j.msea.2024.146231.

32 K. D. Traxel and A. Bandyopadhyay, "Reactive-deposition-based additive manufacturing of Ti-Zr-BN composites", *Addit Manuf*, vol. 24, pp. 353–363, Dec. 2018, doi: 10.1016/j.addma.2018.10.005.

33 K. Benarji, Y. R. Kumar, C. P. Paul, A. N. Jinoop, and K. S. Bindra, "Parametric investigation and characterization on SS316 built by laser-assisted directed energy deposition", *Proc. Inst. Mech. Eng. Part L*, vol. 234, no. 3, pp. 452–466, 2020, doi: 10.1177/1464420719894718.

34 S. A. Bernard, V. K. Balla, S. Bose, and A. Bandyopadhyay, "Direct laser processing of bulk lead zirconate titanate ceramics", *Mater. Sci. Eng. B*, vol. 172, no. 1, pp. 85–88, 2010, doi: 10.1016/J.MSEB.2010.04.022.

35 Y. Zhang, H. Sahasrabudhe, and A. Bandyopadhyay, "Additive manufacturing of Ti-Si-N ceramic coatings on titanium", *Appl. Surf. Sci.*, vol. 346, pp. 428–437, 2015, doi: 10.1016/J.APSUSC.2015.03.184.

36 M. Roy, A. Bandyopadhyay, and S. Bose, "in vitro antimicrobial and biological properties of laser assisted tricalcium phosphate coating on titanium for load bearing implant", *Mater. Sci. Eng. C*, vol. 29, no. 6, pp. 1965–1968, 2009, 10.1016/J.MSEC.2009.03.009.

37 D. Ke, A. A. Vu, A. Bandyopadhyay, and S. Bose, "Compositionally graded doped hydroxyapatite coating on titanium using laser and plasma spray deposition for bone implants", *Acta Biomater.*, vol. 84, pp. 414–423, 2019, doi: 10.1016/J.ACTBIO.2018.11.041.

38 F. Niu, S. Yan, G. Ma, Dongjiang Wu, Directed-energy deposition for ceramic additive manufacturing, *ASM Handbook-Additive Manufacturing Processes*, Volume 24, Edited by D. L. Bourell; W. Frazier; H. Kuhn; M. Seifi, pp. 131–151 (2020). https://doi.org/10.31399/asm.hb.v24.a0006559.

39 A. Bandyopadhyay and B. Heer "Additive manufacturing of multi-material structures," *Mater. Sci. Eng. R. Rep.* 129, 1–16 (2018).

40 Y. Zhang and A. Bandyopadhyay, "Direct fabrication of compositionally graded Ti-Al_2O_3 multi-material structures using laser engineered net shaping," *Addit. Manuf.*, 21, pp. 104–111 (2018).

41 T. Gualtieri and A. Bandyopadhyay, "Additive manufacturing of compositionally gradient metal-ceramic structures: stainless steel to vanadium carbide," *Mater. Des.*, 139, pp. 419–28 (2018). doi: 10.1016/j.matdes.2017.11.007

42 K. D. Traxel and A. Bandyopadhyay, "Diamond-reinforced cutting tools using laser-based additive manufacturing," *Addit. Manuf.*, 37, 101602 (2021). doi: 10.1016/j.addma.2020.101602.

43 V. K. Balla, L. B. Roberson, G. W. O'Connor, S. Trigwell, S. Bose, and A. Bandyopadhyay, "First demonstration on direct laser fabrication of lunar regolith parts," *Rapid Prototyp. J.*, 186, pp. 451 – 457 (2012).

44 A. Afrouzian, K. D. Traxel, and A. Bandyopadhyay, "Martian regolith – Ti6Al4V composites via additive manufacturing," *Int. J. Appl. Ceram. Technol.*, 19, pp. 2998–3006 (2022). doi: 10.1111/ijac.14136.

7

Hybrid AM of Ceramics and AM of Ceramic-Based Multi-Material Components

Giorgia Franchin[1], *Uwe Scheithauer*[2], *and Tassilo Moritz*[2]

[1] *University of Padova, Department of Industrial Engineering, Via Marzolo 9, 35131 Padova (PD), Italy*
[2] *Department of Processes and Components, Fraunhofer-Institut für Keramische Technologien und Systeme IKTS, Winterbergstr. 28, 01277, Dresden, Germany*

Additive manufacturing (AM) represents a game changer, especially for ceramic materials, as these hard and brittle materials, which are otherwise difficult to process, can now be transformed into much more complex geometries much more easily than with conventional technologies.

However, the functionality of components can be further enhanced by hybridizing AM technologies or combining different materials in an AM component. This chapter summarizes the current state of research in these areas.

7.1 Hybrid Additive Manufacturing of Ceramics

Following the CIRP Annals definition published in 2014 [1], a hybrid manufacturing process is described as the result of "simultaneous and controlled interaction of mechanisms and/or energy sources having a significant effect on the process performance." Hybridization is a term used in the field of AM to indicate the combination/integration (either in parallel or in series) of different technologies, where at least one is additive, to generate components with features typically not obtainable when using a single approach, or to provide additional manufacturing benefits [2].

Indeed, the goal of a hybrid manufacturing technique is to exploit the combined advantages of the singular methods, unlocking novel capabilities, thus validating the so-called "1 + 1 = 3" effect [3].

As shown in Figure 7.1 and following what is proposed by Lauwers et al. [1], the hybrid manufacturing techniques can be divided into two groups, consisting of either concurrent processes, which synergistically act at the same time, or sequential techniques, where the separate action of different processes happens on the same processing zone. Furthermore, the combined processes could either be part of the same family (i.e. subtractive, formative, or additive) or belong to different ones.

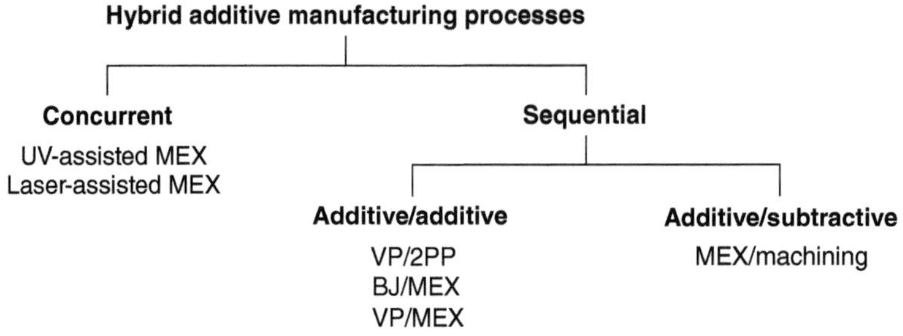

Figure 7.1 Classification of hybrid additive manufacturing processes with relevant examples.

With traditional AM approaches, severe overhangs, small details, or inner channels/pores are not always possible with traditional AM approaches and, if so, require sacrificial support structures; moreover, the layer-by-layer construction produces multiple interlayer interfaces and a stair-stepping effect on the surface, which can act as stress concentrators and reduce the mechanical strength of the printed parts. The common goal behind all hybrid technologies is to overcome these limitations and enhance process capabilities by extending the geometrical complexity and increasing the surface quality of printed parts to reach (near) net-shape components.

Finally, hybrid AM has risen concurrently with multi-material technologies as the two developments slightly overlap: in some cases, the different materials or geometries are better suited for different fabrication technologies. This is particularly relevant in the electronic and information industry, as the fabrication of sensors or embedded circuits requires continuous, thin conductive traces to be embedded in bulk, wide components.

The following paragraphs explore hybrid technologies from the different groups that are relevant to the fabrication of ceramic components.

7.1.1 Concurrent Hybrid Processes

The two processes discussed in this section are both characterized by feedstock material deposition associated with an energy source able to quickly transform the material into a solid on the spot; such transformation eliminates the stringent rheological requirements of conventional material extrusion (MEX) techniques (in particular, direct ink writing (DIW)) and increases shape retention also in case of severe overhangs, opening the door to free-forming fabrication.

7.1.1.1 UV-Assisted MEX

This technology exploits the benefits of both photopolymerization and DIW; in UV-assisted direct ink writing (UV-DIW), a photocurable suspension is extruded through a nozzle and cured by an external UV source. The combination of direct deposition and photocuring capabilities allows overcoming most of the limits of DIW and VP, taken separately. The consolidation of the ink relies on the photocuring process, which proceeds faster and develops higher stiffness compared to gel reconstruction. The process is seldom influenced by light-particle interactions, that is, light scattering or absorption, which are limiting the resolution and penetration depth in standard VP processes to the point where dark materials are not printable at all. In UV-DIW, a limited penetration depth still results in the consolidation of a thin outer shell, which is sufficient to guarantee shape retention.

Most reported examples of UV-DIW with ceramic suspensions focus on conventional layer-by-layer structures, yet with increased shape stability and pronounced overhangs; materials include oxides such as alumina, hydroxyapatite, zirconia, and polymer-derived ceramics such as SiOC and Si_3N_4/SiC [4–7]. A more challenging yet enabling use of UV-DIW aims at overcoming the traditional stacking of 2D layers; in fact, the fast solidification rate coupled with the unconstrained mobility of the nozzle enables the fabrication of self-supporting truss-based structures, a process referred to as freeform fabrication. Such an approach was proposed by Maeng et al., who produced porous biphasic calcium phosphate components with unsupported features [8], and further validated by De Marzi et al. by producing silica glass lattices with smooth obliquus trusses maintaining high transparency as they were the result of a single extrusion [9]. Control on the nozzle inclination other than its position, guaranteed by the employment of a six-axis robotic arm as DIW platform, allowed to produce more complex geometries with higher fidelity, based both on SiOC and Si_3N_4, a material normally difficult to process via conventional VP techniques [10, 11]. Figure 7.2 shows a moment of the printing process and the resulting Si_3N_4 component after sintering.

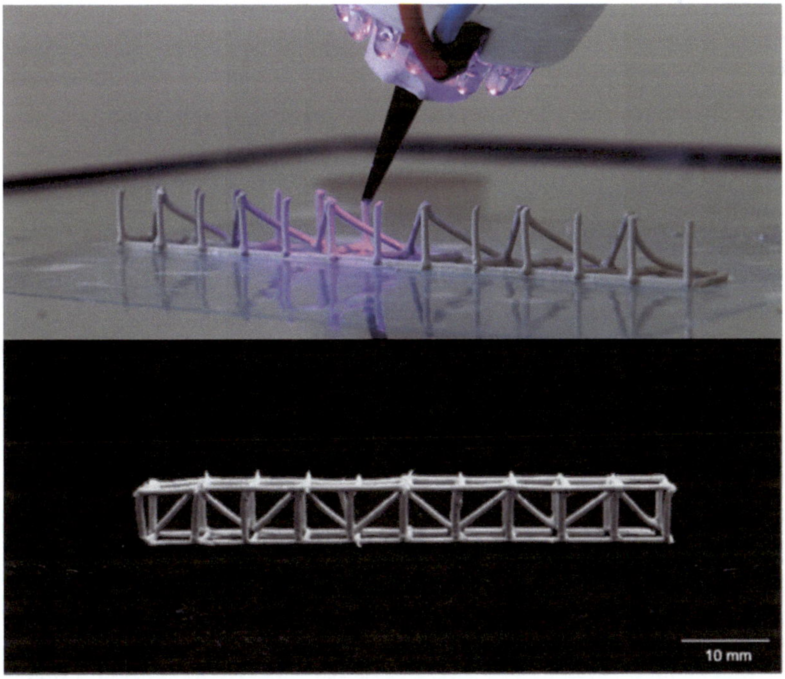

Figure 7.2 UV-assisted DIW of Si_3N_4: (top) printing process and (bottom) sintered component. Adapted from [10] / Elsevier / CC BY-NC-ND 4.0.

7.1.1.2 Laser-Assisted MEX

The process consists of the selective deposition of the feedstock material followed by a laser able to convert it into the final ceramic material on the spot. It has been reported by Sarkar et al. [12] with the name direct laser writing (DLW) in the context of producing pure silica and silica-titania glass in various shapes, forms, and sizes. The powder-based pastes were deposited via DIW, and a CO_2 laser followed the nozzle during the layer deposition to convert such pastes to sintered, transparent glass. Such topical sintering of powders is beneficial for multiple reasons: on one end, it transforms the deposited material from a paste into a solid with higher shape retention capabilities; on the other hand, it eliminates the need for subsequent thermal debinding and sintering treatments. Various curved thin-walled structures were demonstrated; with appropriate laser parameters, a transmittance of about 90% was reached, which is peculiar for glass components obtained from powder-based feedstock.

7.1.2 Sequential Hybrid Processes

The processes discussed in this section are characterized by the application of multiple technologies to produce a single part, which are taking place one after the other within each layer, except for VP/2PP. The general aim is to enhance the resolution of standard technologies, providing more complex geometries (e.g. embedding hollow channels), higher surface quality, or patterning at small scales.

7.1.2.1 Additive/Additive

BJ/MEX The combination of BJ, milling, and material deposition was proposed by de Melo Bernardino et al. [13] to obtain dense ceramic parts with embedded hollow channels. Their proposed workflow consists of the deposition of a ceramic slurry (an alternative step to a powder bed deposition, known as layerwise slurry deposition, LSD), drying, CNC machining, and extrusion of a sacrificial ink within the machined traces. After debinding and sintering, dense porcelain components could be produced with embedded hollow 3D channels resulting from the removal (combustion) of the sacrificial graphite ink. The process promises to be adapted for the manufacturing of ceramic–metal parts employing low- or high-temperature co-fired ceramics (LTCCs or HTCCs) and conductive pastes in place of the sacrificial ink.

VP/MEX Raynaud et al. [14] propose a similar, yet more integrated system, employing SL for the substrate fabrication; here, layers of a photocurable HTCC ceramic slurry are selectively cured leaving open paths, which are then thoroughly cleaned and selectively filled with a tungsten-based ink deposited via DIW; an example of embedded microstrip produced with this technology is shown in Figure 7.3. Finally, the ceramic and metal are co-sintered to dense ceramic components with various complex embedded circuits. As discussed for multi-material technologies, proper cleaning of the uncured slurry and co-sintering are the crucial steps in order to obtain high-quality parts.

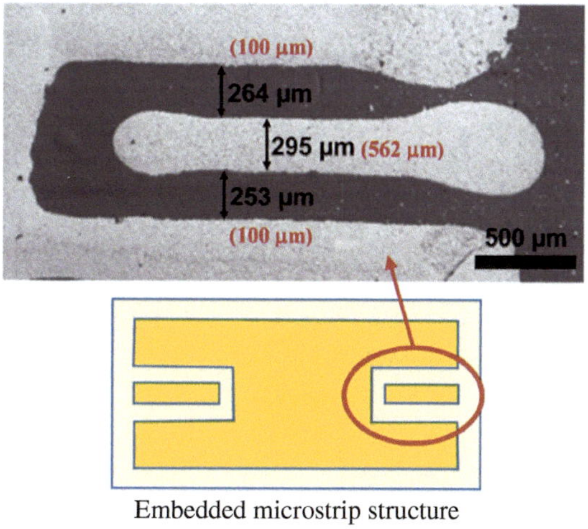

Figure 7.3 Micrograph of the upper side of a structure comprising an HTCC substrate and an embedded tungsten microstrip. The dimensions after manufacture, in black, are compared to the dimensions used by the simulation, in red. [14] / with permission of Elsevier.

VP/2PP The aim of coupling VP techniques (either digital light processing, DLP, or stereolithography, SL) with two-photon polymerization (2PP) is to generate parts with details ranging from the submicron to the mm scale. 2PP can reach submicron resolution, yet it is a tremendously slow process when it comes to producing parts even in the cm scale; VP techniques are much faster but have a resolution of several tens of microns. Their sequential use allows for printing a relatively big scaffold with VP, on top of which high-resolution features can be added where required. This is not a trivial combination, as it should account for the different shrinkage of the scaffold and feature parts during the subsequent shrinkage; it works best when the same ink is employed for both technologies. It should be noted, though, that volumetric AM techniques such as 2PP possess stringent requirements in terms of ink transmittance as the light should be able to penetrate the entire vat. For this reason, the reported examples so far only include SiOC polymer-derived ceramics from powder-free feedstock and YSZ from a well-dispersed, nanoparticle-based photocurable suspension, reported in Figure 7.4 [15, 16].

7.1.2.2 Additive/Subtractive

MEX/Machining The previous examples have shown that hollow channels can be embedded into bulk ceramics by milling and subsequent sacrificial material deposition. Channels with smaller widths (sizes from ~30 to ~400 µm) could also be fabricated by an integrated AM, namely DIW, and in-plane laser machining method. In this case, the paste viscosity had to be thoroughly optimized to ensure that the cover layer did not sag into the microchannels. Sinuous channels, networks, patterns, and vertical channels were embedded into bulk alumina following this approach [17].

More commonly, machining is performed by integrating a milling tool and spindle directly in an AM printer; this kind of architecture is now commercially available and promises to produce parts with the same characteristics as those finished by milling during the green state [18]. Larger nozzle diameters can be selected since the imperfections of the printing process will be corrected by the milling operation. The larger the nozzle diameter, the higher the productivity, since only a

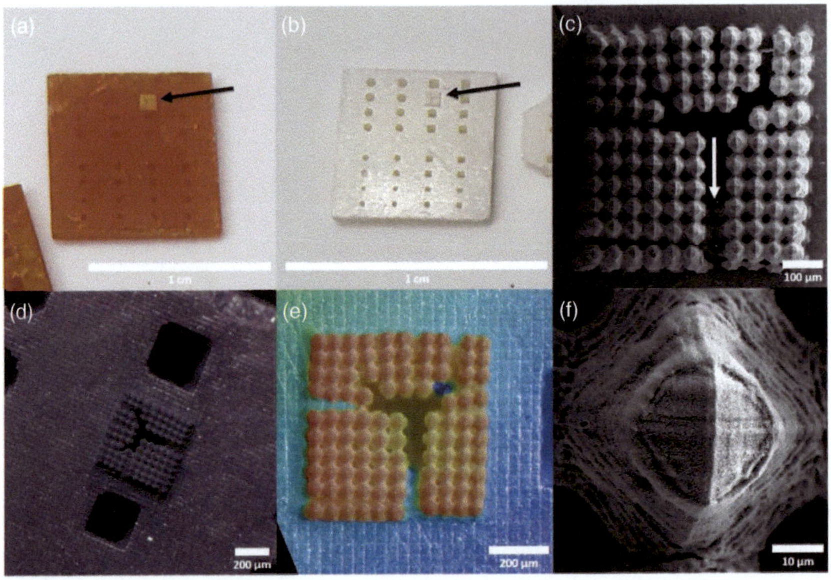

Figure 7.4 YSZ plate printed via SL: (a) non-sintered; (b) sintered, with a structure printed via 2PP on top of it bridging a void; (c, f) sintered, SEM images; (d) sintered, light microscopy image; (e) sintered topological view. [16] / MDPI / CC BY 4.0.

near-net-shape part has to be produced. Most interestingly, the hybrid architecture enables surfaces to be finished progressively, layer by layer, including internal ones when they are still reachable, guaranteeing better and more homogeneous surface quality.

Finally, the combination of additive and subtractive manufacturing could introduce a correction procedure to eliminate printing defects such as under- and over-extrusion. The promising research by Masters et al. [19] couples this concept with deep learning, demonstrating the possibility of a selective layer reworking in a ceramic hybrid AM platform. The experimental setup and process flowchart are reported in Figure 7.5. Each layer was evaluated in situ for under- and over-extrusion using a vision-based monitoring system and a deep learning model, which informed the decision to repair defective layers via subtractive operations, followed by reprinting. Reworking achieved a 68% reduction in void content compared to uncorrected parts.

7.1.3 Conclusions for Hybrid Manufacturing

In conclusion, the integration of AM methods with additional energy sources or other (additive or subtractive) manufacturing techniques signifies a pivotal shift in how ceramic components are conceived and produced. The concurrent hybrid processes like UV- or laser-assisted MEX enable freeform fabrication of ceramics with complex geometries, severe overhangs, and high shape retention capabilities. The sequential hybrid processes can produce ceramic parts with

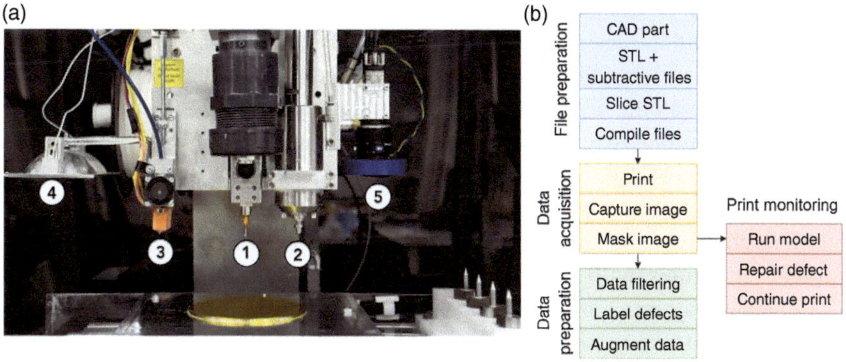

Figure 7.5 (a) Overview of the MEX/machining experimental apparatus. 1) Ceramic paste extruder, 2) micromachining spindle with automatic tool changer, 3) polymer extruder, 4) IR lamp, and 5) in situ monitoring system. (b) Flowchart demonstrating the stages of file generation, data acquisition and training data preparation, and the steps in the print monitoring. [16] / Elsevier / CC BY 4.0.

embedded hollow channels, complex internal features, enhanced surface finish, and multi-material integration. Overall, hybrid AM overcomes the limitations of individual AM processes and unlocks new possibilities in ceramic component design and manufacturing. However, challenges remain, such as optimizing process parameters, developing specialized feedstocks, and integrating different manufacturing platforms seamlessly.

7.2 Additive Manufacturing of Ceramic-Based Multi-Material Components

7.2.1 Introduction Concerning AM of Multi-Material Ceramic Components

AM offers countless degrees of freedom in the geometric design of components and has literally revolutionized conventional manufacturing. To reach their full potential, AM technologies should be able to combine different materials in a single processing step. Multicomponent production is very attractive for the manufacture of multifunctional components, as additional expensive joining steps are no longer necessary [20]. A review by MacDonald et al. [21] highlights the potential of multi-process (or hybrid) 3D printing, in which complementary technologies, both novel and traditional, are "collaboratively" combined to advance the future of manufacturing. The authors state that the next generation of 3D printing will enable the integration of dissimilar materials and allow the integration of active components to deliver functionality that was not possible previously.

The development of multi-material printing is in its infancy; however, it could have a massive impact, for example, in energy applications, electronic devices, and composites. Combining different materials in a single manufacturing process will reduce waste and production time by eliminating the need to assemble components. It will also expand the design space, enabling new products, composites and devices with structural and functional properties [22]. Products that have multi-material structures possess enhanced performance [23] and are utilized in various fields, including the automotive industry, aerospace engineering, biomedicine, and the defense industry [24–26].

Functionality in materials can be enhanced by internal composition gradients and hierarchical architectures. Unlike nature's resource-efficient designs, creating synthetic structures with organized phases across multiple hierarchical levels in a scalable way remains a challenge [27]. However, most AM techniques are ill-suited to create composition gradients or hierarchical structures, as they impose stringent constraints on the materials used. Most of the devices for AM currently on the market are only suitable for single-material applications. The reasons for this are mostly process- or system-specific. The latter is particularly the case when contamination of one material by the other due to carryover cannot be avoided. In the field of ceramic components, the situation is even more complicated, as AM of ceramics, like all other conventional ceramic shaping processes, remains a powder technological shaping method. After the shaping process, whether additive or conventional, the so-called ceramic green parts must be debinded and sintered to obtain their final desired properties. However, additively manufactured components do not always achieve bulk material properties, due to processing limitations [28]. These thermal treatment steps complicate the processing chain and can be seen as the main reason why AM of ceramics lags behind the processes for metals or polymers in terms of market penetration. Moreover, the use of bulk ceramics is often compromised by their inherent brittleness, which is still one of the largest limiting factors for using ceramic components in high-temperature applications. In particular, during rapid temperature changes, thermal shock cracks may initiate at the surface, leading in most cases to catastrophic failure of the ceramic component [29].

These challenges become even more complex when two or more ceramic materials are combined to produce components with multifunctional properties, such as different colors, porous and dense structures, electrical conductivity and insulation, or different mechanical properties. The selection of a suitable material pairing is a decisive factor for a successful co-sintering process, as the chemical and physical material properties cannot simply be changed by process adjustments [30]. First of all, the materials must match chemically to enable bonding at the atomic level [31]. Since they are joined by co-sintering, they must also be compatible in terms of sintering and thermal expansion behavior [32, 33]. The differential dimensional change of the dissimilar materials (strain mismatch) during cooling down from the sintering temperature may induce significant residual stresses [34]. They may be intrinsically generated due to (i) strain mismatch in materials with different coefficients of thermal expansion (CTEs), (ii) through phase transformations [35, 36], or (iii) chemical reactions [37, 38]. In the case of metal/ceramic components, they also suffer from distortion and warpage due to the mismatch of the coefficient of thermal expansion (CTE) during the conventional sintering [28].

On a technological scale, the shrinkage of the starting materials must be adapted in order to precisely match the onset of shrinkage, the shrinkage rate, and the absolute shrinkage of the composite partners. This can be achieved, for example, via tailoring of the particle size distributions and the packing density of the powders in the green parts [39–41].

Since 2018, multi-material AM of ceramics has been gaining more and more attention from the scientific community, as is reflected by the increasing number of publications per year [42]. A second increase could have been realized since 2023, which means that the technologies and the devices have reached a degree of development worthy of closer consideration in this chapter. Based on the seven process categories of AM according to DIN EN ISO/ASTM 52900:2022-03 [43], resp., three manufacturing categories can be considered for multi-material applications in the field of ceramics: (i) Vat Photopolymerization (VPP), (ii) MEX, and (iii) Material Jetting (MTJ).

A less explored area is the use of AM to manipulate the material's structure at the nano- to micro-scale in order to optimize performance. AM of ceramics is appealing as they are widely used in structural and functional applications due to their strength, high-temperature stability, oxidation resistance, and biocompatibility [44]. Functionally graded materials (FGMs) are a generation of technical materials, the final properties of which are determined by a given gradient (volume distribution) of the chemical composition, material microstructure, or structural characteristics during design. This feature allows FGMs to achieve material properties that are impossible to obtain using homogeneous alloys and composites and allows distributing properties in areas where needed [45]. Over the past two decades, advances in AM have proven that gradient design is technologically and economically feasible, especially for prototypes and small-scale production [46–49], even with components that are not entirely compatible or homogeneous. AM suggests additional benefits of precision and repeatability to achieve desired gradients and properties. In this context, FGM transitions are particularly interesting for AM methods, which, in turn, allow point-wise material deposition for stress-optimized design on the basis of finite element simulations.

The multi-material methods described below have been selected as examples from the work of international research groups, without claiming to be exhaustive, and can be assigned to the three process categories mentioned. The chosen methods are suited for ceramic/ceramic and/or for ceramic/metal material compounds.

7.2.2 Multi-Material AM Methods for Ceramics

7.2.2.1 Process Category Vat Photopolymerization

VPP has already reached industrial readiness for series production, enabling the high-precision production of complex and individualized ceramic components [50]. In multi-material VPP, at least two materials are combined in a single component to adjust its properties locally. For example, hard and ductile, heat/electrically conductive and insulating, porous and dense, or bioresorbable and bioinert materials can be combined [51]. The recent progress of the LCM-technology allows the combination of a two-vat multi-material printing system, developed by Lithoz GmbH [52], for the fabrication of porosity-graded as well as multi-material ceramic parts [53]. To do so, at least two slurries are used in a special multi-material VPP machine, for example, a 2M30 (Lithoz, Vienna, Austria). The slurries must initially fulfill the requirements and restrictions of the monolithic VPP process [54], especially in terms of the printing process, for example, a sufficiently high curing depth and a suitable viscosity. Besides the printing process, the sintering of composite parts, the so-called co-sintering, is a major challenge, especially in terms of the chemical compatibility, the coefficient of thermal expansion, and the preferred sintering conditions of the individual materials, which have to match in a co-sintering process. A frequently underestimated step along the VPP process chain is the thermal debinding step, in which the binder required for shaping is eliminated by heating the green part in an oven. In this stage, the majority of the components' defects, especially delaminations and cracks, are initiated [55]. The slurry composition heavily influences the debinding process [56]. On the one hand, the ceramic content of the slurry should be as high as possible in order to keep the amount of binder to be removed as low as possible [57]. On the other hand, a high solid loading generally leads to higher viscosities. Therefore, the interactions along the process chain have to be considered, and the temperature profile for debinding has to be precisely adjusted [55, 58]. Regarding multi-material VPP, an identical or at least very similar binder system for different ceramics has enormous advantages. This enables the knowledge gained from monolithic manufacturing to be largely transferred and the process parameters to be set precisely.

Sutejo et al. [59] developed a novel DLP system to accommodate ceramics slurry feed, which can possibly utilize multiple materials concurrently, as shown in [60–63]. Two lines of film conveyor were installed as feed carriers. Moreover, to tackle the problem of ceramic particles in the feed settling down during the course of the fabrication process, the vessel with the doctor blade slit installed on one side was utilized instead of the static vat. The rotation of the film rotor induces a force that pulls the ceramic slurry through the gaps between the doctor blade, producing a fine layer of ceramic slurry. This layer is then cured with UV light to create the desired model. Meanwhile, the remaining slurry colliding with the blade wall gets knocked back, triggering the mixing mechanism, ensuring the ceramic particles disperse homogeneously throughout the fabrication process. The building plate attached to the axis bar in the middle rotates between two fabrication lines during multi-material printing, enabling extra material to be used on the parallel level. Undoubtedly, the cleaning system was also integrated before line exchanges, introducing a stamping cross-contamination prevention system. A specialized platform equipped with paper cloth, consisting of entangled fibers, functioned as a contaminant hooking and entrapment agent. The pressing and wiping mechanisms of this paper cloth were controlled to gently remove any residual slurry from the built structure. A sequence of dry, ethanol- or diluent-dampened, and dry cloth was applied to maximize cleaning potential. Initial dry stamping removed major contaminants, dampened cloth re-dispersed and rinsed remaining particles, while final dry cloth stamping removed the remaining liquid phase, preventing contamination and foreign compounds from being entrapped between layers.

Multifunctionality or regional functionality is occasionally essential in a structure that sustains multiple types of load or environment. Sutejo et al. [59] selected bone structure as the display of proof of the VPP system described above. In this case, even opposite properties might be needed to balance the structure performance such as rigid–flexible and bioinert–bioactive. The materials used in this study include beta-tricalcium phosphate (β-TCP), hydroxyapatite (HA), and bioglass (BG) to maintain chemical similarity with bone. This combined mechanism of contaminant removal, involving hooking and redispersion, effectively cleaned the residual slurry adhered to the built structure, resulting in a successful removal of 99.6% of the undesired contaminant, as verified by weight loss evaluation and the detection of 0.32% Si atomically.

Other authors described cleaning systems, including ethanol spraying, vacuum swapping, centrifugal spin, or sonication [64–68]. However, these cleaning methods might generate redundant strain to the built structure during the cleaning process; thus, the application is restricted to polymer-based and low-content ceramic feed.

In Schwarzer-Fischer et al. [69], a collaborative work is described aiming to develop a material combination for a multi-material approach via the additive VPP method, which allows for combining electrical conductivity and electrical insulation, plus high mechanical strength in co-sintered ceramic components. As conductive component, titanium suboxides are chosen, whereas zirconia forms the mechanically stable and insulating part. Since titanium suboxides cannot be used for VPP due to their light-absorbing behavior, titania is used instead. After AM, the two-component parts are co-sintered in a reducing atmosphere to transform the titania into its suboxides and, thus, attaining the desired property combination.

7.2.2.2 Process Category Material Extrusion

MEX 3D printing includes, among other processes, fused filament fabrication (FFF) or fused deposition modeling (FDM). Both denote a widespread 3D-printing technique for polymeric thermoplastic materials. The printing filament is fed through a heated extrusion nozzle, and complicated parts are built-up line-wise by moving the extruder head or the printing bed in horizontal and vertical directions. Due to the low prices and simple handling of commercial printers, FFF is the most popular 3D-printing technique for private users, and a wide range of thermoplastic materials in different colors are available commercially. FFF has also been demonstrated for a wide range of ceramics. MEX can be used for a very wide range of materials, while other AM methods may have problems with changing the chemical composition of the material within a single layer (for instance, binder jetting and VPP). Compared to VPP, there are no restrictions in MEX on the optical properties of the materials. Additionally, MEX pastes have higher viscosity, allowing more ceramic powder to be added to the paste, thus resulting in the production of products with higher densities [26].

One limiting step for ceramic FFF is the intricate filament fabrication, which decreases the degree of freedom in choosing the binder material drastically. Main issues of using high solid load filaments are brittleness, insufficient flexibility, buckling, clogging of nozzles, reduced tackiness, and others. The main critical parameters of FFF are filament-specific (e.g. thermal, mechanical, and rheological properties and diameter), operation-specific (e.g. temperature, speed, and structure of 3D-printed object), and apparatus-specific (e.g. number of extrusion heads, nozzle diameter, and gear force) [70, 71].

To circumvent such problems, Wick et al. [72] developed a 3D printer that is based on the microextrusion of granulated feedstocks. The printing still follows the basic principle of controlled deposition of a filament of molten material. Here, however, a granulate feed containing a surface-modified ceramic powder dispersed in a thermoplastic polymer is used instead of a prefabricated filament [73–78]. A study of Wick et al. [72] contributes to the advancement of FFF metal–ceramic composites for space environments using both local laser debinding and induction sintering. A custom-built system that integrates a low-intensity infrared (IR) diode laser and an induction heater has been engineered to locally process metal and ceramic parts. The method and the apparatus involve continuously printing, and in parallel, thermally debinding the main polymeric matrix by vaporizing the polymer binder mass, with the debinded parts being subjected to a subsequent indirect induction sintering treatment. Specifically, this work presents a multi-material stainless steel 316L geometry combined with zirconia ZrO_2, whereby these materials have been extruded with a print head, thermally debinded with the IR laser source, and subsequently induction sintered. A novel hybrid device has been developed to allow local debinding and sintering of metal–ceramic parts, which are shaped via FFF.

In Abel et al. [79], an attempt is made to combine zirconia and modified stainless steel 17-4PH since both materials have a comparable CTE (approx. 11×10^{-6}/K) and can be sintered under the same conditions (reducing hydrogen atmosphere, sintering temperature: 1350–1400 °C). However, for adjusting the shrinkage behavior, a special milling procedure for the metallic powder is required. Within this work, heating devices had been manufactured by FFF containing an electrically conductive, meandric path made of stainless steel in a nonconductive ZrO_2 housing. Figure 7.6

Figure 7.6 Sintered heating elements made of zirconia and stainless steel (17-4PH). The white color of zirconia turns to black during sintering under a hydrogen atmosphere. [79] / MyJoVE Corporation / CC BY-NC-ND 3.0.

shows the sintered samples. These multi-material components have been tested successfully by cyclic heating and cooling.

Multi-material 3D printing with FFF (multi-material FFF) can be easily implemented using multiple extrusion nozzles, and the method has drawn increasing interest. However, some inherent disadvantages still exist when multi-material FFF is used to construct complex structures, such as low resolution accompanied by poor surface finish, insufficient bonding between adjacent sections, and slow build velocity. Additionally, thermoplastic polymer filaments with high solid content are necessary for the fabrication of metal and ceramic parts because of the low shrinkage after sintering [80].

Paste coextrusion using filaments with core–shell structures is an alternative approach of MEX to fabricate multi-material composites through a single printing step. Recent works have focused on the combination of a brittle shell and a tough or hollow core [81–86]. Printing systems for coextrusion usually consist of a modified coaxial printhead equipped with multiple feedstock materials. In some cases, the printhead has a block in the center to form hollow filaments. Carbon–ceramic and hollow ceramic filaments with core–shell structures were extruded from modified coaxial printheads [81, 83, 87, 88], and the mechanical performance of the printed scaffolds was improved when polymer was impregnated in the hollow core [88]. Toughening has been observed in polymer systems by introducing a compliant interfacial layer [82]. Multi-material composites were also prepared by formulating ink cartridges with core–shell structures [84, 86, 89]. After assembling cylindrical core and shell cartridges in a single cartridge as a feed rod, the filaments were pushed through a nozzle tip. By adjusting the rheology, the filaments retained the coaxial structure of the cartridge. Ceramics reinforced with a compliant metal, nickel, or short carbon fibers were coextruded in this way to fabricate composites with elevated mechanical performance through crack deflection and fiber bridging [88, 89]. The above coextrusion methods have been mostly applied to hollow ceramic scaffolds, polymer–polymer lattices and polymer–ceramic scaffolds, with only a few reports on ceramic dense parts [88]. Zhou et al. [44] propose a method to engineer a network of weak interlayers in ceramic bodies by extruding filaments with a core–shell structure and a porous interfacial layer between core and shell. The goal is to design a network of interlayers that will direct crack propagation to improve fracture resistance. Alumina was used for both core and shell material, and the interlayer was composed of a mixture of Al_2O_3 and graphite with different volume ratios. The graphite flakes with high aspect ratio burn out during sintering, leaving a weak porous layer with plate-shaped pores aligned tangentially to the core.

MEX technology enables the successful production of ceramic and multi-ceramic products. Promising materials produced via MEX technology include TiO_2 and ZrO_2, which can be used in electrical and electronic engineering [26].

Pelz et al. [90] utilized the MEX to manufacture functional gradient material from boron carbide (B_4C) and silicon carbide (SiC). The fabrication of this gradient material was achieved by mixing two pastes with different compositions. Additionally, various ceramic-layered samples were made. These samples exhibit cracks induced through residual stresses caused by thermodynamic factors.

Robocasting, also called Direct Ink Writing (DIW), was initially developed to create complex ceramic structures using any material by formulating a paste with controlled rheology and is one further of the MEX technologies enabling to build mono or multi-material ceramics layer by layer [91–94]. It consists of the continuous extrusion of a ceramic paste through a thin nozzle (generally 0.2–0.9 mm), the position of which is controlled in 3D. According to [95], the quality of the printed part will depend on three families of parameters. First, similar to all other ceramic processes, the composition of pastes and their homogeneity are important factors to guarantee the least possible defects. For example, increasing the volume fraction of ceramic particles will lead to rheological modifications that could be harmful for a safe homogeneity during the entire process. Second, environmental parameters, such as temperature and humidity, are also essential to avoid cracks during the printing process. The homogeneity of the filament can be mainly maintained under high humidity to avoid cracks due to too rapid drying [96–98]. Third, the choice of printing parameters will influence shape fidelity and final geometries, so indirectly, final properties, for example, printing speed will influence the overall aspect of the printing filament and its shape [99, 100]; the degree of overlap and size of the nozzle will influence the shape fidelity of ceramics due to the topography of the printed part related to layer stacking. Another advantage of DIW is that it enables a combination of different formulations into complex structures by using multiple extrusion nozzles. Modifications of the extrusion system using microfluidic printheads allow switching [99, 100] or mixing [101, 102]. The main difference between FFF and DIW is that in FFF, polymeric filaments must be melted during extrusion, while DIW systems rely purely on pastes with very specific rheology that are stored in a syringe cartridge [22].

According to [27], one of the most promising strategies is support bath-assisted or embedded 3D printing. In this technique, inks are extruded into a high-yield-stress fluid known as the "support bath." The support bath is deformable at appreciably high shear, allowing the nozzle to translate freely, but it is also recoverable, allowing the fluid to encase printed structures and support them against gravity. After the printed structure has been solidified, the support bath can be removed in a post-processing step. This method enhances DIW capabilities by enabling omnidirectional printing and allowing for the use of inks with reduced viscosity, storage modulus, and yield stress.

One common approach to print structures composed of distinct materials involves using multiple DIW nozzles to sequentially extrude different inks. This method significantly advanced multi-material AM, but it has drawbacks such as prolonged printing times due to frequent nozzle switching. Additionally, there is often poor adhesion both between layers of the same material and between different materials, leading to compromised mechanical properties and performance, such as electrical contact loss in sensors or actuators.

An alternative approach to building 3D multi-material structures via DIW involves customizing the printing nozzle to shape structures within the extruded flow. These extrusion-based AM methods have gained attention for their ability to precisely control multi-material arrangements at scales below the nozzle size (subvoxel), to print multiple filaments in parallel, or to quickly switch in-line between different types of ink being extruded. Printheads designed to coextrude structured filaments mainly focused on biphasic and coaxial configurations. Recently, research has shifted toward rotational DIW printing methods where specialized printheads coextrude multi-material filaments during rotation of the printhead itself. This approach provides precise control over azimuthal variations within helical multi-material filaments, enabling the creation of intricate and versatile structures [27].

Pleij et al. [27] developed a new 3D-printing configuration that combines multi-material 3D printing, advective assembly, support bath 3D printing, and rotational axes to produce complex architectures that are impractical to produce via other means. The advective assembler nozzles build structures in laminar flows and extrude strong, multi-material filaments, for example, internally structured as to be made to be more resistant to potential delamination. The support bath allows the structured filaments to be arranged in curvilinear macroscopic shapes. Combining these distinct advances in AM methods provides new spatial axes for design.

Current approaches for DIW of disparate materials with high spatial resolution are based on precision motion control systems. This is normally an XYZ gantry robot with a few microns (1–3 mm) accuracy and repeatability in the three directions. At the laboratory scale, this robot is normally equipped with linear motors that can reach maximum speeds between 0.1 and 2 m/s. The gantry robot is coupled with a printer head device with multiple syringe pumps or pressure-driven reservoirs for different formulations. Disparate materials can be combined in the same object by synchronizing XYZ motion with the plunger displacement or applied pressure. The precise movements of the XYZ gantry robot and syringe pumps are controlled through a user interface or software that translates a predefined 3D design into G-code commands that guide the printing of the desired object.

In Zhou et al. [44], a straightforward method to obtain ceramics with a network of continuous weak interlayers designed to increase fracture resistance using alumina as a model system is demonstrated. This is achieved by combining DIW with the coextrusion of multi-material pastes with carefully matched rheology based on thermally reversible hydrogels and inorganic powders. The printed Al_2O_3 bars with and without weak interlayers exhibit strengths ranging between 180 and 360 MPa and $K_{IC} \sim 3\,\text{Mpa}\sqrt{m}$. The introduction of weak interlayers using different raster patterns, such as length-wise and Bouligand alignments, can be used to direct crack propagation and promote gradual failure. The result is an improvement in the fracture energy up to $230\,\text{J/m}^2$ and KJ up to $9\,\text{Mpa}\sqrt{m}$.

7.2.2.3 Process Category Material Jetting

In MJT, droplets of the feedstock material are selectively deposited and solidified in successive layers [103]. Several MJT techniques are recognized by the mechanism of droplet generation, in which the most widely used are inkjet printing (IJP), aerosol jet printing (AJP), and electrohydrodynamic inkjet printing (EHDP). Over the past few years, these MJT techniques have attracted interest in the on-demand fabrication of highly customizable electronics because of their abilities to precisely and smartly deliver high-resolution patterns with a diversified structure (<1 μm) from design files in a noncontact manner. Multi-material 3D printing of supercapacitors [104], filters [105], tapered optical waveguides [106], solar cells [107], microsensors [108], flexible circuits [109], and high-density redistribution layers (RDLs) of silicon interposers [110] has been easily implemented by following the same approach of conventional 2D ink jetting by sequential deposition of functional inks via an MJT system equipped with multiple jetting heads [80].

Besides AJP, multi-material jetting (MMJ) will be described here in more detail.

Five core stages (atomization, aerosol transport, collimation, focusing, and impaction) are involved when an AJP system is working. Initially, fine mists of numerous aerosol droplets are generated by atomizers with pneumatic or ultrasonic approaches. Afterward, the carrier gas flow applied to the ink reservoir drives the aerosol droplets through a virtual impactor, where the aerosol droplets are filtered for a more uniform size distribution and ultimately directed to the printhead. At the printhead, the coaxial sheath gas flow plays an important role, which collimates the aerosol droplets into a tight beam and forms an interlayer between the aerosol stream and physical components to reduce nozzle clogging. Finally, a high-velocity stream of collimated aerosol droplets, together with the sheath gas, exits from the printhead and is directed toward the substrate with a standoff distance of 1–5 mm [80]. One of the requirements for ink formulation is that the solid particles do not clog the atomizer nozzle, which is usually achieved using submicron solid particles capable of deformation whose fragments or fines that become amorphized upon impact may facilitate consolidation [111]. Because the atomization process selectively introduces smaller particles into the aerosol stream, controlling a narrow particle size distribution is essential; otherwise, the particles in the ink available for printing will change as a function of printing time [112].

Multi-material jetting (CerAM MMJ) technology (originally thermoplastic 3D printing – T3DP [113]) is an AM method based on the selective application of droplets onto the build platform, which was developed especially for AM of ceramic-based multi-material components [114, 115]. Thermoplastic particle-filled suspensions are used, which are applied in a molten state and then immediately solidify upon cooling. The droplet size, shape, and spacing can be precisely adjusted using numerous process parameters (e.g. needle stroke, pressure, temperature and droplet frequency) [116]. Multiple print heads allow different materials to be combined in a single component. This makes it possible to produce ceramic components with locally varying material properties (e.g. electrically conductive and insulating). After shaping, the green components must be thermally debinded and sintered to achieve the ceramic properties (density, strength). It is crucial to coordinate the shrinkage and thermal expansion characteristics of the various materials in order to prevent cracking and delamination. Compared to other AM technologies, MMJ is characterized by a virtually unlimited material portfolio. Ceramic (including Al_2O_3 [117], ZrO_2 [113, 115], Si_3N_4 [118, 119], SiC, and AlN), metallic [114], glass [120], and hard metal powders [121] have already been processed in thermoplastic binder systems with high solid content (up to 23 approx. 65 vol.%) [113]. By depositing individual droplets (typical droplet diameters between

23 approx. 200 and 1000 μm) [116], fine details and complex geometric degrees of freedom can be achieved. The material is only applied where it is needed, which reduces waste and cleaning effort. MMJ technology is ideal for AM of multi-material components because, unlike other AM processes in which material is applied over a large area and selectively cured (e.g. binder jetting and vat polymerization), MMJ deposits material with pinpoint accuracy, drop by drop, so that no cleaning step is necessary when changing materials.

7.3 Applications

In Schlacher et al. [122], the thermal shock resistance of an AM multi-material design was investigated. The combination of alumina-zirconia (ZTA) with embedded alumina (A) layers with in-plane compressive residual stresses, associated with the different shrinkage of the combined materials during cooling down from sintering, was employed as a strategy to provide resistance against the propagation of (surface) cracks. In a first step, a simple five-layered multi-material design (ZTA/A/ZTA/A/ZTA) with a disc geometry was investigated and compared to the monolithic reference samples (A, ZTA). The corresponding retained (biaxial) strength after thermal shock at critical temperature differences of the multi-material sample is twice as high as that of the reference counterpart. To assess the feasibility of 3D-printing complex alumina-based parts with enhanced damage tolerance, a turbine blade (ZTA/A/ZTA/A/ZTA) was designed by using the multi-material design approach. This work demonstrated for the first time that the damage-tolerant design approach may also be applicable to 3D-printed complex-shaped parts, which could open novel application fields in the future.

The advanced properties of dielectric ceramics can be used in multi-material structures, which are suitable for the radiofrequency (RF) components (e.g. filters and capacitors) in wireless communication applications. These properties include suitable relative permittivity, low dielectric loss, temperature-stable dielectric properties, low sintering temperature below that of the electrode materials' melting point, chemical compatibility with electrode material, and so on. In a study of Gheisari et al. [123], $Bi_2Mo_2O_9$ (BMO)-Ag artificial dielectric structures with varying geometries have been fabricated by DIW and characterized. The printed structures would be very difficult to make with conventional casting methods due to the requirement to have a high solid loading to reduce shrinkage and cracking during co-sintering. The printing resolution used for this investigation was approximately 200 μm. Reduction of the sintering temperature of the ceramic material to a level such that it can be co-fired with silver is essential. The potential use of the (BMO) for the fabrication of low-temperature co-fired ceramics (LTCC) modules and novel RF substrates is worth considering due to its low sintering temperature and outstanding dielectric properties.

Microwave (MW) dielectric ceramics have for a long time been established as the go-to materials for applications related to mobile and wireless telecommunications, such as dielectric substrates, dielectric resonators, filters, capacitors, global positioning, and Wi-Fi components. Their properties, namely relative permittivity (εr, whose

Figure 7.7 AM of ceramic-based Wi-Fi components by MEX. [28] / Elsevier / CC BY 4.0.

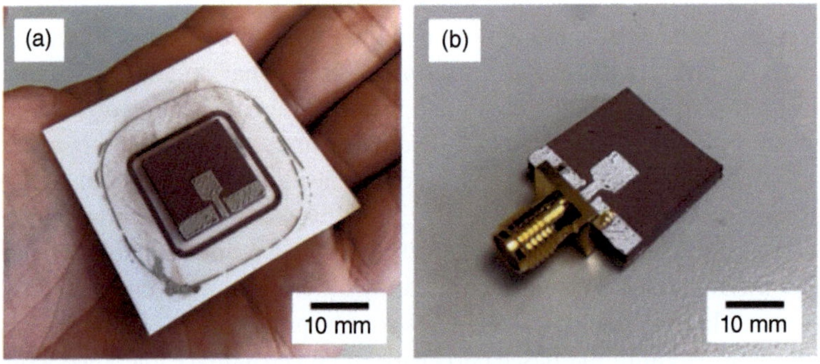

Figure 7.8 Additively manufactured ceramic antenna structure in (a) green and (b) sintered state with an attached SMA connector. [28] / Elsevier / CC BY 4.0.

optimum value is specific to the application), low dielectric loss tangent (tanδ), and high-quality factor (Q × f) are of particular interest to modern applications in 5 G and 6 G telecommunication systems. A study of [28] investigates the MW dielectric properties of Ag_2MoO_4 fabricated through MEX together with the use of MW-assisted sintering of metal/ceramic structures (Figure 7.7). Applying 2 kW microwave energy at 2.45 GHz for 60 minutes was proven sufficient to densify the metallic Ag infilling electrodes, without causing any macroscopic defects. Figure 7.8 shows a fully functional multi-layered antenna structure with a metamaterial artificial magnetic conductor, which had been manufactured in this way.

Most advances in this area have been made to create devices for electrochemical energy storage. This is one of the fields that is pushing the multi-material challenge, given the nature of the application, in which the combination of very disparate materials from polymers, ceramics, and metals in one single device is key for performance [22]. Recently, Shen et al. [124] used DIW to make a multi-material asymmetric micro-supercapacitor (MSC) with ultrahigh areal energy density. Cathode and anode inks of vanadium pentoxide (V_2O_5) and graphene–vanadium nitride quantum dots (G-VNQDs) with highly concentrated graphene oxide (GO) dispersions, as well as a LiCl–PVA gel electrolyte, were designed and formulated. The 3D-printed asymmetric MSC with interdigitated electrodes exhibited excellent structural integrity, a large areal mass loading of 3.1 mg/cm^2, and a wide electrochemical potential window of 1.6 V. Moreover, there were substantial open macropores in the 3D-printed electrodes, which provided numerous channels to facilitate mass transportation. Cheng et al. [125] also tackled the challenge of adding complexity and printed an interdigitated full cell battery. Lithium iron phosphate (LFP) and lithium titanium oxide (LTO) materials were formulated with poly(vinylidene-fluoride) PVDF, deposited onto preprinted polydimethylsiloxane (PDMS) substrates, and used as the cathode and the anode, respectively.

In recent years, ceramic-based AM technologies have been actively developed to implement their unique capabilities, such as a high level of design freedom and shape flexibility, to increase solid oxide cells (SOCs) specific performance per unit mass and volume. The pioneering use of these technologies in functional ceramics has created new opportunities for energy devices. The use of AM enables the fabrication of self-supported ceramic devices, such as SOCs and stacks, with embedded functionality, which represents a competitive advantage, since AM prevents the loss of valuable materials while promoting the automatization of the fabrication process and reducing the total number of manufacturing stages. The presented development of single-step multi-material AM is still required to manufacture complete SOC devices, being the next step to full automatization of the SOC production cycle. The proposed approach implies that multiple thermal treatments applied to each functional ceramic layer are further substituted by a single co-sintering process [126]. In particular, the manufacturing of NiO-YSZ (RC)|8YSZ (SLA)|LSM-CGO (RC) button cells with a total diameter of 2 cm via hybrid multi-material AM technology based on VPP and robocasting is demonstrated, while the use of any commercial substrate is avoided. Planar self-supported solid oxide fuel cells (SOFCs) were produced in a single AM process based on the VPP printing method using a ceramic 3D printer (CERAMAKER C900, 3DCERAM-Sinto, France), which was further customized through hybridization with a robotic arm with 4 robocasting dispensing tips (Figure 7.9).

Since MMJ was developed directly for AM of multi-material components, it is not surprising that a wide range of demonstrators already exist that showcase its potential. The qualification of a material combination based on Si_3N_4-$MoSi_2$, which enables the realization of ceramic components that combine both electrically insulating and conductive areas, has made it possible to produce various electric heaters and igniters [119].

The manufacturing of ceramic heating elements with integrated conductive structures, which are integrated into the ceramic component in a spiral or meander shape, for example, allows rapid and uniform temperature development in combination with outstanding high-temperature resistance and chemical and mechanical resistance. Application scenarios that have already been addressed include chemical reactors and mixers [127], forming tools (Figure 7.10) [128], and glow igniters [118]. If the electrical conductor tracks are separated by an air gap, they can be used as electrodes to generate a spark or plasma and, for example, as an ignition source for combustion processes, such as in engines for space applications [118]. The ignition sources can be placed exactly where they are needed for an optimal combustion process. Work has also been ongoing for some time on combining different colors

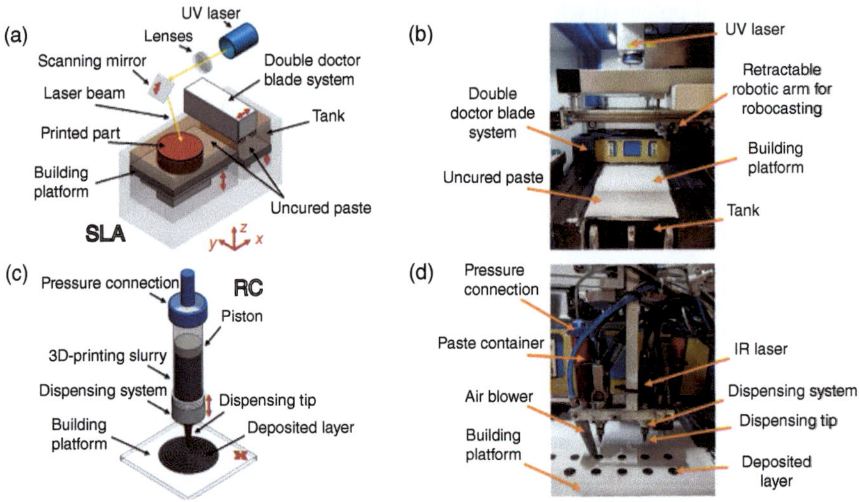

Figure 7.9 AM of ceramic multi-material components by a combination of VPP and MEX: (a) SLA (VPP) working principle, (b) SLA functional part of the CERAMAKER C900 system, (c) MEX working principle, and (d) retractable robotic arm of the CERAMAKER C900 system. [126] / with permission of Royal Society of Chemistry.

Figure 7.10 Glowing, electrically heated forming tool manufactured using MMJ based on two different Si_3N_4-$MoSi_2$ mixtures to integrate an electric heater into an electrically insulated ceramic component. [128] / Inovar Communications Ltd.

in a ceramic (Figure 7.11) [129] or glass-based component [120]. Due to the high resolution of MMJ, this is particularly interesting for the jewelry and watch industry.

The combination of dense and porous areas enables the optimization of heat transfer, for example, in chemical reactors (increased heat transfer) or in heatable tools (reduced heat transfer) [117].

Figure 7.11 Black-and-white zirconia components manufactured with MMJ.

7.4 Conclusion and Outlook for AM of Ceramic Multi-Material Components

AM of multi-material components based on ceramics further expands the degree of freedom in the design of ceramic components. In addition to the outstanding geometric freedom offered by AM technologies, the combination of different materials allows different properties to be integrated into a single component. This enables different functions to be combined in a single component, further advancing the functionalization and miniaturization of components. Compared to current manufacturing strategies, complex and defect-prone joining processes can be eliminated, reducing manufacturing costs and lead times. Despite the progress made in recent years in the AM of ceramic-based multi-material components, there are still many challenges to overcome. The main challenges are the necessary co-processing and the placement of the different materials during shaping.

In general, two different types of AM technologies can be distinguished, which differ in the way the material is applied. [130] introduced the terms direct and indirect AM technologies for this purpose. While indirect AM technologies apply the material over the entire building platform and consolidate it selectively (e.g. VPP and BJT), direct AM technologies only apply the material where it is needed (e.g. MEX and MJT). Direct AM technologies are better suited for the manufacturing of multi-material components, as no cleaning step is required when changing materials to prevent cross-contamination.

When co-processing different materials, co-sintering is particularly challenging. Critical stresses, deformations, and other defects during the sintering process can only be avoided if all materials exhibit comparable shrinkage behavior. To achieve this, the shrinkage behavior can be adjusted by specifically modifying the particle shape and size of the powders used and the feedstock formulation. In addition, all materials must be processable under the same sintering conditions (including temperature profile and atmosphere). However, a major limitation of co-processable material pairs results from the cooling process after sintering. If the different materials do not have a comparable thermal expansion coefficient, this also results in critical stresses, deformations, and other defects (e.g. cracks).

Various material combinations are known that have already been processed and successfully co-processed using conventional shaping methods [128]. These now need to be adapted for AM technologies. In addition, the ability to combine different materials in a single component offers the opportunity to create graded material transitions that can prevent critical stresses. The targeted geometric design of the interface between different materials, taking advantage of the possibilities offered by AM technologies, also reduces the risk of defects. Current developments, such as the MMJ ProX series systems from AMAREA Technology GmbH [131], with up to six print heads working in parallel, offer previously unknown possibilities for combining materials and properties and for designing the interfaces between different materials.

The necessary co-sintering limits the available portfolio of co-sinterable material combinations for multi-material components. However, even material pairs that cannot be co-processed can be combined to shape multi-material components by shaping the materials sequentially and processing them separately using different thermal processes. However, the geometric arrangement of the materials and thus the properties in the component are limited [33, 132].

References

1 Lauwers, B., Klocke, F., Klink, A. et al. (2014). Hybrid processes in manufacturing. *CIRP Ann.* 63: 561–583. https://doi.org/10.1016/J.CIRP.2014.05.003.

2 Gao, W., Zhang, Y., Ramanujan, D. et al. (2015). The status, challenges, and future of additive manufacturing in engineering. *Comput. Aided Des.* 69: 65–89. https://doi.org/10.1016/j.cad.2015.04.001.

3 Schuh, G., Kreysa, J., and Orilski, S. (2009). Roadmap "Hybride Produktion". *Zeitschrift Für Wirtschaftlichen Fabrikbetrieb* 104: 385–391. https://doi.org/10.3139/104.110072.

4 de Hazan, Y., Thänert, M., Trunec, M., and Misak, J. (2012). Robotic deposition of 3d nanocomposite and ceramic fiber architectures via UV curable colloidal inks. *J. Eur. Ceram. Soc.* 32: 1187–1198. https://doi.org/10.1016/J.JEURCERAMSOC.2011.12.007.

5 Faes, M., Vleugels, J., Vogeler, F., and Ferraris, E. (2016). Extrusion-based additive manufacturing of ZrO_2 using photoinitiated polymerization. *CIRP J. Manuf. Sci. Technol.* 14: 28–34. https://doi.org/10.1016/J.CIRPJ.2016.05.002.

6 Wei, L., Li, J., Zhang, S. et al. (2020). Fabrication of SiOC ceramic with cellular structure via UV-Assisted direct ink writing. *Ceram. Int.* 46: 3637–3643. https://doi.org/10.1016/J.CERAMINT.2019.10.083.

7 Clarkson, C.M., Wyckoff, C., Parvulescu, M.J.S. et al. (2022). UV-assisted direct ink writing of Si3N4/SiC preceramic polymer suspensions. *J. Eur. Ceram. Soc.* 42: 3374–3382. https://doi.org/10.1016/J.JEURCERAMSOC.2022.03.001.

8 Maeng, W.Y., Jeon, J.W., Bin Lee, J. et al. (2020). Photocurable ceramic/monomer feedstocks containing terpene crystals as sublimable porogen for UV curing-assisted 3D plotting. *J. Eur. Ceram. Soc.* 40: 3469–3477. https://doi.org/10.1016/J.JEURCERAMSOC.2020.03.047.

9 De Marzi, A., Giometti, G., Erler, J. et al. (2022). Hybrid additive manufacturing for the fabrication of freeform transparent silica glass components. *Addit. Manuf.* 54: 102727. https://doi.org/10.1016/J.ADDMA.2022.102727.

10 De Marzi, A., Diener, S., Campagnolo, A. et al. (2024). Ultra-lightweight silicon nitride truss-based structures fabricated via UV-assisted robot direct ink writing. *Mater. Des.* 244: 113092. https://doi.org/10.1016/J.MATDES.2024.113092.

11 Huang, K., De Marzi, A., Franchin, G., and Colombo, P. (2024). UV-assisted robotic arm freeforming of SiOC ceramics from a preceramic polymer. *Addit. Manuf.* 83: 104051. https://doi.org/10.1016/J.ADDMA.2024.104051.

12 Sarkar, S., Tomlinson, N., Lei, J. et al. (2025). Direct-laser-writing of transparent silica and silica-titania glasses. *Addit. Manuf.* 110: 104934. https://doi.org/10.1016/j.addma.2025.104934.

13 de Melo Bernardino, R., Valentino, S., Franchin, G. et al. (2020). Manufacturing of ceramic components with internal channels by a novel additive/subtractive hybridization process. *Open Ceram.* 2: 100010. https://doi.org/10.1016/J.OCERAM.2020.100010.

14 Raynaud, J., Pateloup, V., Bernard, M. et al. (2021). Hybridization of additive manufacturing processes to build ceramic/metal parts: example of HTCC. *J. Eur. Ceram. Soc.* 41: 2023–2033. https://doi.org/10.1016/J.JEURCERAMSOC.2020.10.032.

15 Sänger, J.C., Schwentenwein, M., Bermejo, R., and Günster, J. (2023). Hybridizing lithography-based ceramic additive manufacturing with two-photon-polymerization. *Appl. Sci.* 13: 3974. https://doi.org/10.3390/APP13063974.

16 Schmidt, J., Brigo, L., Gandin, A. et al. (2019). Multiscale ceramic components from preceramic polymers by hybridization of vat polymerization-based technologies. *Addit. Manuf.* 30: 100913. https://doi.org/10.1016/J.ADDMA.2019.100913.

17 Hong, Y., Lei, J., Heim, M. et al. (2019). Fabricating ceramics with embedded microchannels using an integrated additive manufacturing and laser machining method. *J. Am. Ceram. Soc.* 102: 1071–1082. https://doi.org/10.1111/jace.15982.

18 3DCERAM (2024). Introduces a new pellet head for 3D-printing of ceramic and metal feedstock. *Ceram. Appl.* 12: 31–33.

19 Masters, L., Davie, D., Cevallos, P.J. et al. (2025). Strategic layer reworking using hybrid additive manufacturing for defect-free ceramic parts. *Addit. Manuf.* 102: 104752. https://doi.org/10.1016/J.ADDMA.2025.104752.

20 Schwarzer, E., Götz, M., Markova, D. et al. (2017). Lithography-based ceramic manufacturing (LCM) – viscosity and cleaning as two quality influencing steps in the process chain of printing green parts. *J. Eur. Ceram. Soc.* 37: 5329–5338.

21 MacDonald, E. and Wicker, R. (2016). Multiprocess 3D printing for increasing component functionality. Science 353 (6307): aaf2093.

22 Rocha, V.G., Saiz, E., Tirichenko, I.S., and García-Tuñón, E. (2020). Direct ink writing advances in multi-material structures for a sustainable future. *J. Mater. Chem. A* 8: 15646–15657.

23 Onuike, B., Heer, B., and Bandyopadhyay, A. (2018). Additive manufacturing of Inconel 718 – copper alloy bimetallic structure using laser engineered net shaping (LENSTM). *Addit. Manuf.* 21: 133–140.

24 Bandyopadhyay, A., Traxel, K.D., Lang, M. et al. (2022). Alloy design via additive manufacturing: advantages, challenges, applications and perspectives. *Mater. Today* 52: 207–224.

25 El-Galy, I.M., Saleh, B.I., and Ahmed, M.H. (2019). Functionally graded materials classifications and development trends from industrial point of view. *SN Appl. Sci.* 1: 1378.

26 Repnin, A., Sotov, A., Popovich, A., and Masaylo, D. (2023). Development of TiO_2/ZrO_2 multi-material obtained from ceramic pastes for material extrusion. *Micromachines* 14: 2177. https://doi.org/10.3390/mi14122177.

27 Pleij, T., Bayles, A.V., and Vermant, J. (2024). Advective assembler-enhanced support bath rotational direct ink writing. *Adv. Mater. Technol.* 9: 2400005. https://doi.org/10.1002/admt.202400005.

28 Goulas, A., Whittaker, T., Chi-Tangyie, G. et al. (2023). Multi-material additive manufacture and microwave-assisted sintering of a metal/ceramic metamaterial antenna structure. *Appl. Mater. Today* 33: 101878. https://doi.org/10.1016/j.apmt.2023.101878.

29 Danzer, R., Lube, T., Supancic, P., and Damani, R. (2008), https://doi.org/10.1002/adem.200700347). Fracture of ceramics. *Adv. Eng. Mater.* 10 (4): 275–298.

30 Scheithauer, U., Schwarzer, E., Moritz, T., and Michaelis, A. (2018). Additive manufacturing of ceramic heat exchanger: opportunities and limits of the lithography-based ceramic manufacturing (LCM). *J. Mater. Eng. Perform.* 27: 14–20. https://doi.org/10.1007/s11665-017-3087-1.

31 Zakeri, S., Vippola, M., and Levänen, E. (2020). A comprehensive review of the photopolymerization of ceramic resins used in stereolithography. *Addit. Manuf.* 35: 101177. https://doi.org/10.1016/j.addma.2020.101177.

32 Nazir, A., Gokcekaya, O., Billah, K.M. et al. (2023). Multi-material additive manufacturing: a systematic review of design, properties, applications, challenges, and 3D printing of materials and cellular metamaterials. *Mater. Des.* 226: 111661. https://doi.org/10.1016/j.matdes.2023.111661.

33 Michaelis, A., Scheithauer, U., Moritz, T. et al. (2020). Advanced manufacturing for advanced ceramics. *Procedia CIRP* 95: 18–22. https://doi.org/10.1016/j.procir.2020.03.028.

34 Cheng, L., Sun, M., Ye, F. et al. (2018). Structure design, fabrication, properties of laminated ceramics: a review. *Int. J. Lightweight Mater. Manuf.* 1: 126–141. https://doi.org/10.1016/j.ijlmm.2018.08.002.

35 Virkar, A.V., Huang, J.L.W., and Cutler, R.A. (1987). Strengthening of oxide ceramics by transformation-induced stress. *J. Am. Ceram. Soc.* 70 (3): 164–170. https://doi.org/10.1111/j.1151-2916.1987.tb04952.x.

36 Pontin, M.G., Rao, M.P., ŚAnchez-Herencia, A.J., and Lange, F.F. (2002). Laminar ceramics utilizing the zirconia tetragonal-to-monoclinic phase transformation to obtain a threshold strength. *J. Am. Ceram. Soc.* 85 (12): 3041–3048. https://doi.org/10.1111/j.1151-2916.2002.tb00576.x.

37 Nordberg, M.E., Mochel, E.L., Garfinkel, H.M., and Olcott, J.S. (1964). Strengthening by ion exchange. *J. Am. Ceram. Soc.* 47 (5): 215–219. https://doi.org/10.1111/j.1151-2916.1964.tb14399.x.

38 Krishnamurthy, R. and Sheldon, B.W. (2004). Stresses due to oxygen potential gradients in non-stoichiometric oxides. *Acta Mater.* 52: 1807–1822. 10.1016/j.actamat.2003.12.020.

39 Moritz, T., Scheithauer, U., Mannschatz, A. et al. (2017). Material-and process hybridization for multifunctional ceramic and glass components. *Ceram. Appl.* 5: 66–71.

40 Baumann, A.; Moritz, T.; Lenk, R. Multi component powder injection moulding of metal–ceramic composites. In Proceedings of the International Powder Metallurgy Congress & Exhibition EuroPM 2009, Copenhagen, Denmark, 12–14 October 2009.

41 Suri, P., Heaney, D.F., and German, R. (2003). Defect-free sintering of two material powder injection molded components-Part 1+Part 1. *J. Mater. Sci.* 38: 4869–4881.

42 SCOPUS Analyze Search Results (2025). https://www.scopus.com/term/analyzer.uri?sort=plf -f&src=s&sid=eff77a1853f9ccf26e58406d13e80e91&sot=a&sdt=a&sl=75&s=%28KEY%28additive+ AND+manufacturing%29+AND+KEY%28ceramics%29+AND+KEY%28multi+material%29%29&origin=resultslist&count=10&analyzeResults=Analyze+results (accessed 21st September 2025).

43 EN ISO/ASTM (2021). EN ISO/ASTM 52900:2021: Additive manufacturing – general principles – fundamentals and vocabulary (ISO/ASTM 52900:2021); German version EN ISO/ASTM 52900:2021, international standard.

44 Zhou, S., Cai, Q., Tirichenko, I.S. et al. (2023). Additive manufacturing of Al_2O_3 with engineered interlayers and high toughness through multi-material co-extrusion. *Acta Mater.* 246: 118704. https://doi.org/10.1016/j.actamat.2023.118704.

45 Sotov, A., Kantyukov, A., Popovich, A., and Sufiiarov, V. (2022). A review on additive manufacturing of functional gradient piezoceramic. *Micromachines* 13: 1129. https://doi.org/10.3390/ mi13071129.

46 MRu, H., Nazir, A., Lin, S.-C., and Jeng, J.-Y. (2022). Parametric investigation of functionally gradient wave springs designed for additive manufacturing. *Int. J. Adv. Manuf. Technol.* 119: 1673–1691.

47 Agapovichev, A.V., Sotov, A.V., Kyarimov, R.R. et al. (2018). The investigation of microstructure and mechanical properties of tool steel produced by selective laser melting technology. *IOP Conf. Ser. Mater. Sci. Eng.* 441: 012003.

48 Ren, L., Wang, Z., Ren, L. et al. (2022). Graded biological materials and additive manufacturing technologies for producing bioinspired graded materials: an overview. *Compos. Part B Eng.* 242: 110086.

49 Williams, G.M., Akhavan, H., Dupuy, C., and Harmon, P. (2021). Additive manufacturing of freeform optics for defense applications. *Proceedings of the 2021 IEEE Research and Applications of Photonics in Defense Conference (RAPID)*, Miramar Beach, FL, USA (2–4 August 2021), pp. 1–2.

50 Schubert, J., Lehmann, C.L., and Zanger, F. (2024). Versatile binder system as enabler for multi-material additive manufacturing of ceramics by vat photopolymerization. *Ceram. Int.* 50: 50948–50954. https://doi.org/10.1016/j.ceramint.2024.10.006.

51 Geier, S. and Potestio, I. (2020). 3D-printing: from multi-material to functionally-graded ceramic. *Ceram. Appl.* 8 (2): 32–35.

52 Homa, J. (2025). The multimaterial 3D printer. https://www.lithoz.com/ (accessed 21st September 2025).

53 Nohut, S., Geier, S., Kraleva, I. et al. (2022). Lithography-based additive manufacturing of porosity graded alumina. *Addit. Manuf. Lett.* 3: 100060. https://doi.org/10.1016/j.addlet.2022.100060.

54 Schubert, J., Lehmann, C.-L., Schulze, V., and Zanger, F. (2023). Potentials and challenges of multi-material additive manufacturing of ceramics by vat photopolymerization. *Ceram. Appl.* 11 (3): 12–16.

55 Bae, C.-J. and Halloran, J.W. (2010). Influence of residual monomer on cracking in ceramics fabricated by stereolithography. *Int. J. Appl. Ceram. Technol.* 8 (6): 1289–1295. https://doi.org/10.1111/j.1744-7402.2010.02559.x.

56 Schwarzer-Fischer, E., Abel, J., Sieder-Katzmann, J. et al. (2022). Study on ceramfacturing of novel alumina aerospike nozzels by lithography-based ceramic vat photopolymerization (CerAM VPP). *Dent. Mater.* 15 (9): 3279. https://doi.org/10.3390/ma15093279.

57 Zimbeck, W. and Rice, R. (1997). Stereolithography of ceramics and metals. IS&T Annual Conference, The Society for Imaging Science and Technology, pp. 649–655.

58 Wu, H., Cheng, Y., Liu, W. et al. (2016). Effect of the particle size and the debinding process on the density of alumina ceramics fabricated by 3D printing based on stereolithography. *Ceram. Int.* 42 (15): 17290–17294. https://doi.org/10.1016/j.ceramint.2016.08.132.

59 Sutejo, I.A., Zhang, S., Gal, C.W. et al. (2024). Stamping cleaning conception on cross-contamination free multi-ceramic bioglass-tricalcium phosphate core-shell structure fabrication. *Mater. Des.* 246: 113304. https://doi.org/10.1016/j.matdes.2024.113304.

60 Zhang, S., Sutejo, I.A., Kim, J. et al. (2022). Three-dimensional complex construct fabrication of illite by digital light processing-based additive manufacturing technology. *J. Am. Ceram. Soc.* 105: 3827–3837. https://doi.org/10.1111/jace.18369.

61 Zhang, S., Sutejo, I.A., Kim, J. et al. (2022). Fabrication of complex three-dimensional structures of mica through digital light processing-based additive manufacturing. *Ceramics* 5: 562–574. https://doi.org/10.3390/ ceramics5030042.

62 Zhang, S., Sutejo, I.A., Gal, C.W. et al. (2024). Fabrication of highly transparent yttria by DLP-based additive manufacturing. *J. Eur. Ceram. Soc.* https://doi.org/10.1016/j.jeurceramsoc.2024.03.048.

63 Sutejo, I.A., Kim, J., Zhang, S. et al. (2023). Fabrication of color-graded feldspathic dental prosthetics for aesthetic and restorative dentistry. *Dent. Mater.* 39: 568–576. https://doi.org/10.1016/j.dental.2023.03.021.

64 Han, D., Yang, C., Fang, N.X., and Lee, H. (2019). Rapid multi-material 3D printing with projection micro-stereolithography using dynamic fluidic control. *Addit. Manuf.* 27: 606–615. https://doi.org/10.1016/j.addma.2019.03.031.

65 Choi, J.W., Kim, H.C., and Wicker, R. (2011). Multi-material stereolithography. *J. Mater. Process. Technol.* 211: 318–328. https://doi.org/10.1016/j.jmatprotec.2010.10.003.

66 Hu, K., Zhao, P., Li, J., and Lu, Z. (2022). High-resolution multiceramic additive manufacturing based on digital light processing. *Addit. Manuf.* 54: 102732. https://doi.org/10.1016/j.addma.2022.102732.

67 Cheng, J., Wang, R., Sun, Z. et al. (2022). Centrifugal multimaterial 3D printing of multifunctional heterogeneous objects. *Nat. Commun.* 13. https://doi.org/10.1038/ s41467-022-35622-6.

68 Zhou, C., Chen, Y., Yang, Z., and Khoshnevis, B. (2013). Digital material fabrication using mask- image-projection-based stereolithography. *Rapid Prototyp. J.* 19: 153–165. https://doi.org/10.1108/13552541311312148.

69 Schwarzer-Fischer, E., Günther, A., Roszeitis, S., and Moritz, T. (2021). Combining zirconia and titanium suboxides by vat photopolymerization. *Materials* 14: 2394. https://doi.org/10.3390/ ma14092394.

70 Parulski, C., Jennotte, O., Lechanteur, A., and Evrard, B. (2021). Challenges of fused deposition modeling 3D printing in pharmaceutical applications: where are we now? *Adv. Drug Deliv. Rev.* 175: 113810.

71 Mohamed, O.A., Masood, S., and Bhowmik, J.L. (2015). Optimization of fused deposition modeling process parameters: a review of current research and future prospects. *Adv. Manuf.* 3: 42–53.

72 Wick-Joliat, R., Schroffenegger, M., and Penner, D. (2023). Multi-material ceramic material extrusion 3D printing with granulated injection molding feedstocks. *Ceram. Int.* 49: 6361–6367. https://doi.org/10.1016/j.ceramint.2022.10.170.

73 Bellini, A., Shor, L., and Guceri, S.I. (2005). New developments in fused deposition modeling of ceramics. *Rapid Prototyp. J.* 11: 214–220. https://doi.org/10.1108/13552540510612901.

74 He, Q., Jiang, J., Yang, X. et al. (2021). Additive manufacturing of dense zirconia ceramics by fused deposition modeling via screw extrusion. *J. Eur. Ceram. Soc.* 41: 1033–1040. https://doi.org/10.1016/j.jeurceramsoc.2020.09.018.

75 Shen, T., Xiong, H., Li, Z. et al. (2021). Fused deposition fabrication of high-quality zirconia ceramics using granular feedstock. *Ceram. Int.* 47: 34352–34360. https://doi.org/10.1016/j.ceramint.2021.08.348.

76 Hadian, A., Fricke, M., Liersch, A., and Clemens, F. (2022). Material extrusion additive manufacturing of zirconia parts using powder injection molding feedstock compositions. *Addit. Manuf.* 57: 102966. https://doi.org/10.1016/j.addma.2022.102966.

77 Penner, D., Fassbind, A., Henke, R. et al. (2017). Additive manufacturing of ceramics by composite microextrusion. *Ceram. Forum Int.* 94: 62–65.

78 Lengauer, W., Duretek, I., Fürst, M. et al. (2019). Fabrication and properties of extrusion-based 3D-printed hardmetal and cermet components. *Int. J. Refract. Met. Hard Mater.* 82: 141–149. https://doi.org/10.1016/j.ijrmhm.2019.04.011.

79 Abel, J., Scheithauer, U., Janics, T. et al. (2019). Fused filament fabrication (FFF) of metal-ceramic components. *J. Vis. Exp.* 143: e57693.

80 Chen, H., Guo, L., Zhu, W., and Li, C. (2022). Recent advances in multi-material 3D printing of functional ceramic devices. *Polymers* 14: 4635. https://doi.org/10.3390/polym14214635.

81 Luo, Y., Zhai, D., Huan, Z. et al. (2015). Three-dimensional printing of hollow-struts-packed bioceramic scaffolds for bone regeneration. *ACS Appl. Mater. Interfaces* 7 (43): 24377–24383.

82 Mueller, J., Raney, J.R., Shea, K., and Lewis, J.A. (2018). Architected lattices with high stiffness and toughness via multicore-shell 3D printing. *Adv. Mater.* 30 (12): e1705001.

83 Schlordt, T., Schwanke, S., Keppner, F. et al. (2013). Robocasting of alumina hollow filament lattice structures. *J. Eur. Ceram. Soc.* 33 (15–16): 3243–3248.

84 Moon, Y.W., Choi, I.J., Koh, Y.H., and Kim, H.E. (2015). Macroporous alumina scaffolds consisting of highly microporous hollow filaments using three-dimensional ceramic/camphene-based co-extrusion. *J. Eur. Ceram. Soc.* 35 (16): 4623–4627.

85 Cai, Q. (2020). Robocasting of complex structural ceramics, department of materials, Imperial College London.

86 Cai, Q., Meille, S., Chevalier, J. et al. (2023). 3D-printing of ceramic filaments with ductile metallic cores. *Mater. Des.* 225: 111463.

87 Fu, Z., Freihart, M., Schlordt, T. et al. (2017). Robocasting of carbon-alumina core-shell composites using co-extrusion. *Rapid Prototyp. J.* 23 (2): 423–433.

88 Paredes, C., Martínez-Vázquez, F.J., Pajares, A., and Miranda, P. (2019). Novel strategy for toughening robocast bioceramic scaffolds using polymeric cores. *Ceram. Int.* 45 (15): 19572–19576.

89 Wilkerson, R.P., Gludovatz, B., Watts, J. et al. (2016). A novel approach to developing biomimetic ("Nacre-Like") metal-compliant-phase (Nickel-Alumina) ceramics through coextrusion. *Adv. Mater.* 28 (45): 10061–10067.

90 Pelz, J.S., Ku, N., Shoulders, W.T. et al. (2021). Multi-material additive manufacturing of functionally graded carbide ceramics via active, in-line mixing. *Addit. Manuf.* 37: 101647.

91 Chen, Z., Li, Z., Li, J. et al. (2019). 3D printing of ceramics: a review. *J. Eur. Ceram. Soc.* 39: 661–687.

92 Peng, E., Zhang, D., and Ding, J. (2018). Ceramic robocasting: recent achievements, potential, and future developments. *Adv. Mater.* 30: e1802404.

93 Chen, Z., Sun, X., Shang, Y. et al. (2021). Dense ceramics with complex shape fabricated by 3D printing: a review. *J. Adv. Ceram.* 10: 195–218.

94 Lakhdar, Y., Tuck, C., Binner, J. et al. (2021). Additive manufacturing of advanced ceramic materials. *Prog. Mater. Sci.* 116: 100736.

95 Maillard, M., Chevalier, J., Gremillard, L. et al. (2023). Optimization of mechanical properties of robocast alumina parts through control of the paste rheology. *J. Eur. Ceram. Soc.* 43: 2805–2817. https://doi.org/10.1016/j.jeurceramsoc.2022.12.008.

96 Feilden, E. (2017). Additive manufacturing of ceramics and ceramic composites via robocasting. (Imperial College London, 2017). https://doi.org/10.13140/RG.2.2.29343.25765.

97 Tabard, L. (2020). Elaboration de ćeramiques architecturées pour le stockage d'énergie thermique. Material Chemistry. Université de Lyon. NNT2020LYSEI007, 198 pages.

98 Lorenz, M., Dietemann, B., Wahl, L. et al. (2020). Influence of platelet content on the fabrication of colloidal gels for robocasting: experimental analysis and numerical simulation. *J. Eur. Ceram. Soc.* 40: 811–825.

99 Trincat, T., Saner, M., Schaufelbühl, S. et al. (2022). Influence of the printing process on the traces produced by the discharge of 3D-printed Liberators. *Forensic Sci. Int.* 331: 111144.

100 Allum, J., Moetazedian, A., Gleadall, A., and Silberschmidt, V.V. (2020). Interlayer bonding has bulk-material strength in extrusion additive manufacturing: new understanding of anisotropy. *Addit. Manuf.* 34: 101297.

101 Franchin, G., Maden, H., Wahl, L. et al. (2018). Optimization and characterization of preceramic inks for direct InkWriting of ceramic matrix composite structures. *Materials* 11: 515.

102 Garnault, T. (2020). Frittage par micro-ondes de céramiques mises en forme par impression 3D. Génie des procédés. Université Grenoble Alpes, Français.

103 Dilag, J., Chen, T., Li, S., and Bateman, S.A. (2019). Design and direct additive manufacturing of three-dimensional surface micro-structures using material jetting technologies. *Addit. Manuf.* 27: 167–174.

104 Choi, K.-H., Yoo, J., Lee, C.K., and Lee, S.-Y. (2016). All-inkjet-printed, solid-state flexible supercapacitors on paper. *Energy Environ. Sci.* 9: 2812–2821.

105 Craton, M.T., Sorocki, J., Piekarz, I. et al. (2018). Realization of fully 3D printed W-band bandpass filters using aerosol jet printing technology. *Proceedings of* *the 2018 48th European Microwave Conference (EuMC)*, Madrid, Spain (25–27 September 2018), pp. 1013–1016.

106 Theiler, P.M., Lütolf, F., and Ferrini, R. (2018). Non-contact printing of optical waveguides using capillary bridges. *Opt. Express* 26: 11934–11939.

107 Eggenhuisen, T.M., Galagan, Y., Biezemans, A.F.K.V. et al. (2015). High efficiency, fully inkjet printed organic solar cells with freedom of design. *J. Mater. Chem. A* 3: 7255–7262.

108 Qin, H., Dong, J., and Lee, Y.-S. (2017). Fabrication and electrical characterization of multi-layer capacitive touch sensors on flexible substrates by additive e-jet printing. *J. Manuf. Process.* 28: 479–485.

109 Khan, A., Rahman, K., Ali, S. et al. (2021). Fabrication of circuits by multi-nozzle electrohydrodynamic inkjet printing for soft wearable electronics. *J. Mater. Res.* 36: 3568–3578.

110 Laurila, M.-M., Khorramdel, B., and Mantysalo, M. (2017). Combination of E-jet and inkjet printing for additive fabrication of multilayer high-density RDL of silicon interposer. *IEEE Trans. Electron Devices* 64: 1217–1224.

111 Sarobol, P., Cook, A., Clem, P.G. et al. (2016). Additive manufacturing of hybrid circuits. *Annu. Rev. Mater. Sci.* 46: 41–62.

112 Deiner, L.J. and Reitz, T.L. (2017). Inkjet and aerosol jet printing of electrochemical devices for energy conversion and storage. *Adv. Eng. Mater.* 19: 1600878.

113 Scheithauer, U., Schwarzer, E., Richter, H.-J., and Moritz, T. (2015). Thermoplastic 3D printing-an additive manufacturing method for producing dense ceramics. *Int. J. Appl. Ceram. Technol.* 12 (1): 26–31. https://doi.org/10.1111/ijac.12306.

114 Scheithauer, U., Slawik, T., Schwarzer, E. et al. (2015). Additive manufacturing of metal-ceramic-composites by thermoplastic 3D-printing (3DTP). *J. Ceram. Sci. Tech.* 6 (2): 125–131. https://doi.org/10.4416/JCST2014-00045.

115 Scheithauer, U., Schwarzer, E., Haertel, A. et al. (2016). Processing of thermoplastic suspensions for additive manufacturing of ceramic-and metal-ceramic-composites by thermoplastic 3D-printing (T3DP). In: *Additive Manufacturing and Strategic Technologies in Advanced Ceramics, 216e tome*, Ceramic Transactions Series (ed. K. Shimamura, S. Kirihara, J. Akedo, et al.), 19–30. Hoboken, NJ: Wiley.

116 Scheithauer, U., Johne, R., Weingarten, S. et al. (2018). Investigation of droplet deposition for suspensions usable for thermoplastic 3D printing (T3DP). *J. Mater. Eng. Perform.* 27 (1): 44–51. https://doi.org/10.1007/s11665-017-2875-4.

117 Scheithauer, U., Kerber, F., Füssel, A. et al. (2019). Alternative process routes to manufacture porous ceramics-opportunities and challenges. *Materials (Basel, Switzerland)* 12 (4). https://doi.org/10.3390/ma12040663.

118 Ziener, J., Scheithauer, U., Gottlieb, L. et al. (2024). Additive manufacturing of ceramic multi-material heating and ignition elements for a sustainable space access. *Acta Astronaut.* 219: 609–617. https://doi.org/10.1016/j.actaastro.2024.03.038.

119 Zschippang, E., Kunz, W., Weingarten, S. et al. (2024). Sintering of Si_3N_4 –SiC–$MoSi_2$ composites additively manufactured by multi material jetting. *Int. J. Appl. Ceram. Technol.* 21 (4): 2613–2622. https://doi.org/10.1111/ijac.14715.

120 Moritz, T., Schilm, J., Rost, A. et al. (2019). Ceramic additive manufacturing methods applied to sintered glass components with novel properties. *Ceram. Mod. Technol.* 1 (2): 111–119. https://doi.org/10.29272/cmt.2019.0008.

121 Scheithauer, U., Pötschke, J., Weingarten, S. et al. (2017). Droplet-based additive manufacturing of hard metal components by thermoplastic 3D printing (T3DP). *J. Ceram. Sci. Technol.* 8 (1): 155–160. https://doi.org/10.4416/JCST2016-00104.

122 Schlacher, J., Hofer, A.-K., Geier, S. et al. (2021). Additive manufacturing of high-strength alumina through a multi-material approach. *Open Ceram.* 5: 100082. https://doi.org/10.1016/j. oceram.2021.100082.

123 Gheisari, R., Zhang, S., Whittow, W., and Engstrøm, D.S. (2023). Laser assisted direct ink writing of multi-material bismuth molybdate – silver artificial dielectrics for radio frequency applications. *Eng. Rep.* 5: e12749. https://doi.org/10.1002/eng2.12749.

124 Shen, K., Ding, J., and Yang, S. (2018). 3D Printing Quasi-solid-state asymmetric micro-supercapacitors with ultrahigh areal energy density. *Adv. Energy Mater.* 8: 1800408.

125 Cheng, M., Jiang, Y., Yao, W. et al. (2018). Elevated-temperature 3D printing of hybrid solid-state electrolyte for li-ion batteries. *Adv. Mater.* 30: e1800615.

126 Kostretsova, N., Pesce, A., Anelli, S. et al. (2024). Single-step fully 3D printed and co-sintered solid oxide fuel cells. *J. Mater. Chem. A* 12: 22960. https://doi.org/10.1039/d4ta02490grsc.li/materials-a.

127 Scheithauer, U., Johne, R., Gottlieb, L. et al. (2025). Recent developments in the field of additive manufacturing of ceramic-based multi-material components open the door to highly functionalized components. *Prog. Addit. Manuf.* 10: 4051–4059. https://doi.org/10.1007/s40964-025-01112-6.

128 Scheithauer, U. and Johne, R. (Spring 2024). The additive manufacturing of multi-material and multi-functional ceramic components, PIM International, Vol. 18, No. 1. https://www.pim-international.com/articles/the-additive-manufacturing-of-multi-material-and-multi-functional-ceramic-components/?utm_source=copy&utm_medium=website&utm_campaign=SocialSnap (accessed 8th April 2026).

129 Weingarten, S., Scheithauer, U., Johne, R. et al. (2019). Multi-material ceramic-based components - Additive manufacturing of black-and-white zirconia components by thermoplastic 3D-printing (CerAM - T3DP). *J. Vis. Exp.* 143. https://doi.org/10.3791/57538.

130 Zocca, A., Colombo, P., Gomes, C.M., and Günster, J. (2015). Additive manufacturing of ceramics: issues, potentialities, and opportunities. *J. Am. Ceram. Soc.* 98: 1983–2001.

131 AMAREA Technology GmbH (2025). https://www.amarea.de/3d-printers (accessed 21st September 2025).

132 Scheithauer, U., Schwarzer-Fischer, E., Sieder-Katzmann, J. et al. (2024). Additive manufacturing of ceramic single and multi-material components – a groundbreaking innovation for space applications too? *Acta Astronaut.* 221: 155–162. https://doi.org/10.1016/j.actaastro.2024.05.003.

8

Design and Characterization of Additively Manufactured Ceramics

Riccardo Balzarotti, Marco Pelanconi, Giovanni Bianchi, and Alberto Ortona

University of Applied Sciences and Arts of Southern Switzerland (SUPSI), Department of Innovative Technologies (DTI), Via la Santa 1, 6962, Lugano, Switzerland

8.1 Introduction

This chapter reports an up-to-date overview of some core topics in the field of additive manufacturing (AM) of ceramic materials. First, a brief introduction concerning the computational design of 3D models for AM and the most recent approach to topological optimization is reported, aiming at shredding a light on how data-driven functional design can be implemented in AM. Then, a cross-sectoral report of the most relevant feedstock parameters for each AM process category is provided, with a dedicated focus on the most relevant characterization techniques. Finally, an overview of the key features of the parts before and after heat processing, together with the corresponding characterization techniques, is reported.

8.2 Model Design – General Rules for Ceramic AM

The advent of computational design has revolutionized the creation of complex geometries for ceramic AM. In ceramic AM, the interplay of material properties, geometric complexity, and production constraints requires a unique approach to model design. Unlike traditional subtractive methods, computational design leverages algorithms and simulation tools to develop structures optimized for performance, manufacturability, and material efficiency. By integrating these tools with the capabilities of ceramic AM, engineers and designers can achieve unprecedented precision and functionality in their components [1].

One of the fundamental principles in designing models for ceramic AM is understanding the material's behavior during the printing and post-processing stages. Ceramics are inherently brittle and exhibit minimal tolerance for defects, making it critical to design parts with stress-distribution strategies. This is particularly important in components with thin walls or intricate lattices, as these are prone to cracking during drying or sintering. Another critical consideration in ceramic

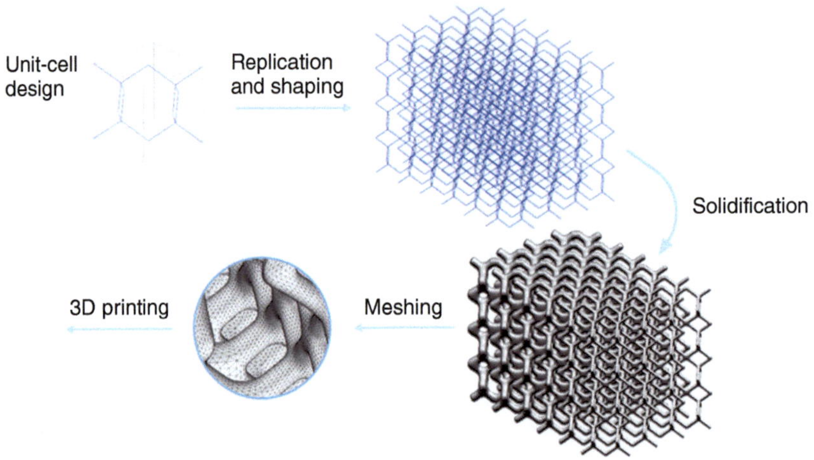

Figure 8.1 Schematic representation of the model design in additive manufacturing. [7] / with permission of Springer Nature.

AM is the need for support structures during printing. Computational design tools can address this challenge by integrating self-supporting features into the model. Overhangs and unsupported spans are minimized through the strategic design of angles and bridging elements [2], reducing the need for additional supports that might otherwise compromise the part's integrity during removal. For intricate designs [3], tools like nTopology or Autodesk Fusion 360 can automatically analyze geometries and recommend modifications to ensure printability without sacrificing functionality. Parametric modeling is also invaluable in ceramic AM [4]. By defining geometric relationships and constraints through parametric equations, designers can create adaptable models that respond to specific design criteria or environmental conditions. For example, a designer can use a parametric approach to vary pore sizes in a filter component based on flow rate requirements, ensuring performance optimization while maintaining structural integrity [5]. These models can be easily iterated upon, allowing rapid prototyping and refinement to meet functional and manufacturing needs [6] (Figure 8.1).

8.3 Topological Optimization and Numerical Simulation

Computational design forms the backbone of model creation for ceramic AM. By leveraging advanced tools and techniques such as topology optimization [8], parametric modeling, and simulation-driven design [9, 10], engineers can unlock the full potential of ceramic AM to produce lightweight, highly functional, and manufacturable components [11, 12].

Topology optimization plays a central role in computational design for ceramic AM [13]. This technique identifies areas of a part where material can be removed or redistributed to achieve the desired mechanical properties while minimizing weight. For example, lightweight structures with high stiffness-to-weight ratios are often achieved using generative design algorithms that produce lattice or porous geometries. These complex forms are impossible to fabricate using conventional techniques but are well-suited for ceramic AM due to its layer-by-layer fabrication capability. Moreover, these optimized structures not only reduce material usage but also enhance thermal and fluidic performance, making them ideal for heat exchangers, filters, or biocompatible implants. Moreover, computational design integrates simulation-driven insights with real-world constraints. Simulations such as computational fluid dynamics (CFD [14, 15]) and finite element analysis (FEA) allow designers to predict how a part will behave under operational conditions, ensuring that the final design meets both performance and durability criteria. For ceramic AM, such simulations are particularly crucial in mitigating risks associated with thermal gradients, shrinkage, and warping during sintering, allowing designers to incorporate features such as fillets, chamfers, and rounded transitions to reduce stress concentrations.

8.4 Characterization of the Feedstocks Prior to Printing

Regardless of the complexity in the classification of AM technology [16], processes for the AM of ceramic materials can be grouped into three categories depending on the used feedstock, namely powder-based, bulk solid-based, and slurry/liquid-based, as shown in Figure 8.2 [17, 18].

With respect to traditional production processes, the advent of AM has introduced new standards in the feedstock properties, which are strictly connected with the requirements of each manufacturing technique.

8.4.1 Powder Based

AM technologies involving a bed of powder are common and very well-established production strategies. In such approaches, a thin layer of powder is spread and consolidated in a layer-by-layer approach either by local melting or gluing. The printed layer is then slightly lowered into the job box, so that more powder is spread over the top to repeat the procedure and to have the object grow.

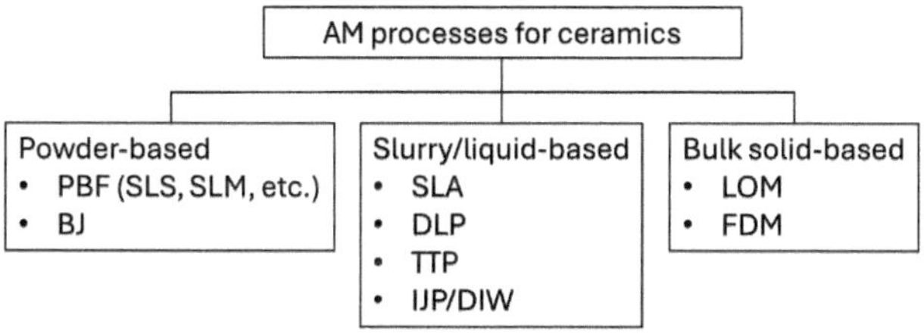

Figure 8.2 Classification of AM technologies for ceramics by the type of feedstock used. [17] / MDPI / CC BY 4.0. BJ, Binder Jetting, DIW, Direct Ink Writing, DLP, Digital Light Processing, FDM, Fused Deposition Modeling, IJP, Ink Jet printing, LOM, Laminated Object Manufacturing, PBF, Powder Bed Fusion, SLA, Stereolithography, SLM, Selective Laser Melting, SLS, Selective Laser Sintering, TPP, Two-Photon Polymerization.

In powder-based AM, the characteristics of the powder feedstock have a strong influence not only on many aspects of the printing operations but also on parts requirements and economic aspects. Thus, it is of paramount importance to have a clear understanding of the requirements of each powder-based AM technology, to prevent undesired part characteristics and keep a control on cost. The most influential powder characteristics toward powder-based AM are:

- powder morphology,
- particle size distribution (PSD),
- packing density,
- powder flow and spreadability.

In the following, each one of these properties will be analyzed, with a specific focus on the most relevant techniques and procedures that are suitable for the characterization.

Powder morphology: The shape of the powder feedstock has a strong influence on powder processability, which affects the process parameters of many powder-based AM technologies [19, 20]. In most cases, powder-based AM processes have beneficial effects in the use of spherical particles, due to the higher powder bed density that can be achieved. Spray-based processes are widely recognized among the most reliable and reproducible techniques for powder aggregation, in which particles' shape, morphology, and surface features can be controlled [21–23].

Many different characterization techniques are available for the assessment of powder morphology, such as dynamic image analysis (DIA), laser diffraction, and optical/electron microscopy [24]. In an increasing complexity scale, techniques can be grouped into four categories, namely (i) light microscopy (LM), (ii) static image analysis (SIA), (iii) DIA, and (iv) micro X-ray-computed tomography (μXCT) [25]. LM is a nonautomatic SIA technique that requires separate processing stages, including imaging. DIA, SIA, and LM can only provide 2D images, while μXCT is able to provide full 3D morphology characterization of the powder sample. While the advantages of using μXCT to obtain 3D characterization are evident, this approach has higher computational demand both in terms of costs and time, in comparison to 2D shape analysis, which provides an automated process of morphology characterization from taking particle images to their processing and quantification [26]. Accordingly, the investigation of potential correlations between 3D and 2D shapes is of high interest, as it could provide a reasonable estimation of a 3D shape from a 2D particle image [27]. For example, the lightness method has been reported as a possible approach to derive three-dimensional features (pseudo-3D, or sometimes referred to as 2.5D) by using a number of particle projections in different orientations for the same particle [28, 29].

Particle size distribution (PSD): As mentioned above, much relevant information on powder processability can be drawn from the analysis of powder morphology. Regardless of the powder shape, PSD is another key parameter for powder-bed-based AM, which is typically used to address the powder size in terms of volume or count percentile. The role of PSD is crucial at different AM stages, namely in the preparation (e.g. flowability and packing behaviors), in the processing (e.g. interaction with liquid binder, saturation, etc.), and in the post-processing (e.g. sintering properties). Therefore, optimizing the PSD as well as the binder selection can help reducing numerous issues related to printing [30]. The pore size distribution and uniformity of the powder bed are affected by PSD, as it may influence, for example, the drop penetration of liquid binders in BJ. Additionally, agglomeration phenomena can occur in case of the use of small particles, which results in the generation of macro-voids within the powder and in an increase of drop penetration time [31]. Smaller particles typically initiate sintering more quickly than larger particles, and thus reach densification faster, due to the higher surface area and the higher energy state.

A variety of well-established approaches are available for the measurement of the PSD [32]; among others, sieve analysis [33], laser diffraction [34], and dynamic light scattering (DLS) [35] are the most used techniques (Figure 8.3).

Sieve analysis uses several stacked sieves with an increasing mesh size to force the particles' distribution from the uppermost sieve to different fractions according to the particle sizes. Despite the low cost and the simple analytical principle, sieving has some intrinsic limitations, including a restricted resolution, the inability to analyze wet or sticky particles, and a minimum measurement limit in the range of tens of microns.

DLS is based on the interaction of light as it travels through the sample, and it measures the PSD, determining the time-dependent intensity scattering fluctuation due to the Brownian motion of suspended particles [36].

Using a different approach, laser diffraction determines the PSD by measuring the angle of diffraction of scattered light as a laser beam passes through the particle dispersion. The technique relies on the fact that the diffraction angle is inversely proportional to particle size. As opposed to DLS, laser diffraction should be used for

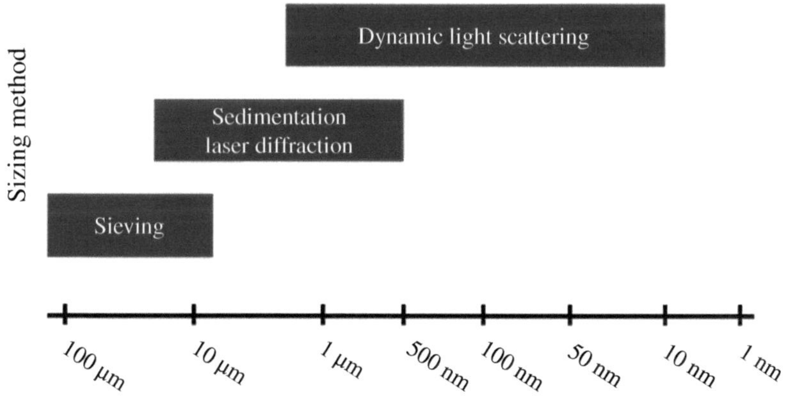

Figure 8.3 Qualitative classification of the PSD analysis techniques depending on particle size. Adapted from [36].

particles ranging from the submicron (approximately 500 nm) to the millimeter size range and, in a more general way, in the range of particle size where sedimentation velocities exceed Brownian motion [36].

Powder packing density: Powder packing density plays a crucial role in the determination of the arrangement of the particle network and, thus, in their maximum contact. Using the simplest approach to the topic, spheres with the same size in a closed-packing arrangement can reach 74% packing, while random closed-packing provides about 63% space filling [37]. International standards specify the procedures to determine the tap density of granulated or ungranulated ceramic powders by a constant volume measuring method [38]. Tap volume and density do not result from a loose powder bed, as they consist of the conditions of the bed (i.e. pores and inter-particular voids) after a defined number of tapping steps. Measuring instruments for the determination of tap densities, also known as tap volumeters, are commercially available instruments.
Under actual printing conditions, powder packing density is influenced by many parameters, which can be grouped into (i) powder properties, (ii) process properties, and (iii) powder flowability (Figure 8.4).

Given the remarkable number of variables, powder packing density under actual printing conditions is typically measured using mock devices mimicking the effective spreading dynamics in binder jet system [30]. Using a slightly different approach, small-sized cups can be printed and used to determine the mass of powder that the printer spreads in a defined volume, thus allowing for determining the packing density of the powder bed [40].

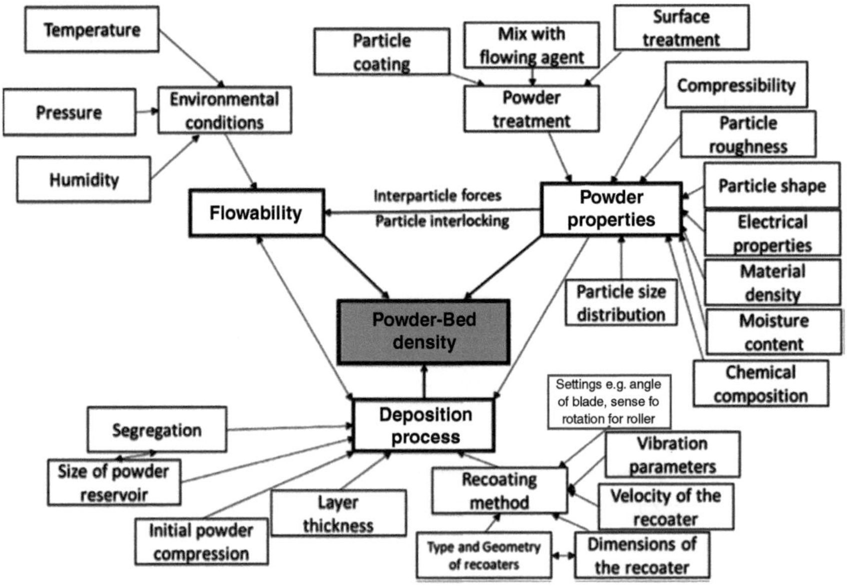

Figure 8.4 Parameters influencing powder bed density. [39] / with permission of Elsevier.

Table 8.1 List of methods and procedures for the characterization of powder flowability.

Method	Procedure
Angle of response	Measurement of the angle between the horizontal plane and the powder heap at the bottom of a funnel.
Time of discharge	Assessment of the elapsed time during powder flowing out of a funnel.
Hausner ratio	Calculation of the ratio between tap and bulk density.
Shear cell	Use of a shear cell to measure the shear resistance and the yield loci of a consolidated bulk.
Rotating drum	Dynamic angle of response, cohesive index and first avalanche angle are measured by image processing of powder flowing into a rotating container.
Powder rheometry	A stirring, rotating blade is used to measure the powder's resistance to flow.

Adapted from ref. [41].

Powder flow and spreadability: Despite the complexity of its characterization, the assessment of powder flow behavior plays a crucial role in many industrial processes, including AM [41]. Given the importance of the topic and its relevance to the final properties of the 3D-printed parts, reviews are available in the literature [42, 43]. From a general perspective, powder flowability is crucial to have the possibility to transfer the feedstock through the process, and it enables many operations at different printing stages (e.g. powder movement from the hopper to the bed, consistent deposition onto the powder bed, and layer spreading) [39]. High flowability is a desired property for the powder feedstock, as it enables high printing resolution, dimensional accuracy, and density of a 3D-printed part. The most widely used methodologies for the characterization of powder flowability are listed in Table 8.1.

In the angle of response (AoR) method, the powder flows through a special funnel placed at a fixed height above a completely flat and level plate. The AoR is the base angle of the resulting cone, which is calculated from the measured height and radius of the formed powder cone [44, 45]. Low AoR values ($<30°$) indicate low interparticle friction and forces and, therefore, very free flowing [41].

The time of discharge method can be used only to powders that flow freely through the test orifice. The ISO 14629:2012 standard specifies a test method to determine the flowability of ceramic powders using a specified funnel.

The Hausner ratio (HR) is calculated as the ratio of the tap density to the apparent density. As mentioned in the previous section, both tap and apparent density can be measured according to international standards (ASTM standard B213-17 and B212-17, respectively). The HR provides a description of the packing behavior of powders, where powders with HR between 1.00 and 1.11 are representative of excellent flow behavior, while HR values >1.4 are typical of cohesive and non-flowing powders [41].

Shear cells are widely utilized to assess the flow behavior of powders during applications that involve powder discharge. In these tests, the feedstock powder is sheared across a single plane to determine its shear strength properties [46]. The use of shear cells allows determining interaction forces both among the particles and between the particles and the walls, thus providing insights into the many feedstock properties, such as compressive strength, powder compressibility, consolidation time, and bulk density [43]. An ASTM standard reports the standard test method for shear testing of bulk solids using the Jenike shear cell [47].

In the rotating drum method, a digital camera is used to acquire images of the powder moving into a partially filled drum. By image and video processing, characteristics of the powder such as avalanche angle, volume expansion ratio, fluidized volume slope, and fluidized height slope can be obtained [30, 48].

Finally, in powder rheometry, different tools (e.g. blades, pistons, and shear heads) are used in the powder bed under simultaneous rotation and axial movement, aiming at measuring the axial and rotational forces. In powder rheometry tests, a variety of control modes are available, including velocity, force, and torque [46].

8.4.2 Liquid/Slurry Based

Liquid feedstock is used in a wide range of AM technologies (Figure 8.2), and reviews on their effective use are available in the literature [49]. From the point of view of characterization, some specific techniques are available, which focus on the assessment of the features that are most critical to printing. In the case of photopolymerization-based techniques, an attempt to provide an international standard for the characterization of ceramic slurries in vat photopolymerization is ongoing at the time of this publication. The characterization protocol includes the assessment of the composition and properties of the slurry, such as solids content, dynamic viscosity, PSD, chemical composition, and solid dispersion stability. This standard also provides available methods for sampling and preparing slurry samples for testing [50].

Regardless of the AM technique, the rheological properties of the liquid feedstock play a significant role in determining the printing performance [51]. The rheological behavior of the liquid feedstock is strongly influenced by the physical characteristics of the powder particles (e.g. PSD, shape, and morphology) [52], which can be characterized using the techniques reported in Section 8.4.1. The rotational rheometer is widely considered the most robust instrument to obtain a wide range of information regarding the liquid feedstock [53]. Viscosimeters are a cost-effective alternative to rheometers, but some intrinsic limitations arise, both in terms of analysis quality and range of operation.

Another prominent issue that concerns specifically the liquid feedstocks that are used for SLA is their high refractive index when compared to the base resin. The difference in the refractive index affects the curing depth (C_d) and the resolution of the final component, due to scattering phenomena. In case of laboratory-developed liquid feedstocks, experimental strategies are reported in literature for the assessment of all the relevant parameters of the system, which rely on the measurement of C_d [mm] at different values of light exposure [mJ·mm^{-1}] [54].

The stability of the liquid feedstock is another crucial parameter to be assessed, as it has a strong influence on the overall quality of the 3D-printed parts [55]. In case of colloidal suspensions, zeta potential is an indicator of the magnitude of electrostatic forces, either repulsive or attractive, that exist between particles within a dispersion [51]. Commercial instruments are available for the measurement of the zeta potential as a function of the pH; based on this characterization, it is possible to determine the range of stable composition of a liquid feedstock (i.e. no settling, aggregation, or flocculation), which is typically reached when particles are highly charged (i.e. more than 30 mV or less than 30 mV).

In case of concentrated liquid feedstocks, settling studies are conducted to evaluate the stability over time. Test tubes are used to determine sedimentation kinetics, based on light diffusion, and by visually measuring sedimentation heights [52].

8.4.3 Bulk Solid Based

Fused deposition modeling (FDM) is an extrusion-based AM process that consists of softening a material and pushing it through an orifice in order to deposit that material in layers to build a 3D structure [56, 57]. To produce ceramic parts by FDM, filaments made of a combination of thermoplastic polymers and ceramic materials are typically used, as the basic principle of material extrusion for AM involves the melting of the material and its movement throughout a nozzle onto the printing area [58, 59]. Depending on the type of extruder used, three main FDM techniques are available, namely extrusion with plungers, extrusion with filaments, and extrusion with screws. While the overall composition of the material is similar (i.e. a polymer-ceramic composite), each technique uses slightly differently shaped feedstock materials (i.e. soft material in a reservoir, filament, and granules, respectively).

Different techniques are commonly used for the characterization of the feedstock materials for FDM. In addition to the traditional microscopy analyses (i.e. optical and scanning electron) and density measurements (He-pycnometry), rheology tests are crucial to evaluate the possibility of obtaining extruded filaments for 3D printing. Rotational rheometers with a "plane-plane" working unit geometry are typically used, and many different measurements (e.g. oscillatory measurements and creep tests at a constant load) can be made to assess the viscosity of the feedstock. Measurements must be performed in the recommended temperature range for the materials [60].

From a mechanical perspective, the filament must be flexible enough to be wound around a spool for continuous feeding, and, thus, it is typically characterized mechanically by using the tensile test. The geometry of tensile test samples and the general principles for determining the tensile properties of plastics and plastic composites under defined conditions are reported in international standards [61]. In addition to this, the flexibility of filaments can be evaluated qualitatively by measuring the minimum bending radius to which a filament can be bent before it fractures [62].

In case of commercially available filaments, a detailed knowledge of the chemical properties of the thermoplastic composite material could be beneficial for the refinement of printing operative conditions. In this view, elemental analysis can be performed on the fracture surface of the filaments using a scanning electron microscope equipped with an energy dispersive spectrometer (SEM-EDS).

Thermal analysis comprises various measuring techniques (e.g. thermogravimetric analysis (TGA), differential scanning calorimetry (DSC), and differential thermal analysis (DTA)), with a common feature, namely the measure of the material response when heated or cooled (or, in some cases, held isothermally) [63]. As an example, TGA is typically used for the assessment of the most suitable debinding procedure, which strongly depends on the chemical composition of the polymeric part of the FDM feedstock.

8.5 Characterization of the Parts Before and After Heat Processing

Additively manufactured (AM) ceramics possess distinctive properties compared to conventional ceramics due to differences in shaping techniques, consolidation steps, and resulting microstructures. Therefore, their characterization must address conventional mechanical and physical metrics, unique microstructural features, and processing-induced phenomena. Ceramic AM processes typically produce a green body consisting of solid preceramic polymers or ceramic particles embedded in an organic or aqueous binder. This intermediate product requires post-processing steps, such as debinding and sintering or pyrolysis, to achieve the desired final density and functional properties. However, consolidation introduces challenges like anisotropic shrinkage, crack formation, and porosity evolution, all of which critically influence the final component's integrity and performance [64]. This chapter presents a condensed overview, without any claim of comprehensiveness, of the characterization methods employed to evaluate AM ceramics' specific properties both before and after thermal consolidation.

8.5.1 Mechanical Properties

Mechanical performance in additively manufactured (AM) ceramics is typically characterized by flexural strength, compressive strength, fracture toughness, microhardness, and thermomechanical stability. These properties are closely influenced by fabrication parameters, microstructure, and post-processing conditions, and are often anisotropic due to the layer-by-layer deposition process [65, 66].

Mechanical integrity before and after consolidation: Mechanical integrity before consolidation plays a critical role in handling, post-processing, and defect evolution. Flexural and compressive strength tests of green bodies offer early insight into structural stability and help detect weak zones prior to consolidation. These properties are dependent on feedstock composition, binder content, and print orientation. Zhao et al. demonstrated how green strength analysis can inform sintering parameters and optimize microstructural evolution [67, 68].

Post-consolidation mechanical strength is assessed through flexural, compressive, and tensile strength tests, which provide quantitative data on the material's load-bearing capability. These tests are critical for evaluating the success of sintering and the influence of porosity, grain structure, and layer interfaces. For example, flexural strength testing is especially useful in predicting the performance of components under bending loads [69]. To address variability, statistical tools like Weibull analysis are often applied to relate strength to flaw distribution and quantify reliability in real-world use. Moster and Mitchell highlighted how Weibull statistics can identify failure probability and defect sensitivity in AM ceramics [70].

Anisotropy in mechanical properties: AM ceramics display distinct anisotropic mechanical properties due to their layer-by-layer fabrication process. This structural anisotropy significantly affects performance and requires systematic mechanical characterization across different building orientations. Standard methods such as flexural, compressive, and hardness testing are used to quantify these directional variations and correlate them with process parameters and microstructural features. The mechanical characterization of anisotropy in AM ceramics is a multidimensional effort involving direction-specific testing, detailed microstructural imaging, statistical modeling, and process refinement.

Anisotropy primarily arises from directional layering, leading to differences in mechanical response depending on test orientation. Flexural and compressive strength measurements conducted along longitudinal (Z), transverse (X), and lateral (Y) directions reveal that Z-direction specimens typically exhibit lower mechanical strength, attributed to weaker interlayer bonding. In contrast, lateral directions often show superior strength due to better cohesion between printed layers [71–73]. To address the variability in strength values, Weibull statistical analysis is frequently employed, providing a more accurate description of strength distributions in 3D-printed ceramics [74].

Beyond mechanical testing, the microstructural origins of anisotropy are examined using SEM, which enables high-resolution imaging of fracture surfaces post-testing. This allows correlation of mechanical behavior with pore distribution, grain orientation, and interlayer bonding characteristics [75, 76].

Surface microhardness testing, particularly Vickers hardness, is also sensitive to anisotropy. Variations in hardness values reflect the effect of print orientation, surface finish, and fabrication parameters like slurry viscosity, printing speed, and layer thickness [77, 78].

Fracture toughness: Fracture toughness defines a material's resistance to crack initiation and propagation, an especially critical property in additively manufactured ceramics, where microstructural defects, residual stresses, and anisotropy may affect failure behavior. Resistance to crack growth under static loading is commonly evaluated using the single edge-notched beam (SENB) method. This technique involves introducing a notch into a beam-shaped specimen and applying a bending load until fracture occurs. Tatami et al. employed SENB to assess local fracture toughness, revealing how microstructural variations influence crack propagation behavior [79].

Stable crack propagation in heterogeneous microstructures is effectively assessed using the chevron-notched beam (CNB) method. This approach creates a chevron-shaped notch that promotes controlled crack growth without requiring a pre-crack. Apparent fracture toughness estimation of additively manufactured alumina from as-printed chevron-notched test specimens was successfully assessed by Snarr et al. [80].

Localized fracture behavior and surface flaw sensitivity are frequently measured using the indentation fracture toughness (IFT) method. This technique estimates toughness from crack lengths generated by a controlled indentation (e.g. Vickers or Knoop). IFT is particularly useful for small, thin, or geometrically intricate AM ceramic parts. Recent work has shown its effectiveness in rapidly estimating toughness in AM samples with minimal preparation [81].

Microscale variations in fracture resistance are investigated using micro-indentation and micro-cantilever bending. Cui et al. demonstrated a modified micro-cantilever technique for measuring fracture toughness in AM ceramics with precision [82].

Crack path visualization and internal flaw analysis are enabled through X-ray-computed tomography (XCT), which provides 3D, nondestructive insight into fracture processes. XCT allows correlation between microstructural inhomogeneities and crack propagation behavior. Cramer et al. utilized XCT to analyze fracture loads and crack development, understanding defect-driven fracture mechanisms [83].

Dynamic fracture resistance, relevant for components subjected to impact or high-strain-rate conditions, is assessed through dynamic fracture toughness testing. This technique captures the effect of loading rate on crack initiation. Chen et al. showed how strain rate significantly affects the measured toughness of piezoelectric ceramics, underscoring the importance of replicating service conditions during testing [84].

Thermomechanical characterization: Viscoelastic behavior and temperature-dependent mechanical response are captured using dynamic mechanical analysis (DMA). This method reveals storage and loss modulus across thermal and dynamic loading conditions, offering insight into stiffness and damping properties. Studies confirm DMA's effectiveness in assessing AM ceramics for applications in aerospace and biomedical fields, where temperature and cyclic loading are critical [85–87]. It also supports optimization of materials subjected to dynamic environments [88, 89].

Thermomechanical stability and elasticity under thermal cycling are evaluated using thermomechanical analysis (TMA). This technique tracks deformation under applied load as a function of temperature and is key to understanding expansion, creep, and stress relaxation. Studies by multiple authors have shown how thermomechanical cycling impacts bond strength and long-term durability in AM ceramics [90–92]. TMA also enables quantification of elasticity and damping over a thermal range, relevant for high-temperature and thermal-shock-prone applications [93–95].

8.5.2 Composition and Microstructure Characterization, from the Green to the Consolidated Part

Understanding the chemical composition and physical evolution of additively manufactured (AM) ceramics is critical for tailoring their mechanical performance, chemical durability, and functional applicability. Key material properties such as phase composition, microstructure, elemental distribution, and thermal behavior are characterized to ensure reliable performance in advanced applications like aerospace, biomedical implants, and electronics.

Crystalline phase composition and phase transitions directly affect the stability, strength, and functional behavior of AM ceramics. These aspects are commonly assessed using X-ray diffraction (XRD), which enables the identification of crystalline phases, evaluation of phase purity, and detection of phase transformations during thermal processing. Xu et al. demonstrated the utility of XRD in examining the structural evolution of 3D-printed ceramics and their impact on final properties [96].

Transmission electron microscopy (TEM) offers insights at the atomic level, allowing for the investigation of fine-scale microstructural features, such as dislocations, nanoprecipitates, and phases that often influence the mechanical behavior of materials. Techniques like selected area electron diffraction (SAED) within TEM can provide detailed crystallographic information for understanding grain orientation and its effects on crack propagation and mechanical properties, especially in laser AM processes [97, 98].

Raman spectroscopy is an optical technique providing detailed insights into the molecular and structural characteristics of ceramic materials, allowing for enhanced identification of phase compositions and crystallinities, essential for optimizing fabrication processes and outcomes [96, 99]. The integration of Raman spectroscopy in AM of ceramics facilitates the monitoring of material transformations that occur during processing, ensuring greater quality control and consistency.

Microstructural features such as particle morphology, pore distribution, and interfacial bonding critically influence mechanical properties like strength and toughness. These are visualized using scanning electron microscopy (SEM), which provides high-resolution imaging of surface and cross-sectional structures. Maier et al. used SEM to investigate the microstructure of stereolithography-printed alumina ceramics, establishing clear correlations between microstructural quality and mechanical performance [69].

Elemental distribution and compositional uniformity are essential for ensuring consistency in ceramic performance, particularly in systems where dopants or multiple phases are involved. These are typically evaluated using energy dispersive spectroscopy (EDS) coupled with SEM. For example, Gupta and Riyad employed EDS to analyze the effects of oxidation-induced sintering on porous ceramics, linking elemental composition to processing-induced changes in material behavior [100].

Thermal stability and phase evolution during heating are crucial for defining appropriate debinding, sintering profiles and predicting material behavior at elevated temperatures. Techniques such as DSC, DTA, and TGA are used to study decomposition processes, phase changes, and overall thermal response.

Michalik et al. applied DTA and TGA to optimize sintering conditions and prevent degradation in AM ceramic systems [101].

8.5.3 Surface Roughness

Surface roughness in additively manufactured (AM) ceramics significantly influences properties such as adhesion, wear resistance, wettability, microhardness, and aesthetic quality. These surface characteristics, shaped by printing resolution, material composition, and post-processing, are critical for optimizing ceramics in biomedical, structural, and optical applications.

Roughness-induced defects like microcracks can reduce microhardness and mechanical reliability. Abdalla et al. reported that increased surface roughness often leads to lower hardness, due to flaw-sensitive stress concentrations [102]. Similarly, Nakai et al. linked rougher zirconia surfaces to higher friction and wear, underlining the importance of surface texture in performance [103].

To evaluate these effects, SEM is widely used for high-resolution imaging of surface morphology and crack formation [104]. Contact profilometry provides roughness metrics (Ra, Rz), with Dong et al. showing strong links between AM process routes and measured surface variation [105].

For noncontact, large-area analysis, optical profilometry via laser scanning or white light interferometry enables rapid surface mapping. Xing et al. demonstrated how this method relates pore structure and surface quality to mechanical performance in ceramic filters [106].

8.5.4 Dimensional and Geometrical Stability

Shrinkage and warpage in additively manufactured (AM) ceramics – resulting from drying, binder burnout, pyrolysis and sintering-affect dimensional accuracy, mechanical integrity, and geometric fidelity. These phenomena are often anisotropic and material – or process-specific, especially in techniques with directional deposition paths such as robocasting or filament extrusion. Green parts may closely match the digital model, but final sintered parts typically undergo shrinkage ranging from 15% to 30%, depending on the formulation and thermal cycle.

High-resolution 3D scanning and digital subtraction techniques are widely used to assess these dimensional deviations. By scanning green and sintered parts using laser or structured-light systems, and digitally comparing their meshes, researchers can precisely measure shrinkage, distortion, and warpage. This approach enables noncontact, full-field analysis with high spatial resolution and is particularly important in precision-critical fields such as dentistry. Mesh comparison workflows have proven effective in detecting localized shrinkage and nonuniform sintering behavior [107] (Figure 8.5).

Thermal shrinkage behavior during sintering is directly measured using dilatometry, which tracks linear dimensional changes as a function of temperature. This method is particularly useful for evaluating shrinkage kinetics and anisotropy throughout the thermal cycle. Manière et al. used dilatometry to analyze directional

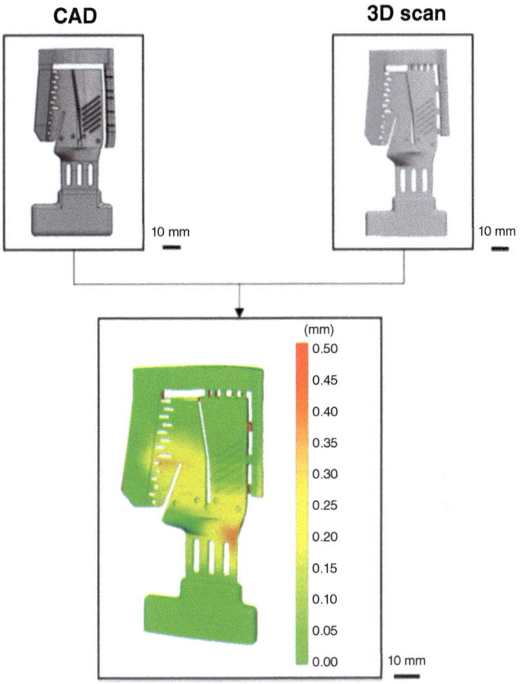

Figure 8.5 Comparison between CAD and 3D-scanned sintered core. [107] / Elsevier / CC BY 4.0.

shrinkage in 3D-printed alumina ceramics, highlighting the influence of printing parameters on dimensional evolution [108].

Optical scanning and image analysis techniques provide spatially resolved shrinkage data by comparing high-resolution images or digital models of parts before and after processing. Matos et al. applied this method to investigate processing-induced shrinkage variations in architected ceramic composites, revealing how localized effects influence overall dimensional stability [109].

Internal shrinkage patterns and densification behavior can be visualized using XCT. This nondestructive 3D-imaging method reveals internal geometry, void evolution, and structural changes. Tiseanu et al. demonstrated the power of XCT in tracking shrinkage and densification in field-assisted sintered ceramics, correlating microstructural changes with dimensional behavior [110].

Finally, manual measurements and statistical analysis remain relevant for batch-level shrinkage evaluation. Using tools like calipers or micrometers to measure key dimensions before and after sintering, followed by regression analysis, enables correlation of processing parameters with dimensional changes. Though low-tech, this method provides practical insights into process repeatability and part-to-part variability.

8.5.5 Density and Porosity

Density and porosity critically influence the mechanical, thermal, and functional performance of additively manufactured (AM) ceramics. Proper characterization

Table 8.2 Techniques for characterizing density and porosity in additively manufactured ceramics.

Technique	Typical accuracy range	Suitable for: Green parts	Suitable for: Sintered parts	Porosity range	Notes
Archimedes' method	±1–3%	No	Yes	<1%	Simple, limited with high porosity or complex geometries
Gas pycnometry	±0.1–0.2%	Yes	Yes	Derived via comparison	Dry, closed-pore samples; true density
Powder-based method	±2–5%	Yes	Yes	<5%	Sensitive to packing and particle shape
Mercury intrusion porosimetry (MIP)	±5–10%	No	Yes	Pore size distribution, 3–4000 nm	Detailed pore distribution; hazardous material
Gas adsorption (BET)	±3–5%	Yes	Yes	Micro and mesoporous	Excellent for fine porosity; assumes ideal geometry
X-ray computed tomography (XCT)	~0.5–5 μm	Yes	Yes	Open and closed porosity	3D pore visualization: resolution depends on sample size

is essential for optimizing processing strategies and ensuring consistency in performance-critical applications such as filters, biomedical devices, and structural components. Each method below provides specific insight into density or porosity, with differences in measurement range, resolution, and sample suitability; the most common methods are summarized in Table 8.2.

Bulk density and open porosity in sintered ceramics are commonly assessed using Archimedes' method, based on buoyancy measurements in air and water. This method is nondestructive and suitable for relatively dense parts, with a typical density measurement range of ~1–6 g/cm^3 and porosity detection above ~1%. It offers an accuracy of ±1–3% for simple geometries but can be less reliable with high-porosity or complex-shaped samples due to trapped air. Yang et al. applied this method to investigate porosity-related leaching in ceramic cores [111], while Choi et al. tracked density increases in ferroelectric ceramics under different sintering regimes [112].

True and envelope densities, particularly important for green or highly porous ceramics, are measured using gas pycnometry (He) and powder-based methods. These techniques provide density values ranging from ~0.1 to 20 g/cm^3 and porosity estimates greater than ~5%, with high accuracy (±0.1–0.2% for gas pycnometry; ±2–5% for powder methods). They are ideal for evaluating densification and overall porosity prior to sintering [113–115].

Pore size distribution and total porosity in sintered ceramics are measured using mercury intrusion porosimetry (MIP), which forces mercury into pores under pressure to measure their volume and size. It covers pore diameters from ~3 nm to 100 μm and detects porosity with ±5–10% accuracy. As a destructive technique, MIP is excellent for characterizing open porosity but cannot detect closed pores and involves handling toxic mercury. Pelanconi et al. used MIP to show how porosity levels in the fabrication of Si-SiC ceramics impacted compression strength and processing [116].

Micro- and mesoporosity, especially in materials with fine-scale porosity, are best evaluated using gas adsorption techniques such as the BET method. This approach measures specific surface area and pore volume, with sensitivity to pore sizes in the ~0.3–100 nm range. While highly effective for microporous ceramics, it assumes ideal surface geometry and is typically destructive. Pechlivani et al. demonstrated how gas adsorption measurements helped correlate porosity levels with the densification process in porous AM ceramics [117].

3D pore morphology and volumetric porosity are nondestructively assessed through XCT [115, 118]. This method provides resolution-dependent pore detection from ~0.5 μm to 500 μm and allows complete internal visualization. While excellent for structural integrity studies and defect mapping, its resolution and equipment cost can be limiting. Diener et al. and Saâdaoui et al. used XCT to study interconnected porosity in ceramics and analyze the mechanical behavior [119, 120].

Layer-by-layer porosity and cross-sectional features in AM ceramics can be imaged using optical coherence tomography (OCT), a high-resolution, nondestructive optical technique. OCT achieves depth resolution of ~1–15 μm and layer thickness analysis up to ~2 mm. Studies have explored its role in evaluating subsurface layer quality [121].

8.6 Summary

In this chapter, an overview of relevant AM topics that are cross-sectoral to all the 3D-printing technologies has been provided. Regardless of the AM technique, the preliminary model design and the following topological optimization stage are crucial to provide the most suitable object design for printing. In this sense, the computational design allows for an optimized management of the interplay of material properties, geometric complexity, and production constraints. Furthermore, the combination of topology optimization and numerical simulations can unlock the full potential of ceramic AM to highly functional and manufacturable components.

In view of matching the properties of the final part to the design specification, an accurate characterization of the feedstock is a mandatory stage in which many parameters (e.g. PSD and feedstock flowability) can be measured and correlated to the resulting features of the green body. Finally, the characterization strategies of some of the phenomena occurring in the conversion of the green body to the final consolidated part (e.g. anisotropic shrinkage, crack formation, and porosity evolution) have been discussed in this chapter, by providing without any claim of comprehensiveness, a condensed overview of the characterization methods employed to evaluate AM ceramics specific properties both before and after thermal consolidation.

References

1 Pelanconi, M., Barbato, M., Zavattoni, S. et al. (2019). Thermal design, optimization and additive manufacturing of ceramic regular structures to maximize the radiative heat transfer. *Mater. Des.* 163: 107539.

2 Pelanconi, M., Rezaei, E., and Ortona, A. (2020). Cellular ceramic architectures produced by hybrid additive manufacturing: a review on the evolution of their design. *J. Ceram. Soc. Japan* 128 (9): 595–604.

3 Gibson, L.J. and Ashby, M.F. (1997). *Cellular Solids: Structure and Properties.* Cambridge: Cambridge University Press.

4 Al-Ketan, O., Pelanconi, M., Ortona, A., and Abu Al-Rub, R.K. (2019). Additive manufacturing of architected catalytic ceramic substrates based on triply periodic minimal surfaces. *J. Am. Ceram. Soc.* 102 (10): 6176–6193.

5 Tedeschi, A. (2011). *Parametric Architecture with Grasshopper: Primer*. Le Penseur.

6 Pelanconi, M. and Ortona, A. (2019). Nature-inspired, ultra-lightweight structures with gyroid cores produced by additive manufacturing and reinforced by unidirectional carbon fiber ribs. *Materials (Basel)* 12 (24): 4134.

7 Pelanconi, M. and Ortona, A. (2021). Review on the design approaches of cellular architectures produced by additive manufacturing. In: *Industrializing Additive Manufacturing* (ed. M. Meboldt and C. Klahn), 52–64. Cham: Springer International Publishing.

8 Pan, C., Han, Y., and Lu, J. (2020). Design and optimization of lattice structures: a review. *Appl. Sci.* 10 (18): 6374.

9 Liu, Z., Xia, L., Xia, Q., and Shi, T. (2020). Data-driven design approach to hierarchical hybrid structures with multiple lattice configurations. *Struct. Multidiscip. Optim.* 61 (6): 2227–2235.

10 Cucinotta, F., Guglielmino, E., Longo, G. et al. (2019). Topology optimization additive manufacturing-oriented for a biomedical application. In: *Advances on Mechanics, Design Engineering and Manufacturing II* (ed. F. Cavas-Martínez, B. Eynard, F.J. Fernández Cañavate, et al.), 184–193. Cham: Springer International Publishing.

11 Song, J., Wang, Y., Zhou, W. et al. (2019). Topology optimization-guided lattice composites and their mechanical characterizations. *Compos. Part B Eng.* 160: 402–411.

12 Pelanconi, M., Zavattoni, S., Cornolti, L. et al. (2021). Application of ceramic lattice structures to design compact, high temperature heat exchangers: material and architecture selection. *Materials (Basel)* 14 (12): 3225.

13 Rezaei, E., Barbato, M., Ortona, A., and Haussener, S. (2020). Design and optimization of a high-temperature latent heat storage unit. *Appl. Energy* 261: 114330.

14 Son, K.N., Weibel, J.A., Kumaresan, V., and Garimella, S.V. (2017). Design of multifunctional lattice-frame materials for compact heat exchangers. *Int. J. Heat Mass Transf.* 115: 619–629.

15 Ortona, A., Yoon, D.H., Fend, T. et al. (2015). Tubular Si-infiltrated SiCf/SiC composites for solar receiver application – part 2: thermal performance analysis and prediction. *Sol. Energy Mater. Sol. Cells* 140: 382–387.

16 Kawalkar, R., Dubey, H.K., and Lokhande, S.P. (2022). A review for advancements in standardization for additive manufacturing. *Mater. Today Proc.* 50: 1983–1990.

17 Solís Pinargote, N.W., Smirnov, A., Peretyagin, N. et al. (2020). Direct Ink writing technology (3D printing) of graphene-based ceramic nanocomposites: a review. *Nanomaterials* 10 (7): 1300.

18 Wong, K.V. and Hernandez, A. (2012). A review of additive manufacturing. *Int. Sch. Res. Not.* 1: 208760.

19 Zhang, M., Fang, B., and Fan, L. (2025). Optimizing powder morphology and scanning strategies to enhance shape accuracy in L-PBF alumina ceramic formation. *Ceram. Int.* 51 (6): 8042–8053.

20 Westbeek, S., van Dommelen, J.A.W., Remmers, J.J.C., and Geers, M.G.D. (2019). Influence of particle shape in the additive manufacturing process for ceramics. *Comput. Math. Appl.* 78 (7): 2360–2376.

21 Stunda-Zujeva, A., Irbe, Z., and Berzina-Cimdina, L. (2017). Controlling the morphology of ceramic and composite powders obtained via spray drying – a review. *Ceram. Int.* 43 (15): 11543–11551.

22 Messing, G.L., Zhang, S.-C., and Jayanthi, G.V. (1993). Ceramic powder synthesis by spray pyrolysis. *J. Am. Ceram. Soc.* 76 (11): 2707–2726.

23 Pratsinis, S.E. (1998). Flame aerosol synthesis of ceramic powders. *Prog. Energy Combust. Sci.* 24 (3): 197–219.

24 Argamosa, A.J., Miciano, S.L., Monis, A. et al. (2023). Recent advances in particle characterization. *Eng. Proc.* 56 (1): 84.

25 Kootahi, K., Kwan Leung, A., Jiang, Z. et al. (2024). Evaluation of the methods of particle morphology characterization: CT scanning, digital imaging and light microscopy. *Comput. Geotech.* 174: 106648.

26 Iskander, M. and Li, L. (2024). Comparison of 2D/3D DIA and μCT BT. In: *Dynamic Image Analysis of Granular Materials* (ed. M. Iskander and L. Li), 249–270. Cham: Springer.

27 Tripathi, P., Lee, S.J., Lee, C.H., and Shin, M. (2025). Towards 3D shape estimation from 2D particle images: a state-of-the-art review and demonstration. *KONA Powder Part. J.* 42: 37–56.

28 Zhang, L. and Liu, G. (2021). Mathematical modeling for ceramic shape 3D image based on deep learning algorithm. *Adv. Theor. Math. Phys.* 1: 4343255.

29 Bagheri, G.H., Bonadonna, C., Manzella, I., and Vonlanthen, P. (2015). On the characterization of size and shape of irregular particles. *Powder Technol.* 270: 141–153.

30 Mostafaei, A., Elliott, A.M., Barnes, J.E. et al. (2021). Binder jet 3D printing – process parameters, materials, properties, modeling, and challenges. *Prog. Mater. Sci.* 119: 100707.

31 Hapgood, K.P., Litster, J.D., Biggs, S.R., and Howes, T. (2002). Drop penetration into porous powder beds. *J. Colloid Interface Sci.* 253 (2): 353–366.

32 Bowen, P. (2002). Particle size distribution measurement from millimeters to nanometers and from rods to platelets. *J. Dispers. Sci. Technol.* 23 (5): 631–662.

33 Allen, T. (1990). Sieving. In: *Particle Size Measurement* (ed. T. Allen), 192–216. Dordrecht: Springer Netherlands.

34 Agrawal, Y.C., McCave, I.N., and Riley, J.B. (2010). 9 – laser diffraction size analysis. In: *Principles, Methods and Application of Particle Size Analysis*, 119–128.

35 Sommer, H.T., Harrison, C.F., and Montague, C.E. (1992). Particle size distribution from light scattering. In: *Particle Size Analysis* (ed. N.G. Stanley-Wood and R.W. Lines). The Royal Society of Chemistry.

36 Kastner, E. and Perrie, Y. (2016). Particle size analysis of micro and nanoparticles. In: *Analytical Techniques in the Pharmaceutical Sciences* (ed. A. Müllertz, Y. Perrie, and T. Rades), 677–699. New York, NY: Springer.

37 Averardi, A., Cola, C., Zeltmann, S.E., and Gupta, N. (2020). Effect of particle size distribution on the packing of powder beds: a critical discussion relevant to additive manufacturing. *Mater. Today Commun.* 24: 100964.

38 ISO 23145 (2007) ISO 23145-1:2007 fine ceramics (advanced ceramics, advanced technical ceramics) – determination of bulk density of ceramic powders. issued 2007.

39 Diener, S., Zocca, A., and Günster, J. (2021). Literature review: methods for achieving high powder bed densities in ceramic powder bed based additive manufacturing. *Open Ceram.* 8: 100191.

40 Li, M., Miao, G., Du, W. et al. (2022). Difference between powder bed density and green density for a free-flowing powder in binder jetting additive manufacturing. *J. Manuf. Process.* 84: 448–456.

41 Baesso, I., Karl, D., Spitzer, A. et al. (2021). Characterization of powder flow behavior for additive manufacturing. *Addit. Manuf.* 47: 102250.

42 Nan, W., Ge, L., He, Z. et al. (2025). Powder spreading and spreadability in powder-based additive manufacturing: state of the art and perspectives. *Powder Technol.* 449: 120393.

43 Talebi, F.A., Haydari, Z., Salehi, H. et al. (2024). Spreadability of powders for additive manufacturing: a critical review of metrics and characterisation methods. *Particuology* 93: 211–234.

44 Moghadasi, M., Du, W., Li, M. et al. (2020). Ceramic binder jetting additive manufacturing: effects of particle size on feedstock powder and final part properties. *Ceram. Int.* 46 (10): 16966–16972.
45 ISO 4324 (1977) ISO 4324:1977 surface active agents – powders and granules – measurement of the angle of repose. issued 1977.
46 Freeman, R. (2007). Measuring the flow properties of consolidated, conditioned and aerated powders – a comparative study using a powder rheometer and a rotational shear cell. *Powder Technol.* 174 (1): 25–33.
47 ASTM D6128-06 (2007). Standard test method for shear testing of bulk solids using the jenike shear cell. issued 2007.
48 Krantz, M., Zhang, H., and Zhu, J. (2009). Characterization of powder flow: static and dynamic testing. *Powder Technol.* 194 (3): 239–245.
49 Franchin, G., Elsayed, H., Botti, R. et al. (2022). Additive manufacturing of ceramics from liquid feedstocks. *Chinese J. Mech. Eng. Addit. Manuf. Front.* 1 (1): 100012.
50 ISO/ASTM DIS 52940 (2025) ISO/ASTM DIS 52940 additive manufacturing of ceramics – feedstock materials – characterization of ceramic slurry in vat photopolymerization.
51 Bhandari, S., Vajpayee, G., da Silva, L.L. et al. (2025). A review on additive manufacturing of piezoelectric ceramics: from feedstock development to properties of sintered parts. *Mater. Sci. Eng. R. Rep.* 162: 100877.
52 Bertrand, G., Roy, P., Filiatre, C., and Coddet, C. (2005). Spray-dried ceramic powders: a quantitative correlation between slurry characteristics and shapes of the granules. *Chem. Eng. Sci.* 60 (1): 95–102.
53 Rau, D.A., Bortner, M.J., and Williams, C.B. (2023). A rheology roadmap for evaluating the printability of material extrusion inks. *Addit. Manuf.* 75: 103745.
54 Halloran, J.W. (2016). Ceramic stereolithography: additive manufacturing for ceramics by photopolymerization. *Annu. Rev. Mater. Res.* 46 (1): 19–40.
55 Bae, C.-J., Ramachandran, A., and Halloran, J.W. (2018). Quantifying particle segregation in sequential layers fabricated by additive manufacturing. *J. Eur. Ceram. Soc.* 38 (11): 4082–4088.
56 Gonzalez-Gutierrez, J., Cano, S., Schuschnigg, S. et al. (2018). Additive manufacturing of metallic and ceramic components by the material extrusion of highly-filled polymers: a review and future perspectives. *Materials (Basel)* 11 (5): 840.
57 Spina, R. and Morfini, L. (2024). Material extrusion additive manufacturing of ceramics: a review on filament-based process. *Materials (Basel)* 17 (11): 2779.
58 Özden, I., Iveković, A., and Kocjan, A. (2021). Additive manufacturing of ceramics from thermoplastic feedstocks. *Open Ceram.* 6: 100129.
59 Buj, I., Vidal, D., Tejo, A. et al. (2021). Recent advances in the extrusion methods for ceramics. *IOP Conf. Ser. Mater. Sci. Eng.* 1193 (1): 12030.
60 Smirnov, A., Terekhina, S., Tarasova, T. et al. (2023). From the development of low-cost filament to 3D printing ceramic parts obtained by fused filament fabrication. *Int. J. Adv. Manuf. Technol.* 128 (1): 511–529.

61 ISO 527-1 (2019). ISO 527-1:2019 plastics – determination of tensile properties part 1: general principles. issued 2019.

62 Clemens, F., Sarraf, F., Borzì, A. et al. (2023). Material extrusion additive manufacturing of advanced ceramics: towards the production of large components. *J. Eur. Ceram. Soc.* 43 (7): 2752–2760.

63 Blanco, I. (2022). A brief review of the applications of selected thermal analysis methods to 3D printing. *Thermo* 2 (1): 74–83.

64 Travitzky, N., Bonet, A., Dermeik, B. et al. (2014). Additive manufacturing of ceramic-based materials. *Adv. Eng. Mater.* 16 (6): 729–754.

65 Olhero, S.M., Ganesh, I., Torres, P.M.C. et al. (2009). Aqueous colloidal processing of ZTA composites. *J. Am. Ceram. Soc.* 92 (1): 9–16.

66 Tijjani, Y. (2022). Quartz, glass, and glass-ceramic matrix nanocomposites containing carbon nanotubes: a review. *Bayero J. Pure Appl. Sci.* 15 (1): 1–10.

67 Zhao, Y., Zhu, J., He, W. et al. (2023). 3D printing of unsupported multi-scale and large-span ceramic via near-infrared assisted direct ink writing. *Nat. Commun.* 14 (1): 2381.

68 Sampaio, F.B.W.R., Özcan, M., Gimenez, T.C. et al. (2019). Effects of manufacturing methods on the survival rate of ceramic and indirect composite restorations: a systematic review and meta-analysis. *J. Esthet. Restor. Dent.* 31 (6): 561–571.

69 Maier, R.A., Allen, A.J., Cox, B., and Levin, I. (2025). Interlaboratory study of flexural strength in additively manufactured alumina. *J. Am. Ceram. Soc.* 108 (2): e20133.

70 Moster, A.L. and Mitchell, B.S. (2009). Hydration and proton conduction in Nafion/ceramic nanocomposite membranes produced by solid-state processing of powders from mechanical attrition. *J. Appl. Polym. Sci.* 113 (1): 243–250.

71 Fan, J., Xu, X., Niu, S. et al. (2022). Anisotropy management on microstructure and mechanical property in 3D printing of silica-based ceramic cores. *J. Eur. Ceram. Soc.* 42 (10): 4388–4395.

72 Brinckmann, S.A., Young, J.C., Fertig, R.S., and Frick, C.P. (2023). Effect of print direction on mechanical properties of 3D printed polymer-derived ceramics and their precursors. *Mater. Lett. X* 17: 100179.

73 Suominen, J.M., Frankberg, E.J., Vallittu, P.K. et al. (2019). Three-dimensional printing of zirconia: characterization of early stage material properties. *Biomater. Investig. Dent.* 6 (1): 23–31.

74 Wu, R., Wang, C., Xu, G. et al. (2025). Preparation and characteristics of porous mullite ceramics by 3D printing and in-situ synthesis. *Materials (Basel)* 18 (5): 956.

75 Mummareddy, B., Burden, E., Carrillo, J.G. et al. (2021). Mechanical perfor mance of lightweight ceramic structures via binder jetting of microspheres. *SN Appl. Sci.* 3 (4): 402.

76 Khakzad, M., Mosadegh, M., Sepasi, Z. et al. (2024). Digital light processing 3D-printed porous yttria-stabilized zirconia with high-thermal shock resistance. *Adv. Eng. Mater.* 26 (21): 2400206.

77 Xu, X., Zhou, S., Wu, J. et al. (2023). Study of alumina ceramic parts fabricated via DLP stereolithography using powders with different sizes and morphologies. *Int. J. Appl. Ceram. Technol.* 20 (2): 1167–1193.

78 Mei, Z., Lu, Y., Lou, Y. et al. (2021). Determination of hardness and fracture toughness of Y-TZP manufactured by digital light processing through the indentation technique. *Biomed. Res. Int.* 1: 6612840.

79 Tatami, J., Katayama, M., Ohnishi, M. et al. (2015). Local fracture toughness of Si_3N_4 ceramics measured using single-edge notched microcantilever beam specimens. *J. Am. Ceram. Soc.* 98 (3): 965–971.

80 Snarr, P.L., Jones, K.V., Ghezawi, E.F. et al. (2025). Apparent fracture toughness estimation of additively manufactured alumina from as-printed chevron-notched test specimens. *J. Am. Ceram. Soc.* 108 (4): e20327.

81 Sang, G., Wang, C., Zhao, Y. et al. (2022). Ni@CNTs/Al_2O_3 ceramic composites with interfacial solder strengthen the segregated network for high toughness and excellent electromagnetic interference shielding. *ACS Appl. Mater. Interfaces* 14 (3): 4443–4455.

82 Cui, J., Gong, Z., Rao, P. et al. (2022). Fracture toughness of 3Y-TZP ceramic determined by modified SCF method based on femtosecond laser. *J. Am. Ceram. Soc.* 105 (1): 614–625.

83 Cramer, C.L., Rossy, A.M., Dehoff, R.R., and Nelson, A.T. (2024). Ability of x-ray computed tomography to resolve critical flaw size in laser-based, paste stereolithography ceramic printing of alumina. *Int. J. Ceram. Eng. Sci.* 6 (5): e10233.

84 Hao-Sen, C., Xia, Y., Ya-Zhou, G., and Dai-Ning, F. (2013). Dynamic fracture toughness of piezoelectric ceramics. *J. Am. Ceram. Soc.* 96 (7): 2036–2038.

85 Tank, M., De Leon, A., Huang, W. et al. (2023). Manufacturing of stereolithographic 3D printed boron nitride nanotube-reinforced ceramic composites with improved thermal and mechanical performance. *Funct. Compos. Struct.* 5 (1): 15001.

86 Markandan, K., Zhang, Z., Chin, J. et al. (2019). Fabrication and preliminary testing of hydroxylammonium nitrate (HAN)-based ceramic microthruster for potential application of nanosatellites in constellation formation flying. *Microsyst. Technol.* 25 (11): 4209–4217.

87 Zhu, W.-J., Tian, G.-Q., Lu, Y. et al. (2019). Leaching improvement of ceramic cores for hollow turbine blades based on additive manufacturing. *Adv. Manuf.* 7 (4): 353–363.

88 Gonzalez, P., Schwarzer, E., Scheithauer, U. et al. (2019). Additive manufacturing of functionally graded ceramic materials by stereolithography. *J. Vis. Exp.* 143: e57943.

89 Yang, Y., Liu, B., Zhang, K. et al. (2024). Correction of manufacturability based on the slice for ceramic vat photopolymerization. *Proc. SPIE* 13226: 1322614.

90 Asaad, R.S. (2019). Effect of thermo-mechanical cycling on the shear bond strength of porcelain to milled and cast cobalt-chromium alloys. *Egypt. Dent. J.* 65 (3): 2709–2722.

91 Nassar, H.S. and Abo El-Mal, E.O. (2024). in vitro evaluation of the marginal adaptation of monolithic ceramic crowns fabricated with different margin designs using novel CAD/CAM material after thermomechanical aging. *Egypt. Dent. J.* 70 (1): 575–585.

92 Soliman, T.A. (2017). Flexural strength and adhesion of zirconia-reinforced lithium silicate glass ceramic to resin cement after thermo-mechanical load cycling. *Egypt. Dent. J.* 63 (3): 2785–2793.

93 Húlan, T., Štubňa, I., Al-Shantir, O., and Trník, A. (2023). The apparatus for thermomechanical analysis of clay-based ceramics. *Meas. Sci. Rev.* 23 (3): 130–135.

94 Colón Quintana, J.L., Soto Medina, S., Hernández, M., and Suárez, O.M. (2018). Study of thermomechanical properties of an Al-Zn-based composite reinforced with dodecaboride particles. *Adv. Mater. Sci. Eng.* 2018 (1): 2975234.

95 Kempen, D., Piccolroaz, A., and Bigoni, D. (2019). Thermomechanical modelling of ceramic pressing and subsequent sintering. *Int. J. Mech. Sci.* 156: 146–158.

96 Xu, X., Li, P., Ge, C. et al. (2019). 3D printing of complex-type SiOC ceramics derived from liquid photosensitive resin. *ChemistrySelect* 4 (23): 6862–6869.

97 Sänger, J.C., Riechers, B., Pauw, B.R. et al. (2024). Microplastic response of 2PP-printed ceramics. *J. Am. Ceram. Soc.* 107 (10): 6636–6645.

98 Saha, M. (2022). *On the Advanced Microstructural Characterisation of Additively Manufactured Alumina-zirconia Based Eutectic Ceramics: Overview and Outlook. ChemRxiv.*

99 Redkov, A.V., Lipovskii, A.A., and Tagantsev, D.K. (2016). Micro-Raman spectroscopy study of glass-ceramics with gradient of volume fraction of crystalline phase. *J. Am. Ceram. Soc.* 99 (8): 2558–2560.

100 Gupta, S. and Riyad, M.F. (2014). Oxidation-induced sintering: an innovative method for manufacturing porous ceramics. *Int. J. Appl. Ceram. Technol.* 11 (5): 817–823.

101 Michalik, D., Pawlik, T., Plewa, J., and Sopicka-Lizer, M. (2016). Influence of homogenization and micro/nano source of starting powders on format ion of the single YAP phase. *Arch. Metall. Mater.* 61 (4): 1753–1760.

102 Abdalla, M.M., Ali, I.A.A., Khan, K. et al. (2021). The influence of surface roughening and polishing on microbial biofilm development on different ceramic materials. *J. Prosthodont.* 30 (5): 447–453.

103 Nakai, H., Inokoshi, M., Nozaki, K. et al. (2022). Osteoblast response of additively manufactured zirconia and alumina-toughened zirconia. *Materials (Basel)* 15 (23): 8685.

104 Kaczmarek, K., Konieczny, B., Siarkiewicz, P. et al. (2022). Surface characterization of current dental ceramics using scanning electron microscopic and atomic force microscopic techniques. *Coatings* 12 (8): 1122.

105 Dong, S., Zheng, K., and Xiao, X.Z. (2014). Ultrasonic vibration assisted grinding of sintered dental zirconia ceramics: an experimental study on surface roughness. *Adv. Mater. Res.* 1017: 800–805.

106 Xing, H., Zou, B., Li, S., and Fu, X. (2017). Study on surface quality, precision and mechanical properties of 3D printed ZrO_2 ceramic components by laser scanning stereolithography. *Ceram. Int.* 43 (18): 16340–16347.

107 Rosa, A., Santoliquido, O., Camerota, F. et al. (2025). Highly complex ceramic cores for investment casting applications made possible by additive manufacturing. *Open Ceram.* 21: 100748.

108 Manière, C., Harnois, C., and Marinel, S. (2023). Sintering behavior of ultra-thin 3D printed alumina lattice structures. *Acta Mater.* 250: 118865.

109 Matos, C., Santos, S., Duarte, I. et al. (2024). Insights into morphology and mechanical properties of architected interpenetrating aluminum-alumina composites. *J. Compos. Mater.* 58 (24): 2623–2637.

110 Tiseanu, I., Craciunescu, T., Aldica, G.V., and Iovea, M. (2008). X-ray micro-tomography as a tool for quantitative characterization of advanced materials manufacturing processes. *Adv. Mater. Res.* 47–50: 698–701.

111 Yang, Z., Li, K., Ma, S. et al. (2021). Preparation, mechanical, and leaching properties of $CaZrO_3$ ceramic cores. *Int. J. Appl. Ceram. Technol.* 18 (5): 1490–1497.

112 Choi, J.H., Kim, J.S., and Cheon, C.I. (2014). Effect of process condition on the ferroelectric properties in $BiFeO_3$-(Bi,K)TiO_3 ceramics. *J. Korean Phys. Soc.* 65 (3): 382–386.

113 Panyata, S., Eitssayeam, S., Rujijanagul, G. et al. (2012). Property development of hydroxyapatite ceramics by two-step sintering. *Adv. Mater. Res.* 506: 190–193.

114 Su, D., Yan, X., and Hou, F. (2014). Fabrication of macroporous SiCN ceramics from mixed polysilazanes. *Key Eng. Mater.* 602: 384–387.

115 Pelanconi, M., Blyweert, P., Bianchi, G. et al. (2023). New, 3D binder-jetted carbons with minimal periodic surface structures. *Carbon N. Y.* 213: 118252.

116 Pelanconi, M., Bottacin, S., Bianchi, G. et al. (2024). High-strength Si–SiC lattices prepared by powder bed fusion, infiltration-pyrolysis, and reactive silicon infiltration. *J. Am. Ceram. Soc.* 107 (7): 4436–4450.

117 Pechlivani, E.M., Kastrinaki, G., Zisis, C. et al. (2023). 3D ceramic structure fabrication by Stereolithography and physicochemical characterization. *Mater. Today Proc.* 93: 54–60.

118 Khosravani, M.R. and Reinicke, T. (2020). On the use of X-ray computed tomography in assessment of 3D-printed components. *J. Nondestruct. Eval.* 39 (4): 75.

119 Diener, S., Franchin, G., Achilles, N. et al. (2021). X-ray microtomography investigations on the residual pore structure in silicon nitride bars manufactured by direct ink writing using different printing patterns. *Open Ceram.* 5: 100042.

120 Saâdaoui, M., Khaldoun, F., Adrien, J. et al. (2020). X-ray tomography of additive-manufactured zirconia: processing defects – strength relations. *J. Eur. Ceram. Soc.* 40 (8): 3200–3207.

121 Lapre, C., Brouczek, D., Schwentenwein, M. et al. (2024). Rapid non-destructive inspection of sub-surface defects in 3D printed alumina through 30 layers with 7 µm depth resolution. *Open Ceram.* 18: 100611.

9

Industrial VPP Equipment for AM of Functional Ceramic Composites

Martin Schwentenwein, Serkan Nohut, and Johannes Homa

Lithoz GmbH, Mollardgasse 85a/2/64-69, 1060, Vienna, Austria

9.1 Introduction

Ceramic additive manufacturing (AM) technologies and in particular vat photopolymerization (VPP) of ceramic materials are split into two worlds that share the same name and underlying physics but not the same expectations and requirements. On the one hand, there are industry-ready platforms that include matched machines, logged calibrations, and qualified and quality-controlled slurries. On the other hand, there are desktop or consumer platforms that significantly differ in the price point but are also focused and limited to more basic requirements. Both can make green parts using light and photocurable ceramic suspensions. Only one is designed to promise the same part next week, next quarter, and at a thousand units.

For industrial platforms, the development and manufacturing of the corresponding feedstocks/slurries follow high standards, typically in the framework of relevant industrial standards such as ISO 9001 or ISO 13485. Powders are sourced with certificates instead of assumptions; particle size distributions are tuned, solids loadings are pushed high to achieve green density, and the slurry's viscosity and cure depth need to be measured, controlled, and recorded in a highly precise manner [1]. Exposure and the corresponding curing are calibrated to yield reproducible results, not only from batch to batch but also uniformly over the whole area of the printing envelope and in between different printers. Even the seemingly simple act of recoating is engineered for ceramics, involving temperature control where needed and agitation strategies to guarantee a uniform and reproducible quality of the coated layer before exposure.

Consumer printers usually stem from the printing of neat photopolymers; they can also print ceramic-filled systems, but they typically do so within a framework that is designed for handling unfilled resins. Thus, the materials printed on these platforms exhibit lower solids loadings to help the resin stay printable, which also makes it easier for the optics and mechanics, but this also means more shrinkage to manage later and a tighter tolerance window in the furnace. Exposure settings are typically less granular, monitoring is lighter, and data capture is not the default. None of this disqualifies them; it simply defines what they are best at: quick exploration, proof-of-concept parts, and straightforward geometries where a good furnace and a careful operator can carry a lot of weight.

9.2 Materials and Rheology

Industrial ceramic VPP emphasizes powder and slurry as controlled materials, not consumables. Powders are specified by particle size distribution, morphology, impurity profile, and surface chemistry to achieve high green density at the required viscosity levels. Dispersant systems, photoinitiating systems, and the monomer backbone are tuned to achieve well-controlled curing kinetics that not only allow processing of simple and light material systems such as alumina, but also guarantee the processability of more complex material systems such as other non-oxides with higher refractive indices (e.g. zirconia), functional ceramics such as piezoceramic materials (e.g. lead-zirconate-titanate or barium titanate), or non-oxide ceramic systems (e.g. aluminum nitride or silicon nitride). Routine characterization – solids content, viscosity versus shear rate, and working pot life – ensures run-to-run stability.

Consumer workflows accept lower solids loading and simpler dispersant strategies to remain printable on general-purpose optics and mechanics. This reduces viscosity but increases sintering shrinkage and the sensitivity of final dimensions to local cure depth and green density also making process control much more difficult.

9.3 Processing

Processing of ceramics in VPP typically follows a two-stage approach across all established technologies, consisting of layer-by-layer shaping followed by thermal post-processing (TPP). The latter can be executed in a single step or can be divided into different segments (e.g. debinding and sintering separately), which mostly depends on the used ceramic, but there can also be other influencing factors such as the geometry (e.g. wall thickness) [2, 3].

Concerning the shaping stage of this process, industrial systems enforce exposure uniformity and stability through calibrated projection systems or laser optics. Layer formation is supported by recoating hardware designed for rather viscous suspensions (doctor blade-based systems, possible thermal conditioning). Exposure strategies can be used to compensate areas near fine features and thin walls to mitigate over-polymerization and preserve channels and delicate features.

Consumer printers adapted for ceramics are not developed to achieve excellent detail; uniformity of the building envelope and environmental control are typically weaker or not implemented. The result is a workable path for simple geometries and short builds, but a narrower process window for long jobs, thin sections, and features approaching the optical resolution limit.

The thermal steps determine whether optical precision translates into ceramic performance. However, the infrastructure used for this process step is not inherently linked to industrial or consumer VPP platforms, so this step should not have an impact. Nonetheless, many defects or flaws only become visible at this stage (e.g. delamination cracks, insufficient densification, and deformation); however, this does not mean that these flaws are caused by TPP, it merely brings to light flaws that were already introduced during the printing stage. Sintering takes place under controlled conditions with mapped furnaces, established loading strategies, and setter materials/coatings chosen to minimize warpage, sticking, and contamination. Anisotropic shrinkage is anticipated via geometry-specific compensation and fixturing.

9.4 Reproducibility

For industrial VPP platforms for ceramics, reproducibility is engineered rather than inferred. Each batch of slurry is tested and qualified; each print job offers the possibility to be seamlessly traced through the whole processing chain, exposure data, environmental conditions, and operator interventions to the 3D-printed green part and, in extension, to the sintered ceramic component. Debinding and sintering profiles are controlled and established protocols, and often distinguish between different geometries to allow for a more time and energy-efficient processing.

In consumer regimes, reproducibility hinges on disciplined local practice: consistent resin agitation, stable exposure settings, repeatable cleaning, and conservative thermal profiles. While teams can stabilize performance, the absence of enforced calibration schedules, automatic data logging, and frozen recipes makes it harder to demonstrate sameness across time, equipment, and operators.

Along the lines of reproducibility, scalability is also a topic that relies on a similar basis, as scaling up production only becomes feasible with a robust and reproducible process and workflow. On top of that, scaling from prototypes to series production requires more than simply scaling printers by number. Industrial platforms offer synchronized printers with harmonized calibrations, shared material pools under lot control, and coordinated green-part logistics. Furnace capacity, load configuration, and setter availability are planned to takt time; inspection and packaging are sized to avoid bottlenecks.

Consumer setups scale by multiplication of what can be done at bench scale, but hidden couplings – resin age, optical drift, green handling variability – emerge as yield loss when throughput rises. Furnaces and metrology, rather than printers, often become the first constraints, and the lack of data makes the integration into workflows difficult.

9.5 Quality Management, Compliance, and Performance Metrics

Industrial ceramic VPP fits within established quality frameworks [4–6]. ISO 9001 provides the backbone for document control, calibration, and corrective actions and is often extended by other relevant standards such as ISO 13485 (medical), AS9100 (aerospace), or automotive frameworks, which extend requirements to risk management, traceability, and validation. Equipment is typically qualified (IQ/OQ/PQ – Installation Qualification, Operational Qualification, Performance Qualification), recipes are version-controlled, and software changes are assessed for process impact. Risk analyses (e.g. FMEA) are conducted and implemented and explicitly consider ceramic-specific failure modes, and corrective/preventive actions (CAPA) are executed within the system.

Consumer environments usually cannot adopt these practices due to the associated costs that arise from this process. Where regulated markets or supplier qualification are in scope, building up these processes becomes necessary and would most likely often outweighing the initial equipment savings – if they are possible on such platforms in the first place.

For a reliable comparison between different platforms, materials, and processes, or even between batches of a given material, standardized methods and metrics are very important. However, current standards do not necessarily reflect all relevant parameters of AM-produced parts, which is why it is important to also take into account and clearly state additional characteristics of the tests:

- **Dimensional accuracy and precision:** Measuring as-printed (green) and as-sintered dimensions, through-channels, and overhangs, and calculating mean error and standard deviations. Evaluating the feature resolution (e.g. minimum resolvable channel/wall thickness); these depend on the quality of the used optical system, exposure strategies, and solids loading of the applied ceramic slurry; assessment of surface quality by quantification of surface roughness.
- **Density and microstructure:** Determining the bulk density (e.g. Archimedes' principle) and, when relevant, closed-porosity via helium pycnometry. Correlate with microstructure (polished sections) to assess grain growth, residual porosity, and defects at interfaces.
- **Mechanical properties and fractography:** Assessing the strength of the sintered specimens; it is of utmost importance not to simply state a value but also to clarify the way the parts were produced (e.g. printed standing upright or lying flat), the testing mode (e.g. three-point-bending, four-point-bending, and ball-on-three-balls), the size of the samples, and also any possible preparation steps (e.g. grinding, polishing, chamfering).

The quality of the sintered ceramic components is, of course, partially predetermined by the quality of the green part and, at that, by the quality of the photocurable ceramic slurry, which serves as feedstock for all ceramic VPP processes. Sedimentation and instability in photocurable slurries can influence the quality of the printed parts and the printability itself. These effects may manifest as property gradients along the printing direction. During printing or in the green state, such instabilities can lead to defects, including layer delamination, detachment from the build platform, or overcuring with increasing exposure time, which may ultimately result in aborted print jobs. During subsequent debinding and sintering, these issues become evident through inconsistent densification, non-uniform shrinkage along the printing axis, and the formation of delamination cracks. Industrial systems mitigate this with validated shelf-life times for the slurries, agitation/homogenization protocols, and controlled idle behaviors. Over-/undercure at fine features leads to dimensional drift or weak lamination; calibrated dose-to-depth curves and feature-aware exposure adjustments reduce risk. Cracking during debinding (e.g. delaminations or interlaminar cracks) arises when stresses due to gas evolution outpace strength; to overcome this, multistage thermal removal, special atmospheres, adequate venting, and mass-scaled ramps are essential. Sintering warpage follows anisotropic shrinkage and setter interaction; apart from that, CAD compensation, fixtures, and/or compatible setters (which might also be directly 3D printed) are standard countermeasures.

These mitigations are in principle available to consumer users too but are seldom formalized, making their effectiveness operator-dependent and less transferable across builds.

9.6 Economic and Regulatory Considerations

Industrial platforms are associated with higher capital and operating costs, but eventually reduce costs by quality through higher first-pass yield, faster process transfer, and fewer engineering iterations during qualification. Consumer platforms minimize entry cost and excel in early-stage learning. As production volumes and compliance demands rise, the balance typically shifts toward industrial systems, where documentation, repeatability, and throughput stop being optional.

In the context of regulated or critical components (medical/dental, aerospace, precision microfluidics, wear-critical ceramics), industrial platforms provide the process and quality infrastructure needed for qualification, while consumer platforms are more appropriate for exploratory designs or educational purposes, where noncritical fixtures and more generous tolerances can be applied.

9.7 Industrial Applications of Lithography-Based Ceramic AM

Industrial applications of lithography-based ceramic additive manufacturing (LCM/VPP) span biomedical, energy and aerospace, chemical processing, and emerging electronics markets. In all of these domains, industrial platforms combine high-solids slurries, calibrated exposure, and validated thermal workflows to deliver reproducible properties at fine feature scales – capabilities that are difficult to reproduce on consumer/prosumer systems at scale [7, 8].

9.7.1 Biomedical Engineering

Ceramic LCM enables patient-specific parts with demanding aesthetics and biocompatibility, including non-prep veneers and crowns in lithium disilicate as well as resorbable bone scaffolds in hydroxyapatite or tricalcium phosphate. Thin-walled veneers and core–shell crowns illustrate how industrial exposure control and shrinkage management preserve thin sections and interfaces after sintering. For resorbable bioceramics, validated debinding and sintering are essential to maintain pore architecture and, consequently, appropriate degradability while maintaining mechanical integrity [7–10].

9.7.2 Energy and Aerospace Engineering

Ceramic casting cores with intricate, multi-wall cooling architectures are a representative industrial use case. Traditional injection molding of such cores reaches geometrical limits; LCM provides smooth surfaces, precise resolution, tight channel control, and stable green-to-sintered shrinkage behavior. Processed on industrial lines, these cores survive debinding without blistering and sinter with minimized warpage, eventually supporting higher turbine blade efficiency via more effective internal cooling [8, 11, 12].

9.7.3 Chemical and Process Engineering

Chemically inert, high-temperature-capable ceramics fabricated by LCM/VPP are increasingly used for nozzle handling abrasive streams, catalyst supports, and microreactors. Additive manufacturing (AM) enables conformal channels and complex lattice or triply-periodic structures for enhanced mass and heat transfer. Industrial printers and qualified furnaces are pivotal for these applications because dimensional tolerances, leak-tightness, and microstructural homogeneity must be met consistently across batches [7–9, 12].

9.7.4 Electronics and Micro-Systems

With continued miniaturization and integration, lithography-based printing of ceramics offers a path to electrically and thermally functional components (e.g. separators, substrates, and sensor bodies) where the coefficient of thermal expansion, dielectric loss, and thermal conductivity must be tailored. Industrial LCM's optical fidelity and validated sintering provide the dimensional stability and material uniformity that microscale applications demand [7–9].

9.8 Market Considerations and Selection Guidance

Applications with small part size, high customizability, and complex internal geometry benefit most from ceramic AM. Industrial lines are preferable when yield, documentation, and throughput are decisive (e.g. regulated medical devices, series production of casting cores or microreactors). Consumer systems remain valuable for early exploration but typically lack the validated materials, exposure control, and furnace infrastructure to assure repeatability and compliance at scale [7, 8].

Multi-material AM, emerging as an advanced level of traditional AM, which typically uses a single material, involves using more than one type of material in a single printing process. This innovative approach may open new possibilities for the widespread use of ceramic-based composites in industries such as aerospace, healthcare, and automotive by providing customized high-performance ceramic components with improved functionality, tailored properties and design flexibility. The following chapter will describe basic principles of the multi-material AM of ceramic-based structures by using lithography-based ceramic manufacturing (LCM) and explain, by referencing prior results published in the literature, how this process can develop innovative products in various fields. This chapter ends with an overview of the state-of-the-art on the multi-material AM of ceramics and a discussion on the future aspects of this technology in terms of potential application areas, material combinations, and design methodologies for multi-material components.

9.9 Multi-Material Additive Manufacturing

Advanced ceramics are known for their high melting point, strength, hardness, and wear resistance, making them ideal for specialized applications. However, these same properties also make it difficult to produce near-net-shaped components using conventional manufacturing methods. Their hardness and brittleness pose significant challenges during machining, making the process both complex and costly [13]. The challenges in the realization of complex designs and high costs associated with conventional manufacturing methods have driven a growing interest in AM for producing defect-free ceramics with properties comparable to those made through conventional processes. AM technologies have been explored in laboratories around the world for approximately 25 years, with substantial efforts dedicated to developing new materials and enhancing scalability, material variety, and printing speeds [14]. While AM adoption in ceramics has lagged polymers and metals, researchers increasingly focused on the development of methods and ceramic feedstocks to produce monolithic, near-net-shape ceramic parts with controlled microstructures to achieve the same technological maturity as metal and plastic AM. Nowadays, a variety of AM technologies can be used to shape ceramics [11, 15, 16]. Choosing the most suitable AM process depends on factors like density, surface finish, size and geometry, as well as the type of ceramic. According to the ASTM, complex shape ceramic part production using the AM techniques can be classified into two groups: direct or single-step and indirect process, also known as multistep [17]. While multistep processes (e.g. sheet lamination, materials extrusion, binder jetting, VPP, and direct inkjet printing) create a green body that requires debinding and sintering to achieve the final ceramic part, the single-step processes (e.g. directed energy deposition and selective laser melting) offer faster production by skipping debinding and sintering. However, due to the thermal stresses that induce delamination and cracks in single-step processes, it is more common to see multistep processes in commercialized ceramic AM applications [18, 19].

As industries increasingly demand advanced, functional, and customized components, multi-material AM is becoming a pivotal innovation. The progression from monolithic to multi-material AM represents a significant leap in 3D-printing technology. Multi-material AM allows the simultaneous use of multiple materials in a single build process, enabling the creation of complex, functional, and highly personalized products with enhanced properties. This transition boosts design flexibility that reduces the need for assembly while allowing for the customization of properties such as electrical, thermal, and mechanical characteristics [20–22]. For example, multi-material AM for metallic biomaterials could enable advanced orthopedic treatments by improving design and functionality [23].

This integration may not completely resolve the complexity of end-use products, as it involves combining materials at different scales, each with varying interfacial energies, and incorporating multiple functionalities at the system level [20]. Unlike traditional manufacturing, multi-material AM creates complex multi-material parts directly from the CAD model, enabling the shift from standardized to customized products [24]. In addition to improving functionality and customization, multi-material AM also supports the shift toward sustainable manufacturing practices. By enabling more efficient designs and minimizing material waste, this technology aligns with the increasing demand for environmentally friendly production methods. As industries work to reduce their carbon footprint, multi-material AM provides a pathway to achieve these sustainability goals while maintaining high levels of innovation and competitiveness [25].

Multi-material components can include a material variant that creates microstructural or morphological transitions, such as porosity, density, grain size, or texturization, or a bi-material structure that combines different materials like metals, ceramics, polymers, composites, and resins. Each multi-material AM technology offers distinct advantages and drawbacks, particularly in areas such as dimensional accuracy, size limitations, material diversity, the need for post-processing, efficiency for large volumes, and specific environmental requirements [26]. Several review papers on the multi-material aspects of AM have been published in recent years, with most focusing on polymer printing [27–31], metal printing [20, 32–35], or the development of methods and applications [20, 25, 36, 37]. There are only a few articles that investigated the multi-material AM of ceramic-based components such as ceramic–ceramic and ceramic–metal material combinations [38–44].

VPP methods, which are compatible with a broad spectrum of materials, machine configurations, and light sources, are well-suited for the cost-effective, near-net-shape production of multi-material components [15, 45, 46]. VPP, a multistep process, facilitates the creation of dense structural parts with strong mechanical properties, high resolution, and smooth surface finish [17]. This chapter explores the use of lithography-based ceramic manufacturing (LCM) for the multi-material AM of ceramic-based composites.

9.10 Multi-Material AM of Ceramic Composites with LCM

9.10.1 Lithography-Based Ceramic Manufacturing (LCM)

LCM is a type of VPP technique that operates on the principle of digital light processing (DLP). It is designed for processing ceramic suspensions with varying solid loadings [47, 48]. A photosensitive ceramic slurry is exposed to blue light with a wavelength of 460 nm in the targeted areas. After each layer is cured, the building platform moves upward, with the gap distance determining the thickness of the subsequent layer. In general, the photoreactivity and the required resolution of the printed part determines the selected layer thickness. In contrast to traditional subtractive methods that carve objects from raw materials, LCM constructs its items layer by layer, utilizing only the required material. Theoretically, the LCM process is applicable to all sinterable powders since it does not necessitate any specialized powder processing methods during the preparation of the slurry. As a result, numerous customer-specific materials have been effectively processed using the LCM technique in recent years [49–54]. In LCM technology, the printed samples – referred to as green bodies – consist of ceramic particles embedded within a polymer network. These samples undergo debinding and sintering cycles, known as TPP, to transform into the final dense ceramic component. During the debinding phase, the organic binder is removed, leaving behind a porous structure. Sintering then facilitates the diffusion of adjacent ceramic grains, leading to densification and shrinkage of the final product. The shrinkage from the green body to the finished part is a crucial factor in the design of ceramic components. The LCM method offers the manufacturing of very complex and net-shaped structures that have up to 100 μm wall thicknesses [55].

9.10.2 LCM Multi-Material Printing

Lithoz has introduced an innovative platform called CeraFab Multi 2M30 (refer to Figure 9.1a), which facilitates multi-material AM. This system features a dual-vat design that allows for horizontal movement to switch between materials, a vertically moving building platform, and an integrated inline cleaning station. The vats utilized in LCM technology are specially designed that significantly minimize over-polymerization, thereby reducing material diffusion at the interfaces between two different materials. The projector optics in this printer provide a lateral resolution of 35 μm per pixel, covering a build area of 2194 × 1234 pixels.

The LCM multi-material printer operates using a bottom-up approach [15] (refer to Figure 9.1b), which is particularly well-suited for multi-material printing when utilizing a dual-vat system. This method requires only a small amount of slurry (approximately 10–15 g) in the vat to initiate the printing process, and since the entire part does not submerge in the slurry, the risk of cross-contamination between materials remains minimal. The CeraFab Multi 2M30 provides two approaches for multi-material printing (see Figure 9.1c).

In the **layer-by-layer approach**, each layer is printed using a single material, which can be either of the two available materials. This method allows for material changes along the printing direction, making it ideal for creating multilayered structures. The thickness of each material in the design can be controlled by adjusting the number of layers printed with each material. Figure 9.1d displays multicolored alumina samples in pink and white, which were produced using the layer-by-layer approach. To achieve the pink color, chromium oxide was mixed with alumina powder.

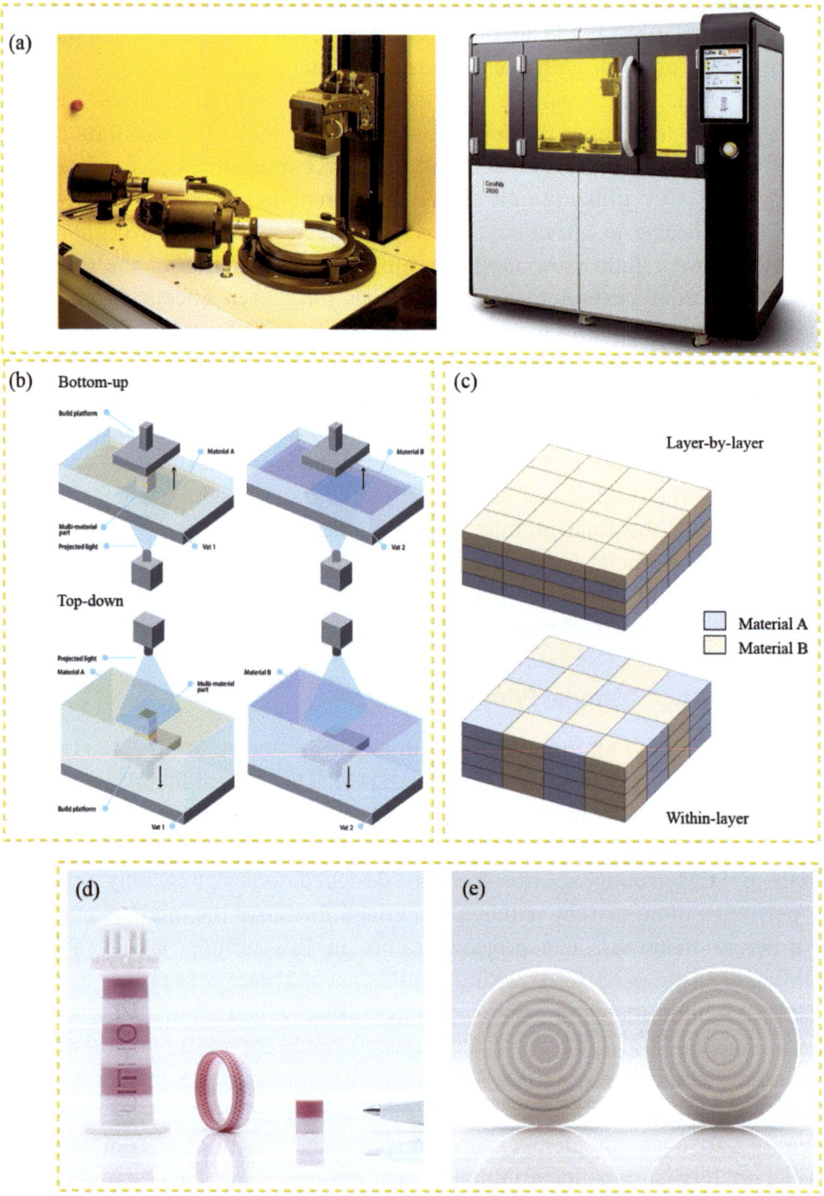

Figure 9.1 (a) CeraFab Multi 2M30 printer, (b) the bottom-up approach of LCM for multi-material printing, (c) multi-material printing approaches, (d) pink/white multi-colored alumina samples, and (e) porous/dense alumina samples. Lithoz.

In the **within-layer approach**, both materials are used within the same layer. In this method, specific pixels are cured with their respective material based on the component design's layer image. This allows for material changes in the horizontal direction across the printing plane. Figure 9.1e shows an alumina sample that combines both porous and dense alumina, produced using the within-layer approach.

Functionally graded material (FGM) is a type of composite material in which a gradient interface is introduced, creating a strong bond between two dissimilar and incompatible materials. The use of AM methods for producing FGMs can address some of the limitations of conventional manufacturing techniques, enabling the creation of complex, near-net-shaped structures with improved control over material gradients in a more cost-effective manner [56, 57]. Each of the proposed printing approaches, as well as their combinations, also facilitates the creation of functionally graded ceramics [46, 58]. Using LCM for producing functionally graded ceramics can help overcome challenges associated with TPP, particularly those caused by mismatches in the thermal expansion coefficients of dissimilar materials.

9.11 Applications for LCM Multi-Material Printing

Reports indicate that the mechanical properties and reliability of ceramic components can be enhanced through the development of multilayered ceramic architectures, which incorporate either weak or strong ceramic interfaces within a ceramic matrix [59, 60].

In the **first approach**, weak interfaces characterized by specific porosity and low strength are positioned between dense and strong ceramic layers. This arrangement allows for a deviation in crack propagation through interface delamination, resulting in improved mechanical properties by enabling a greater amount of energy to be released during the cracking process [61].

In the **second approach**, alternating layers of distinct ceramic materials are bonded with strong interfaces, allowing for the customization of in-plane residual stresses. These stresses result from the differing thermal strains in the layers as they cool from the sintering temperature [62]. Crack deflection and bifurcation are key mechanisms that contribute to the enhanced mechanical properties observed in this type of composite. When designing multilayered structures, strong interfaces are strategically placed either on the surface or embedded within the ceramic matrix, depending on the desired enhancement of properties such as mechanical strength or fracture resistance [63–65].

LCM multi-material printing has effectively facilitated the incorporation of both strong and weak interfaces within multilayered ceramic structures. The following sections will present the key findings. Additionally, complex-shaped multi-material components were printed to showcase these phenomena in practical applications.

9.11.1 Alumina/Zirconia-Toughened Alumina

Multilayered discs were fabricated by incorporating zirconia-toughened alumina (ZTA) layers, which are subjected to compressive residual stress, between alumina

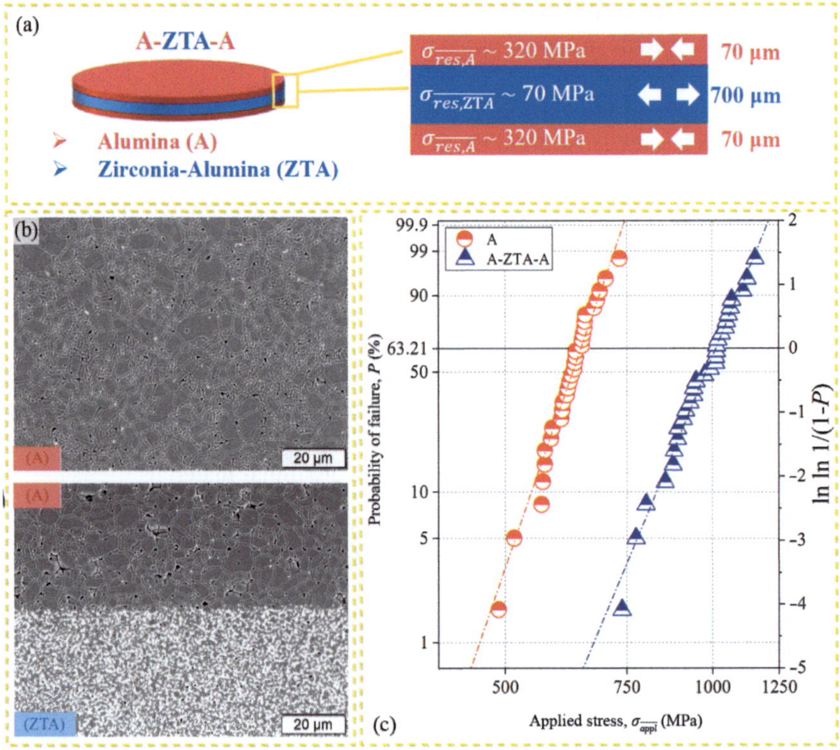

Figure 9.2 (a) A-ZTA-A multilayered discs with designed layer thicknesses and calculated residual stresses, Schlacher et al. (2021) / Elsevier / CC BY 4.0 (b) SEM images of the monolithic alumina and multilayered A-ZTA-A sample, Schlacher et al. (2021) / Elsevier / CC BY 4.0 and (c) Weibull plot of the probability of failure of the monolithic alumina and A/ZTA/A ceramic samples, Schlacher et al. (2021) / Elsevier / CC BY 4.0.

(A) surface layers (refer to Figure 9.2a). This multi-material design aims to enhance the strength of pure alumina [66]. For the printing materials, LithaLox MS548 (Lithoz GmbH, Austria) and LithaLox ZTA20 (Lithoz GmbH, Austria) were used for alumina and zirconia-toughened alumina layers respectively. LithaLox ZTA20 included a total composition of 80% vol. alumina and 20% vol. zirconia In addition to pure alumina, multilayered test samples were printed using the layer-by-layer approach, featuring a design where the A-region layers measured 70 µm on each side and the ZTA-region measured 700 µm. This design resulted in a volume ratio of V_A/V_{ZTA} of approximately 1/5. The samples were sintered at 1600 °C for 2 hours, with a heating rate of 1 °C/min.

Figure 9.2b illustrates the microstructure of the A/ZTA interface in the multi-material sample. The image reveals a distinct interface between the two materials, demonstrating no cross-contamination and highlighting the precision achieved through LCM multi-material printing. All monolithic and multi-material disc-shaped specimens were mechanically tested under biaxial loading by using the ball-on-three-balls (B3B) test [67]. Figure 9.2c presents the probability of failure (P) versus characteristic strength (σ_0) curves for both bulk alumina and the multilayered discs. The A-ZTA-A multi-material system demonstrated a characteristic strength exceeding 1 GPa, while the monolithic alumina reference exhibited a strength of approximately 650 MPa thanks to the shielding effect created by the compressive residual stresses in the alumina region [66].

The following study examined the incorporation of alumina layers, placed under compressive residual stresses, between a ZTA matrix in a multilayered design. This approach aims to enhance the thermal shock resistance of monolithic ceramics [68]. Five-layered multi-material designs (ZTA/A/ZTA/A/ZTA) as well as monolithic discs (A, ZTA) were fabricated using the layer-by-layer method in the CeraFab Multi 2M30 printer (see Figure 9.3a). The multi-material sample was designed with a volume ratio of V_A/V_{ZTA} approximately equal to 1/3 that should optimize the shielding effect in the internal layer [69]. Thermal shock tests were conducted with temperature differences varying from 200 °C to 400 °C, following the EN820–3 standards. In Figure 9.3a, the microscopic images of the cross section of the samples are given that show the subsurface damage of the A-, ZTA- and multi-material samples that are thermally shocked at a temperature difference of 400 °C. In comparison to monolithic samples, thermal shock cracks in the multi-material design are deflected upon encountering the first alumina layer, preventing the cracks from reaching the following A/ZTA interface. This behavior

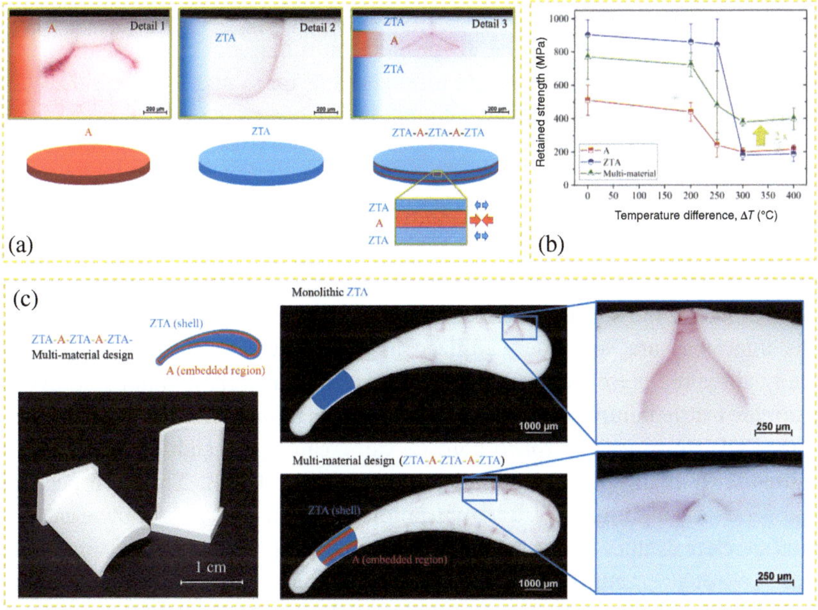

Figure 9.3 (a) Schematic representation of 3D-printed A- and ZTA-monoliths as well as the multi-material designs and their microscope images after thermal-shocked experiments at a temperature difference of $\Delta T = 400$ °C, (b) retained strength of the thermal-shocked samples as a function of the temperature difference, and (c) cross-sections of the thermal-shocked monolith ZTA and ZTA/A/ZTA/A/ZTA multi-material blades. Schlacher et al. 2024a / Elsevier / CC BY 4.0.

showcases the multi-material design's effectiveness in improving the damage tolerance of the ceramic component.

Figure 9.3b illustrates the retained thermal shock biaxial strength of both monolithic and multi-material samples as a function of temperature difference. In the samples that were not subjected to thermal shock, the ZTA exhibited the highest flexural strength. However, the retained strength of the multi-material ceramic after thermal shock is double that of the monolithic samples, which is linked to the crack-arresting properties of the embedded layers. This relatively high retained strength in the multilayer sample can be attributed to the crack-arresting effect provided by the embedded layers, which are under in-plane compressive residual stresses.

To evaluate the potential for 3D-printing of complex multilayered ceramic components with improved damage tolerance, a turbine blade (ZTA/A/ZTA/A/ZTA) was designed using a multi-material approach (refer to Figure 9.3c). Compressive internal stresses were incorporated into alumina layers, by placing them between the ZTA layers so that they can serve as crack barriers. For comparison, pure ZTA blades were also printed under identical printing and sintering conditions. The crack deflection and bifurcation capabilities of the multi-material turbine blade design, following thermal shock, are illustrated in microscopic images of the polished cross-sections of the blades in Figure 9.3c. In the case of the pure ZTA samples, surface cracks propagated freely; however, in the multi-material design, most cracks were stopped within the embedded layers. This demonstrates the "damage tolerance" concept of the 3D-printed multi-material component, which has been validated under temperature differences of up to 400 °C, while the ZTA counterpart exhibited significant degradation due to thermally induced cracks [68].

9.11.2 Porous/Dense Alumina

Incorporating weak interfaces with lower strength can also facilitate crack bifurcation and enhance the reliability of ceramic composites under specific conditions. Porous interfaces are commonly utilized as weak interfaces in this approach. Typically, since the presence of pores often serve as critical flaws that can provoke catastrophic brittle failure, it is avoided to have pores in load-bearing ceramic components. [70]. However, controlled porosity can create lightweight, strain-tolerant structures with high specific strength and reduced stiffness. This is especially advantageous for high-temperature applications, as such materials show excellent thermal shock resistance due to their low elastic moduli [71]. Preliminary investigations into porous ceramics fabrication using LCM involved adding a pore-forming agent (PFA) to the ceramic suspension [72]. Porous alumina samples – designated AP0, AP10, AP20, and AP25 – were created by blending polymethyl methacrylate (PMMA) powders with LithaLox 350 (Lithoz GmbH, Austria), a commercially available photocurable alumina slurry, at weight percentages of 10, 20, and 25 wt.% (with AP0 as the pure slurry). During debinding, PMMA decomposes, leaving voids that form the pores in the alumina structure. In Figure 9.4, the SEM images of the samples AP0, AP10, AP20, and AP25 are shown [73].

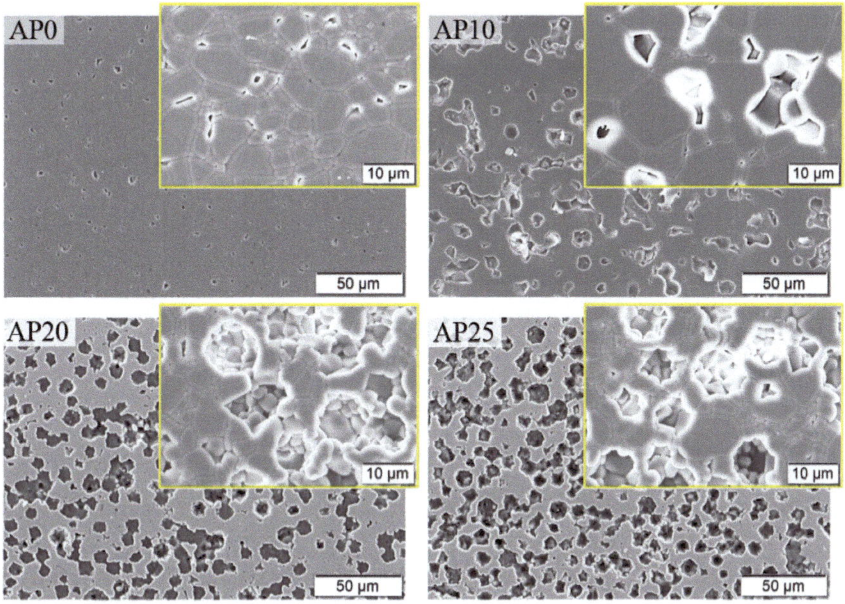

Figure 9.4 SEM images of alumina samples with different PMMA contents: AP0, AP10, AP20 and AP25. Nohut et al., 2023b / John Wiley & Sons / CC BY 4.0.

Table 9.1 The porosity of the alumina samples with different PMMA contents.

Sample	PMMA content (%)	Porosity (%)
AP0	0	2.2 ± 0.1
AP10	10	23.4 ± 1.0
AP20	20	48.9 ± 0.8
AP25	25	51.3 ± 0.9

Adapted from Ref. [73].

The porosity of the samples was measured by using the optical images taken by a light microscope. The measured porosity values are given in Table 9.1. It was observed that in AP10 (see Figure 9.4), the average size of individual pores was approximately 10 μm, comparable to the particle size of PMMA. As the weight fraction of PMMA increased to 20% and 25%, both the porosity and the interconnection between individual pores became more pronounced.

For the fabrication of porous-dense multi-material components, AP0 and AP20 ceramic slurries were utilized as dense and porous alumina, respectively and porosity-graded alumina samples were printed by using layer-by-layer and within-layer approaches that resulted in discrete and continuous gradients in the samples in the printing direction [72].

Incorporating porous interlayers between dense regions in a layered ceramic architecture has proven to be an effective strategy for improving thermal shock

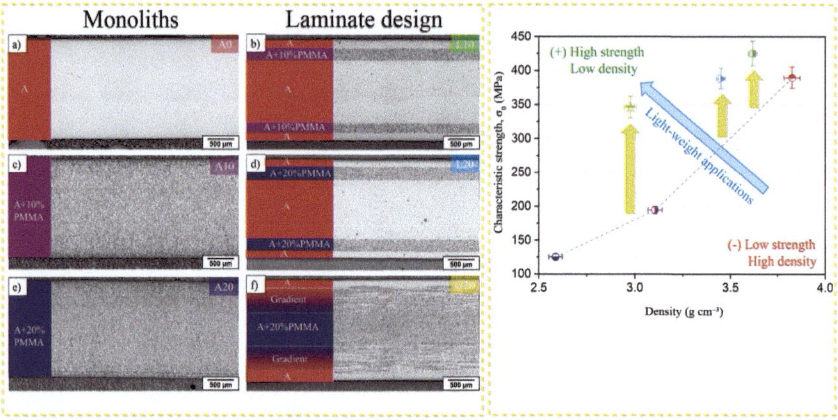

Figure 9.5 (a) SEM images of monolith and laminate disc-shaped specimens; (b) characteristic strength of the non-shocked samples as a function of their density. Schlacher et al. 2024b / Elsevier / CC BY 4.0.

resistance following rapid temperature fluctuations [74]. Disc-shaped samples were created, including monolithic variants such as A0 (pure alumina), A10 (alumina with 10% PMMA), and A20 (alumina with 20% PMMA). Additionally, laminate discs with porous interlayers were produced, namely L10 (dense alumina with an A10 porous interlayer), L20 (dense alumina with an A20 porous interlayer), and G20 (dense alumina with A20 porous gradient regions), as depicted in Figure 9.5a. The cross sections of L10 and L20 designs show sharp interfaces between pure alumina and porous interlayers, while the G20 design exhibits a continuous increase in porosity toward the center.

The biaxial bending strength, measured both before and after thermal shock, was evaluated for monolithic and multi-material designs using the B3B test. Figure 9.5b illustrates the characteristic strength of the non-shocked monolithic samples (A0, A10, A20) and the multi-material samples (L10, L20, G20) as a function of their density. Unlike porous monolithic samples (A10 and A20), the laminated samples exhibited comparable strength to pure alumina. The strength of G20 samples was slightly lower than A0, indicating that fracture in the laminates initiated near the outer pure surface and that the introduction of porous layers did not significantly reduce strength. In Figure 9.5b, the top left corner highlights the trend toward lightweight solutions, and the strength measurements support the notion that a porous-dense layered design could be an excellent option for lightweight ceramic applications. Thermal shock resistance of the monoliths and laminates was evaluated using water quench tests on five specimens per sample, with temperature differences set to approximately 200 °C, 300 °C, and 400 °C, following the EN-820–3 standard. To analyze the thermal shock cracks, samples were immersed in a liquid dye penetrant.

Figure 9.6a illustrates the retained strength as a function of temperature difference for both monolithic and layered designs. Increasing porosity notably reduces the strength of the monolithic samples. The pure A0 monolith's strength drops sharply

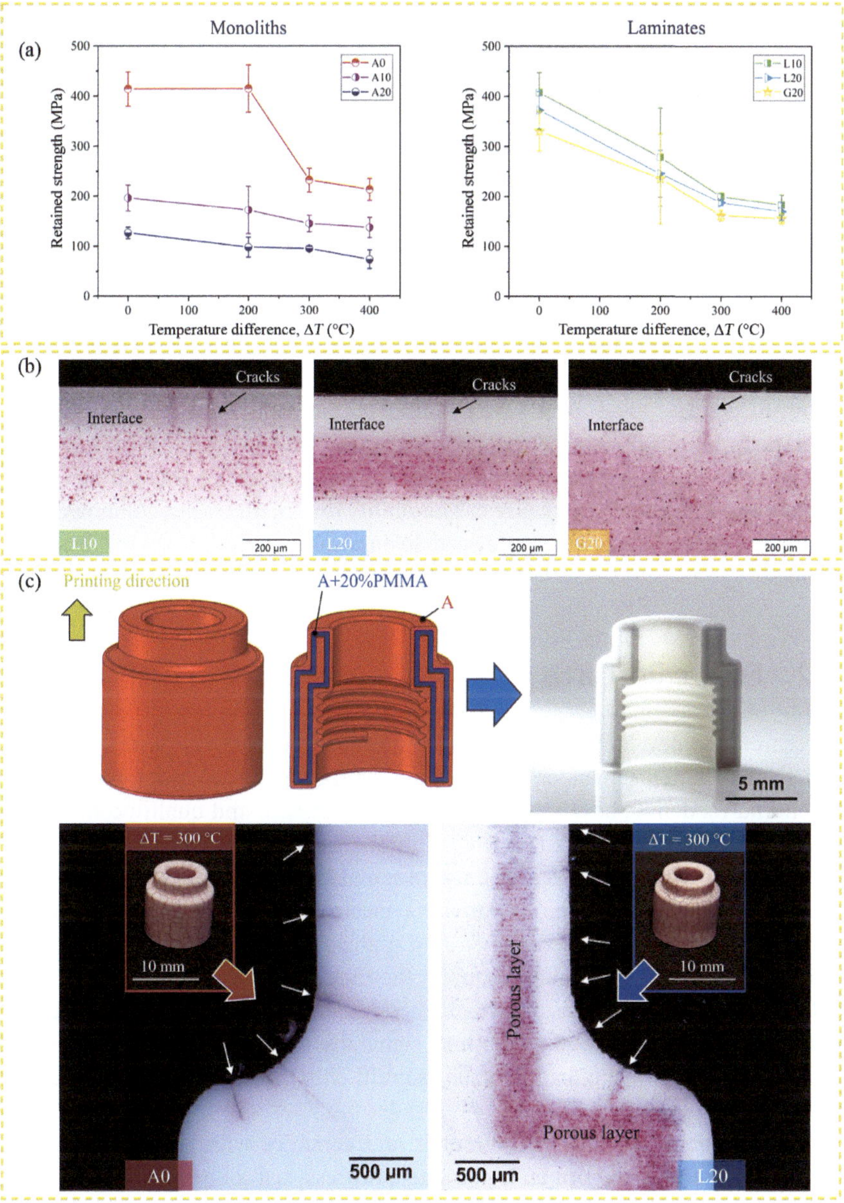

Figure 9.6 (a) Retained strength of the thermal-shocked samples versus temperature difference, (b) cross sections of the layered L10, L20, and G20 designs thermal shocked at 300 °C, and (c) a welding nozzle printed as pure alumina and as multi-material combining both dense and porous alumina regions and cross-sectional view after thermal shocking at 300 °C. Schlacher et al. 2024b / Elsevier / CC BY 4.0.

at a temperature difference of around 300 °C due to thermal shock cracks surpassing the critical temperature difference. This observation aligns with findings reported in the literature [75]. Based on the findings in Figure 9.6a, it can be hypothesized that after reaching a critical thermal shock threshold, cracks propagate into the embedded porous (weak) regions before arresting. To test this, SEM analysis was performed on the cross-sections of thermally shocked laminate designs (L10, L20, and G20) (see Figure 9.6b). The results show that surface cracks interact with the porous regions, deflecting from dense areas. Notably, the L20 design demonstrates significant deflection and bifurcation along the dense/porous interface. To showcase the impact of porous layers on thermal shock behavior, a complex welding nozzle was printed as both a monolith and a multi-material component using a two-vat system (Figure 9.6c). After thermal shock at approximately 300 °C, significant cracks (300–900 μm) were found in the monolithic alumina nozzle. In contrast, the multi-material component showed that surface cracks interacted with weak internal layers, preventing most from surpassing the porous/dense interface and resulting in crack arrest. This ability to contain cracks, combined with lightweight design potential, could lead to innovative applications in structural ceramics.

9.12 Future Research

As research and development in this area continue to evolve, it is anticipated that further enhancements in material development, printing process development, simulation and modeling, advanced thermal post-processing and quality control will lead to even more sophisticated ceramic solutions that meet the growing demands of modern engineering challenges. The use of metals and ceramics in multi-material AM is still developing [38]. The growing electronics market is increasing the demand for ceramics in devices for various applications. Low-temperature co-fired ceramics (LTCC) are a notable example; they are multilayer glass ceramic substrates co-fired with low-resistance metal conductors. LTCCs offer advantages like improved functionality, stability, minimal signal loss, and high component density, making them ideal for applications such as multi-chip modules and high-frequency modules [76, 77].

The development of slurries for metals and ceramics with similar properties, along with the ability to print both in a single process, is a complex task. Figure 9.7 shows glass ceramic – copper multi-material green-body ceramic components printed by CeraFab Multi 2M30. The primary challenge in multi-material AM arises from the differing thermal properties of ceramics and metals. Establishing a strong bond between the two is particularly difficult due to the significant mismatch in melting points and thermal expansion, which can lead to thermal stress. These stresses could have negative stresses on mechanical properties and limit the applications. [44, 78]. Additionally, the TPP of most metal–ceramic multi-material components necessitates various debinding and sintering atmospheres to prevent oxidation. Therefore, increased efforts are needed to focus on the co-sintering of metal–ceramic multi-material components.

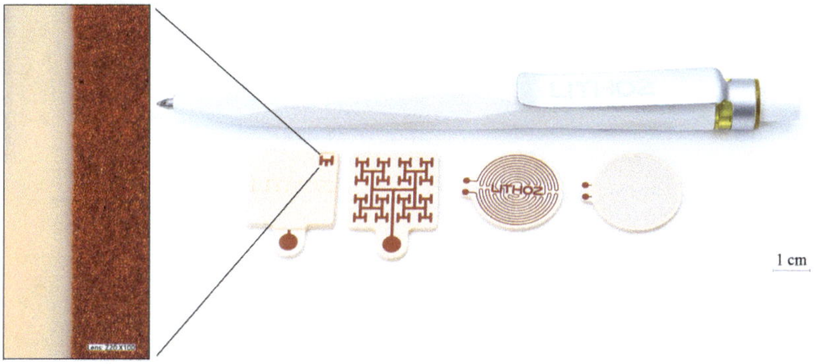

Figure 9.7 Copper-glass ceramic multi-material components in green state produced by LCM.

The multi-vat printing setup enables the assignment of different materials to any predefined layer, including the ability to use multiple materials within the same layer. This flexibility facilitates the creation of complex gradients both parallel and perpendicular to the deposition direction. When high precision is essential, it is crucial to implement rigorous in-line quality control measures throughout the printing process. This includes monitoring factors such as curing depth, over-polymerization, and cross-contamination, both during and after printing. Mid-infrared optical coherence tomography (mid-IR OCT) was used to analyze porous and dense alumina samples in both green and sintered states [79]. This nondestructive technique provides 3D structural imaging of complex specimens. A porous/dense alumina sample was printed using a discrete gradient, as shown in Figure 9.8a, b. The porosity maps of the green and sintered parts (Figure 9.8c, d) clearly distinguish the materials, with the porosity profile illustrating varying density and sufficient contrast. The differences in density allow for the analysis of material transitions. Integrating the OCT measurement head into a 3D ceramic printer for in-line monitoring could help identify defects like air bubbles and adhesion issues, enhancing production efficiency by enabling quicker detection of potential problems and reducing costs.

Multi-material AM has expanded design possibilities and underscored the necessity for more efficient design methodologies and tools. Given that materials with differing properties are printed together, the importance of design in multi-material AM is magnified. There is a growing demand for intelligent tools that facilitate not only the simulation of printing and thermal processing behavior but also the design, preparation of layer images, and slicing. New structures and architectures must be developed to leverage the unique property design space unlocked by these advanced manufacturing capabilities. Currently, STL files are predominantly used for monolithic AM across various materials. However, this file format fails to account for material attributes, color, gradients, or substructures in parts intended for multi-material AM. For certain multi-material AM processes, there is a requirement to utilize two separate STL files and potentially merge

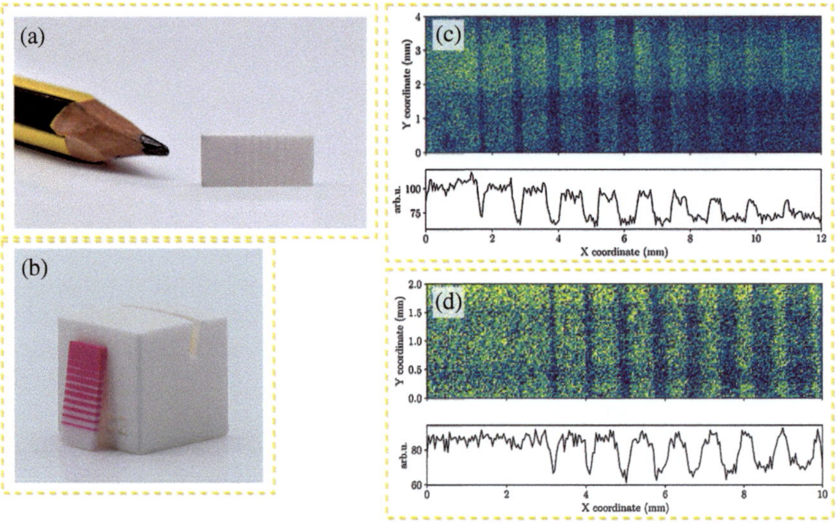

Figure 9.8 (a) Porous/dense alumina green-body sample, (b) sintered sample dipped into crack-infiltration liquid, where the porous regions are colored pink (Lithoz), image of the top interface and the porosity map and porosity profile, (c) of the green-body, and (d) sintered sample. Zorin et al. 2022 / Elsevier / CC BY 4.0.

them. The slicing process should be capable of recognizing and accommodating the assignment of these two STL files to different materials [20]. The development of tools for generating gradients in CAD or STL files has become increasingly important, particularly for FGMs. Recently, the Fraunhofer Institute for Computer Graphics Research IGD introduced a new software called GraMMaCAD, which enables the integration of smooth material gradients into traditional CAD models. GraMMaCAD allows users to add material gradients to CAD models based on surfaces, between subparts, and using auxiliary geometries. This digital technology facilitates the definition of material gradients within virtual products, enabling interactive adjustments of grading characteristics to achieve functional grading on CAD models, independent of the CAD system used. In the framework of a collaboration with Fraunhofer Institute for Computer Graphics Research IGD, the porosity-graded alumina samples were printed as shown in Figure 9.9.

Figure 9.9 Porosity gradient alumina samples. The design of gradient and preparation of layer images were performed by GraMMaCAD, and the samples were printed with A0 and A20 alumina material combination on the CeraFab Multi 2M30 printer.

9.13 Summary

Industrial and consumer ceramic VPP share the underlying physical principles, but the technical execution differs significantly. Industrial systems integrate material control, calibrated optics, validated thermal processing, and comprehensive quality management to deliver reproducibility and scalability. Consumer systems provide low entry barriers through low prices but are limited to a smaller number of accessible ceramics, which are light in color and exhibit low refractive indices. Due to lower solids loadings in the used suspension, full densification is limited, and users typically have to make compromises in the context of the achievable material properties and/or the homogeneity and isotropy of the resulting parts, and consumer platforms often rely heavily on operator discipline and iterative tuning. The appropriate choice follows from the product's intended environment: where sameness must be demonstrated rather than asserted – and where production volumes or audits are expected – industrial ceramic VPP is not merely advantageous; it is enabling.

In this chapter, we have also provided a more detailed overview of the significant capabilities of multi-material AM utilizing LCM technology, particularly in the context of developing ceramic–ceramic multi-material solutions that enhance the mechanical properties and reliability of ceramic components. Through a series of experiments and case studies, it has been demonstrated that the proposed multi-material printing techniques allow for the creation of design architectures that feature both discrete and gradient material transitions. This flexibility in design is applicable not only parallel to the printing direction but also perpendicular, opening a wide range of possibilities for engineers and designers. The CeraFab Multi 2M30 multi-material printing platform has proven to be capable of producing complex-shaped components with exceptional resolution. This advanced technology enables the integration of different materials within a single print job, allowing for the strategic placement of various material properties where they are most needed. The discussions highlighted the importance of understanding the interactions between different materials during the printing process and how these interactions can affect the overall performance of the components. The findings suggest that by carefully designing the interfaces and selecting appropriate material combinations, it is possible to achieve improved damage tolerance and reliability, which are critical factors in the performance of ceramic components in demanding applications. In summary, the future of multi-material AM in ceramics holds immense potential for innovation and transformation. As technology matures, it will not only enhance the performance and functionality of ceramic components but also contribute to more sustainable and efficient manufacturing practices. The ongoing exploration of new materials, design methodologies, and integration with smart technologies will pave the way for groundbreaking applications across diverse industries, ultimately reshaping the landscape of ceramics in manufacturing.

Acknowledgments

We gratefully acknowledge the collaboration with the Technical University of Leoben – Chair of Structural and Functional Ceramics on multi-material LCM 3D printing, and the support of Fraunhofer Institute for Computer Graphics Research IGD for providing the use of GraMMaCAD (software) for the generation of layer images for the fabrication of porosity-graded alumina samples.

References

1 Halloran, J.W. (2016). Ceramic stereolithography: additive manufacturing for ceramics by photopolymerization. *Annu. Rev. Mater. Res.* 46: 19–40. annualreviews.org.

2 Bove, A., Calignano, F., Galati, M. et al. (2022). Photopolymerization of ceramic resins by stereolithography: a review. *Appl. Sci.* 12 (7): 3591.

3 Křížová, M., Safka, J.I., Seidl, M. et al. (2020). Ceramic 3D printing: comparison of SLA and DLP technologies (testing Formlabs Form 2 Ceramic Resin and Tethon3D materials). *MM Sci. J.* 2020: 3905.

4 ISO (2015). ISO 9001:2015 quality management systems – requirements.

5 ISO (2016). ISO, ISO 13485:2016 medical devices – quality management systems – requirements.

6 IAQG (2016). IAQG, 9100:2016 quality management systems – requirements for aviation, space and defense organizations (AS9100D) (overview and guidance).

7 Zocca, A., Colombo, P., Gomes, C.M., and Günster, J. (2015). Additive manufacturing of ceramics: issues, potentialities, and opportunities. *J. Am. Ceram. Soc.* 98 (7): 1983–2001.

8 Stampfl, J., Schwentenwein, M., and Homa, J. (2023). Lithography-based additive manufacturing of ceramics: materials, applications and perspectives. *MRS Commun.* 13: 786–794.

9 Chen, Z., Li, Z., Li, J. et al. (2019). 3D printing of ceramics: a review. *J. Eur. Ceram. Soc.* 39 (2–3): 661–687.

10 Arora, P., Mostafa, K.G., Russell, E. et al. (2023). Shrinkage compensation and effect of building orientation on dimensional accuracy of SLA-AM ceramics. *Materials* 16 (19).

11 Travitzky, N., Bonet, A., Dermeik, B. et al. (2014). Additive manufacturing of ceramic-based materials. *Adv. Eng. Mater.* 16 (6): 729–754.

12 Santoliquido, O., Camerota, F., Ortona, A. et al. (2021). Influence of topology on DLP 3D printing, debinding and sintering behavior of alumina. *Addit. Manuf. Lett.* 1.

13 Lakhdar, Y., Tuck, C., Binner, J. et al. (2021). Additive manufacturing of advanced ceramic materials. *Prog. Mater. Sci.* 116: 100736.

14 Hashemi, S.M.H., Babic, U., Hadikhani, P., and Psaltis, D. (2020). The potentials of additive manufacturing for mass production of electrochemical energy systems. *Curr. Opin. Electrochem.* 20: 54–59.

15 Wang, J.C., Dommati, H., and Hsieh, S.J. (2019). Review of additive manufacturing methods for high-performance ceramic materials. *Int. J. Adv. Manuf. Technol.* 103: 2627–2647.

16 Moritz, T. and Maleksaeedi, S. (2018). 4 – additive manufacturing of ceramic components. In: *Additive Manufacturing: Materials, Processes, Quantifications and Applications* (ed. J. Zhang and Y.G. Jung), 105–161. Butterworth-Heinemann.

17 de Camargo, I.L., Fortulan, C.A., and Colorado, H.A. (2022). A review on the ceramic additive manufacturing technologies and availability of equipment and materials. *Cerâmica* 68: 329–347.

18 Dadkhah, M., Tulliani, J.M., Saboori, A., and Iuliano, L. (2023). Additive manufacturing of ceramics: advances, challenges, and outlook. *J. Eur. Ceram. Soc.* 43 (15): 6635–6664.

19 Galante, R., Figueiredo-Pina, C.G., and Serro, A.P. (2019). Additive manufacturing of ceramics for dental applications: a review. *Dent. Mater.* 35 (6): 825–846.

20 Zheng, X., Williams, C., Spadaccini, C.M., and Shea, K. (2021). Perspectives on multimaterial additive manufacturing. *J. Mater. Res.* 36: 3549–3557.

21 Yuan, C., Wang, F., Qi, B. et al. (2020). 3D printing of multimaterial composites with tunable shape memory behavior. *Mater. Des.* 193: 108785.

22 Pusateri, V., Goulas, C., and Olsen, S.I. (2023). Technical challenges and future environmentally sustainable applications for multi-material additive manufacturing for metals. In: *Advances in 3D Printing* (ed. A. Sharma). IntechOpen.

23 Putra, N.E., Mirzaali, M.J., Apachitei, I. et al. (2020). Multi-material additive manufacturing technologies for Ti-, Mg-, and Fe-based biomaterials for bone substitution. *Acta Biomater.* 109: 1–20.

24 Mahamood, R.M. and Akinlabi, E.T. (2016). Achieving mass customization through additive manufacturing. In: *Advances in Ergonomics of Manufacturing: Managing the Enterprise of the Future. Advances in Intelligent Systems and Computing* (ed. C. Schlick and S. Trzcieliński), 385–390. Springer.

25 Han, D. and Lee, H. (2020). Recent advances in multi-material additive manufacturing: methods and applications. *Curr. Opin. Chem. Eng.* 28: 158–166.

26 Hasanov, S., Alkunte, S., Rajeshirke, M. et al. (2022). Review on additive manufacturing of multi-material parts: progress and challenges. *J. Manuf. Mater. Process.* 6: 4.

27 Zheng, Y., Zhang, W., Baca Lopez, D.M., and Ahmad, R. (1957). Scientometric analysis and systematic review of multi-material additive manufacturing of polymers. *Polymers* 13 (12).

28 Brancewicz-Steinmetz, E. and Sawicki, J. (2022). Bonding and strengthening the PLA biopolymer in multi-material additive manufacturing. *Materials* 15 (16): 5563.

29 Shaukat, U., Rossegger, E., and Schlögl, S. (2022). A review of multi-material 3D printing of functional materials via vat photopolymerization. *Polymers* 14 (12): 2449.

30 Boulaala, M., Elmessaoudi, D., Buj-Corral, I. et al. (2020). Towards design of mechanical part and electronic control of multi-material/multicolor fused deposition modeling 3D printing. *Int. J. Adv. Manuf. Technol.* 110: 45–55.

31 García-Collado, A., Blanco, J.M., Gupta, M.K., and Dorado-Vicente, R. (2022). Advances in polymers based multi-material additive-manufacturing techniques: state-of-art review on properties and applications. *Adv. Manuf.* 50: 102577.

32 Wei, C. and Li, L. (2021). Recent progress and scientific challenges in multi-material additive manufacturing via laser-based powder bed fusion. *Virtual Phys. Prototyp.* 16 (3): 347–371.

33 Schneck, M., Horn, M., Schmitt, M. et al. (2021). Review on additive hybrid- and multi-material manufacturing of metals by powder bed fusion: state of technology and development potential. *Prog. Addit. Manuf.* 6: 881–894.

34 Demir, A.G., Kim, J., Caltanissetta, F. et al. (2022). Enabling multi-material gradient structure in laser powder bed fusion. *J. Mater. Process. Technol.* 301: 117439.

35 Wei, C., Zhang, Z., Cheng, D. et al. (2020). An overview of laser-based multiple metallic material additive manufacturing: from macro-to micro-scales. *Int. J. Extreme Manuf.* 3: 012003.

36 Yao, L., Ramesh, A., Xiao, Z. et al. (2023). Multimetal research in powder bed fusion: a review. *Materials* 16 (12): 4287.

37 Rafiee, M., Farahani, R.D., and Therriault, D. (2020). Multi-material 3D and 4D printing: a survey. *Adv. Sci.* 7 (12): 1902307.

38 Pfeiffer, S., Lorenz, H., Fu, Z. et al. (2018). Al_2O_3/Cu-O composites fabricated by pressureless infiltration of paper-derived Al_2O_3 porous preforms. *Ceram. Int.* 44 (17): 20835–20840.

39 Veron, F., Lanoue, F., Baco-Carles, V. et al. (2022). Selective laser powder bed fusion for manufacturing of 3D metal-ceramic multi-materials assemblies. *Adv. Manuf.* 50: 102550.

40 Koopmann, J., Voigt, J., and Niendorf, T. (2019). Additive manufacturing of a steel–ceramic multi-material by selective laser melting. *Metall. Mater. Trans. B* 50: 1042–1051.

41 Hu, K., Zhao, P., Li, J., and Lu, Z. (2022). High-resolution multiceramic additive manufacturing based on digital light processing. *Adv. Manuf.* 54: 102732.

42 Zhang, D., Jonhson, W., Herng, T.S. et al. (2022). High temperature co-firing of 3D-printed AlZnO/Al_2O_3 multi-material two-phase flow sensor. *J. Mater.* 8 (3): 710–718.

43 Gheisari, R., Chamberlain, H., Chi-Tangyie, G. et al. (2020). Multi-material additive manufacturing of low sintering temperature $Bi_2Mo_2O_9$ ceramics with Ag floating electrodes by selective laser burnout. *Virtual Phys. Prototyp.* 15 (2): 133–147.

44 Verma, A., Kapil, A., Klobčar, D., and Sharma, A. (2023). A review on multiplicity in multi-material additive manufacturing: process, capability, scale, and structure. *Materials* 16 (15): 5246.

45 Zhang, Y., Li, S., Zhao, Y. et al. (2017). Digital light processing 3D printing of AlSi10Mg powder modified by surface coating. *Adv. Manuf.* 39: 101897.

46 Nohut, S. and Schwentenwein, M. (2022). Vat photopolymerization additive manufacturing of functionally graded materials: a review. *J. Manuf. Mater. Process.* 6 (1): 17.

47 Schwentenwein, M. and Homa, J. (2014). Additive manufacturing of dense alumina ceramics. *Int. J. Appl. Ceram. Technol.* 12: 1–7.

48 Schwentenwein, M., Schneider, P., and Homa, J. (2014). Lithography-based ceramic manufacturing: a novel technique for additive manufacturing of high-performance ceramics. *Adv. Sci. Technol.* 88: 60–64.

49 Altun, A.A., Ertl, F., Marechal, M. et al. (2021). Additive manufacturing of lunar regolith structures. *Open Ceram.* 5: 100058.

50 Rauchenecker, J., Rabitsch, J., Schwentenwein, M., and Konegger, T. (2021). Additive manufacturing of aluminum nitride ceramics with high thermal conductivity via digital light processing. *Open Ceram.* 9: 100215.

51 Schwarzer, E., Götz, M., Markova, D. et al. (2017). Lithography-based ceramic manufacturing (LCM) – viscosity and cleaning as two quality influencing steps in the process chain of printing green parts. *J. Eur. Ceram. Soc.* 37 (16): 5329–5338.

52 Solórzano-Requejo, W., Cendrero, A.M., Altun, A.A. et al. (2024). Topology optimisation and lithography-based ceramic manufacturing of short-stem hip prostheses with enhanced biomechanical and mechanobiological performance. *Virtual Phys. Prototyp.* 19 (1): e2387280.

53 Unkovskiy, A., Beuer, F., Metin, D.S. et al. (2022). Additive manufacturing of lithium disilicate with the LCM process for classic and non-prep veneers: preliminary technical and clinical case experience. *Materials* 15 (17): 6034.

54 Harrer, W., Schwentenwein, M., Lube, T., and Danzer, R. (2017). Fractography of zirconia-specimens made using additive manufacturing (LCM) technology. *J. Eur. Ceram. Soc.* 37 (14): 4331–4338.

55 Scheithauer, U., Schwarzer, E., Moritz, T., and Michaelis, A. (2018). Additive manufacturing of ceramic heat exchanger: opportunities and limits of the lithography-based ceramic manufacturing (LCM). *J. Mater. Eng. Perform.* 24: 14–20.

56 Bhavar, V., Kattire, P., Thakare, S. et al. (2017). A review on functionally gradient materials (FGMs) and their applications. *IOP Conf. Ser.: Mater. Sci. Eng.* 229: 012021.

57 Udupa, G., Rao, S.S., and Gangadharan, K.V. (2014). Functionally graded composite materials: an overview. *Procedia Mater. Sci.* 5: 1291–1299.

58 Nohut, S., Geier, S., Kraleva, I. et al. (2023a). Functionally graded ceramics by lithography-based ceramic manufacturing (LCM). *J. Manuf. Mater. Process.* 2 (1): 678.

59 Clegg, W.J., Kendall, K., Alford, N. et al. (1990). A simple way to make tough ceramics. *Nature* 347: 455–457.

60 Minatto, F.D., Milak, P., De Noni, A. et al. (2015). Multilayered ceramic composites – a review. *Adv. Appl. Ceram.* 114 (3): 127–138.

61 Lugovy, M., Slyunyayev, V., Subbotin, V. et al. (2011). Mechanical behavior and failure mechanisms of boron carbide based three-layered laminates with weak interfaces. *Ceram. Int.* 37 (7): 2255–2261.

62 Hillman, C., Suo, Z., and Lange, F.F. (1996). Cracking of laminates subjected to biaxial tensile stresses. *J. Am. Ceram. Soc.* 79 (8): 2127–2133.

63 Pascual, J., Lube, T., and Danzer, R. (2008). Fracture statistics of ceramic laminates strengthened by compressive residual stresses. *J. Eur. Ceram. Soc.* 28 (8): 1551–1556.

64 Bermejo, R. (2017). "Toward seashells under stress": bioinspired concepts to design tough layered ceramic composites. *J. Eur. Ceram. Soc.* 37 (13): 3823–3839.

65 Chang, Y., Bermejo, R., and Messing, G.L. (2014). Improved fracture behavior of alumina microstructural composites with highly textured compressive layers. *J. Am. Ceram. Soc.* 97 (11): 3643–3651.

66 Schlacher, J., Hofer, A.K., Geier, S. et al. (2021). Additive manufacturing of high-strength alumina through a multi-material approach. *Open Ceram.* 5: 100082.

67 Boerger, A., Supancic, P., and Danzer, R. (2002). The ball on three balls test for strength testing of brittle discs: stress distribution in the disc. *J. Eur. Ceram. Soc.* 22 (9): 1425–1436.

68 Schlacher, J., Geier, S., Schwentenwein, M., and Bermejo, R. (2024a). Towards 3D-printed alumina-based multi-material components with enhanced thermal shock resistance. *J. Eur. Ceram. Soc.* 44 (4): 2294–2303.

69 Sestakova, L., Bermejo, R., Chlup, Z., and Danzer, R. (2011). Strategies for fracture toughness, strength and reliability optimisation of ceramic-ceramic laminates. *Int. J. Mater. Res.* 102 (6): 613–626.

70 Green, D.J. and Colombo, P. (2003). Cellular ceramics: intriguing structures, novel properties, and innovative applications. *MRS Bull.* 28: 296–300.

71 Greil, P. (2002). Advanced engineering ceramics. *Adv. Mater.* 14 (10): 709–716.

72 Nohut, S., Geier, S., Kraleva, I. et al. (2022). Lithography-based additive manufacturing of porosity graded alumina. *Addit. Manuf. Lett.* 3: 100060.

73 Nohut, S., Schlacher, J., Kraleva, I. et al. (2023b). 3D-printed alumina-based ceramics with spatially resolved porosity. *Int. J. Appl. Ceram. Technol.* 21 (1): 89–104.

74 Schlacher, J., Mateus, L.B., Nohut, S. et al. (2024b). Thermal shock resistant 3D-printed ceramic components through spatially tailored porosity. *Addit. Manuf.* 96: 104582.

75 Shen, L., Liu, M., Liu, X., and Li, B. (2007). Thermal shock resistance of the porous Al_2O_3/ZrO_2 ceramics prepared by gelcasting. *Mater. Res. Bull.* 42 (12): 2048–2056.

76 Sebastian, M.T. and Jantunen, H. (2008). Low loss dielectric materials for LTCC applications: a review. *Int. Mater. Rev.* 53 (2): 57–90.

77 Ullah, U., Mahyuddin, N., Arifin, Z. et al. (2015). Antenna in LTCC technologies: a review and the current state of the art. *IEEE Antenna Propag. Mag.* 57 (2): 241–260.

78 Sahasrabudhe, H. and Bandyopadhyay, H. (2018). In situ reactive multi-material Ti_6Al_4V calcium phosphate-nitride coatings for bio-tribological applications. *J. Mech. Behav. Biomed. Mater.* 85: 1–11.

79 Zorin, I., Brouczek, D., Geier, S. et al. (2022). Mid-infrared optical coherence tomography as a method for inspection and quality assurance in ceramics additive manufacturing. *Open Ceram.* 12: 100311.

10

Applications and Impacts of AM of Ceramics

Kareen Malsallez[1], Maxence Bourjol[1], Natalia Kostretsova[2], Marc Torrell[2], Albert Tarancón[2,3], Malte Hartmann[4], Johannes Noll[4], Georgia Astbury[5], Matthew Drane[5], and Elsa Antunes[5]

[1] *3DCeram, 4 rue du parc maison rouge, 87280, Limoges, France*
[2] *Catalonia Institute for Energy Research (IREC), Department of Advanced Materials for Energy, Jardins de les Dones de Negre, 1, Sant Adrià de Besòs, 08930, Barcelona, Spain*
[3] *ICREA, Passeig Lluís Companys 23, 08010, Barcelona, Spain*
[4] *Bosch Advanced Ceramic, Sinto Advanced Ceramics Europe GmbH Building B306, Robert-Bosch-Str. 1, 87509, Stuttgart, Germany*
[5] *James Cook University, College of Science and Engineering, 1 James Cook Drive Townsville, 4814, Townsville, Australia*

Additive manufacturing (AM) of ceramics represents a paradigm shift in advanced materials processing, transcending the geometric and functional limitations inherent in conventional ceramic fabrication. This chapter examines how AM technologies, particularly stereolithography (SLA), digital light processing (DLP), and vat photopolymerization (VPP), unlock unprecedented design freedom, enabling complex monolithic structures, integrated multifunctionality, and mass customization unattainable. By eliminating tooling constraints and minimizing material waste, ceramic AM disrupts established manufacturing paradigms while accelerating innovation across advanced sectors. The discussion is structured to demonstrate AM's cross-industry impact. We begin with aerospace and defense, where lightweight, high-strength ceramic components, such as satellite mirrors with internal cooling channels and radiofrequency antennas, enhance performance while reducing system mass. Next, catalytic reactors with microfluidic architectures and semiconductor handling systems consolidate multipart assemblies into single printed units. Biomedical advancements showcase patient-specific implants and surgical tools featuring internal channels for sensors or fluid delivery. Finally, detailed case studies are explored through environmental protection and energy applications, such as 3D-printed coral settlement substrates for ecological restoration and solid oxide cells (SOCs) with corrugated electrolytes that boost active area by 60%. Underpinning these innovations are digital workflows and AI-driven process optimization, which ensure reproducibility and scalability. This chapter provides insights into how ceramic AM redefines material efficiency, supply-chain resilience, and functional design, positioning it as a cornerstone of next-generation industrial and sustainable technologies.

10.1 Aerospace and Defense Applications

Technical ceramics have become essential in high-performance industries – most notably aerospace, defense, and semiconductors – owing to their remarkable properties: high mechanical strength, thermal stability, and chemical resistance. For decades, shaping these materials remained a challenge, restricting their use to basic geometries and low-stress applications. Over the past three years, AM – and particularly SLA – has revolutionized its design and industrial potential, opening the door to complex, functional, and lightweight components that could not be realized otherwise.

Aerospace now represents the most significant revenue opportunities for 3D-printed technical ceramics. Market projections anticipate these sectors will reach about USD 764 million by 2030, driven by the need for lighter, more integrated, and rapidly customizable components. AM's advantages – design freedom, integrated functionalities, reduction in waste, and lead time – are driving adoption for mission-critical, high-value applications. With ongoing developments in printer platforms, process automation, and materials science, 3D-printed ceramics are poised to meet the most demanding industrial challenges of the coming decade.

10.1.1 Aerospace Applications: Large-Scale, Lightweight, and Integrated Structures

In space optics and unmanned aerial vehicle (UAV) payloads, reducing mass and integrating multiple functionalities into a single compact structure are critical. Conventional manufacturing typically involves six steps (blank shaping, machining, grinding, polishing, coating, and interface integration), with machining and grinding generating up to 90% waste of the blank's initial mass – a costly and risky process.

Ceramic AM overcomes these limitations through three key innovations:

- **Lightweight and disruptive designs:** 3D printing enables semi-closed, ribbed, or latticed mirrors, allowing substantial weight reduction and robust, custom-tailored architectures.
- **Functional integration:** AM facilitates internal cooling channels, thermal insulators, and direct integration of optical and structural functions.
- **Size and complexity:** the latest SLA platforms (e.g. 3DCERAM's C3601 Ultimate) print optical substrates up to 600 mm × 600 mm × 300 mm, enabling the manufacture of large, monolithic mirrors for space and scientific instruments (Figure 10.1).

Figure 10.1 3D-printed mirror for satellite.

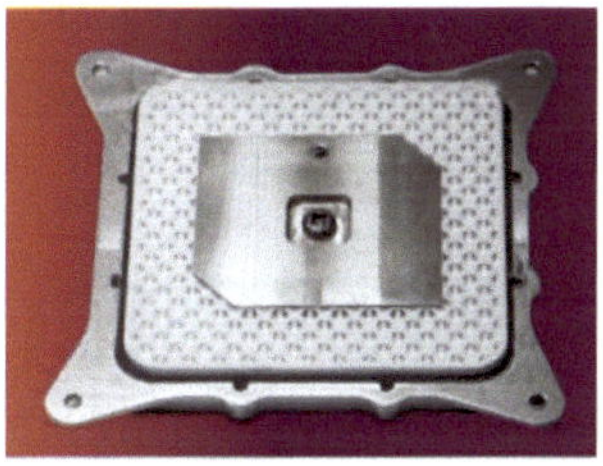

Figure 10.2 3D-printed RF antenna for satellite.

10.1.2 Space Communication: Ceramic Antennas and NewSpace Innovation

AM of ceramic antennas – such as those developed with the French company ANYWAVES (Figure 10.2) – demonstrates the method's strength for miniaturized, high-frequency satellite communications. The integration of lattice structures using high-permittivity ceramics (e.g. zirconia) results in improved radiofrequency performance and robust mechanical properties. Advantages also include:

- **Shorter process chain:** from feasibility, risk analysis, and iterative design to CAD optimization and series production, AM enables rapid development (from concept to part in under 18 months).
- **Industrialization:** manufacturers can internalize production (technology transfer) or outsource small series to specialized suppliers.

10.1.3 Defense: High-Performance, Tailored Components

Defense sectors require materials that combine lightness, extreme robustness, and stability in harsh environments. Technical ceramics meet these demands in applications ranging from ballistic armor to advanced sensors and propulsion systems. Contributions of AM might lie in:

Figure 10.3 3D-printed radome for defense application.

- **Monolithic and complex shapes:** ballistic panels, radomes, and engine components can be produced without joints, maximizing mechanical and thermal integrity (Figure 10.3).
- **Rapid response and sovereignty:** local, on-demand production of critical components; AM contributes to supply-chain autonomy.
- **Material intelligence:** AI-driven process optimization (e.g. 3DCERAM's CERIA) and customizable material formulations (3DMIX) streamline development, reduce risk, and ensure reproducibility.

10.2 Chemical Applications

One of the functional applications in which the benefits of AM, particularly VPP, can be fully leveraged is catalysis. In catalytic processes, according to transition state theory, the activation energy of a chemical reaction is reduced by forming an intermediate transition state complex. Lowering the activation energy means that chemical reactions can take place, for example, at lower temperatures, in a shorter time frame or – when several reaction pathways are possible – the reaction can be guided toward a certain desired pathway [1].

If the substrates and catalyzers exist in different phases, for example, gaseous and solid or liquid and solid, the catalytic process is defined as heterogeneous catalysis. In ceramic applications, the catalyst or the catalyst carrier (where the catalytic reactive species is applied to an inert carrier structure) is usually a dense or porous solid. As the transition state complex during the chemical reaction is then formed by the substrate adsorbing to the solid surface, maximizing the available catalytic surface area increases the efficiency of the catalytic process.

This is where AM comes into play. As AM facilitates the production of complex geometries, including internal channels, wavy structures, thin walls, and very fine holes, catalyst structures with enormous surface area and/or directed flow of fluids can be realized. Figure 10.4 shows an example of a high surface area AM ceramic part: a chemical reactor produced by Bosch Advanced Ceramics for BASF, Ludwigshafen am Rhein, Germany. (a) Shows the CAD cutaway drawing, where

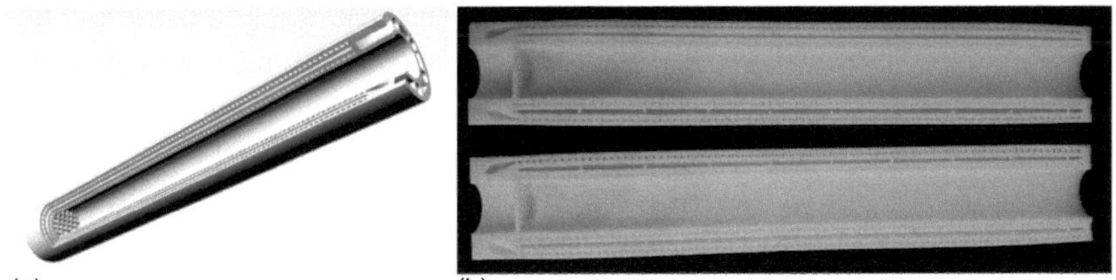

(a) (b)

Figure 10.4 Example of a high surface area ceramic part: a chemical reactor produced for BASF (design and innovation by Andrea Hurtado and Jürgen Brandner, Karlsruher Institut für Technologie, Karlsruhe, Germany). (a) CAD cutaway drawing; (b) two halves of the reactor in sintered state (produced and processed separately).

a quarter of the part has been digitally removed to show the complex internal structure. The reactor features four walls and 700 microcolumns, each having a thickness of ~350 μm over an outer diameter of 16 mm and a total specimen length of 108.1 mm. The material of choice was alumina, due to its chemical inertness and low thermal expansion, as well as mechanical stiffness at elevated temperatures. The sintered specimens can be seen in (b). Two halves of the reactor were printed separately for illustration purposes.

By increasing the variety of ceramic materials available for AM production, the industry can support in realizing complex ceramic geometries for all kinds of functional applications, ranging from dielectrics for 5G and 6G wireless communication systems to transparent ceramics for optical switches, laser amplifiers, or lenses. It is estimated that multi-material printing will gain significance in the future for applications like, for example, solid-state batteries or solid oxide fuel cells.

10.3 Electronic and Semiconductor Applications

AM is particularly well-suited for complex applications. The advantages of AM have been evident since its inception: the production of geometries that are not feasible with any other production technique and the possibility to produce small quantities without the need for enormous starting investments in the form of tooling costs.

These advantages make the technology ideal for prototyping and small batch production, as in those cases, production via AM can result in lower production costs than most conventional production techniques, such as ceramic injection molding (CIM), which require extensive tooling investment. Nevertheless, in recent years, ceramic AM has made a significant leap from prototyping toward industrial series production, by combining high precision, density close to the theoretical maximum, and – often overlooked – reproducibility.

10.3.1 Downholders for Welding

As the quality of many properties of technical ceramics (e.g. thermal conductivity or electrical resistance) stems from a dense and fine microstructure [2], manufacturing technologies capable of enabling such microstructures are essential to open up AM parts for functional applications. An AM category that can realize a dense and fine microstructure – provided processing is performed adequately and in a sophisticated manner – is VPP, due to the small starting grain size (and therefore high sintering activity) of the ceramic powder, being applied in the form of photoreactive slurries with varying viscosity. Beneficially, the small grain size allows for the realization of a low as-fired surface roughness in the range of $R_a \sim 1\,\mu m$, depending on material type and processing parameters.

These characteristics – particularly low surface roughness, thermal conductivity, and mechanical durability – are critical for the application of downholders, which are needed in certain welding processes, such as in the welding of aluminum. On the one hand, the mechanical performance of a ceramic downholder must sustain a significant load applied during processing, as the component is pressed onto the welding substrate to hold it in place. On the other hand, the ceramic downholder must endure elevated thermal stress, which is why a high thermal conductivity is beneficial for its long-term performance. Due to those reasons, alumina is often the material of choice for these kinds of applications.

As the welding process leads to sublimation of aluminum, gaseous reactive species come into contact with the downholder. Therefore, it is beneficial if the material is a chemically inert ceramic with low surface roughness, to minimize the number of solid products forming on its surface and not deteriorating its internal composition and structural integrity.

An example of such a ceramic downholder produced via AM is shown in Figure 10.5. The part was produced for Manz AG, Reutlingen. A rendering of the step-file is shown in (a). The part has a length of 39 mm, a width of 20 mm, and a depth of 30.26 mm. It features a 3.4 mm hole for assembly purposes. Two "fingers" serve the purpose of fixating the welding substrate, while processing gases are distributed locally to the welding area via two internal channels, each with a diameter of 1.3 mm, one of which can be seen in the half-section view of the model (b). VPP is the technology of choice when it comes to realizing thin, lengthy, and curved internal channels like these. Using traditional production techniques, such features would only be possible with the utmost of effort – if at all. The sintered functional part can be seen on c. It is distinguishable from the CAD model mostly by its color impression.

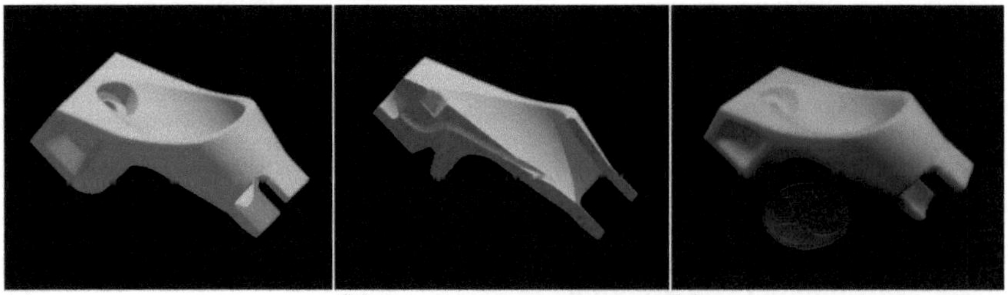

Figure 10.5 (a) Virtual CAD model, (b) half-section view of the CAD model showing the internal channel, and (c) sintered functional part. Bosch Advanced Ceramics, Immenstadt, Germany.

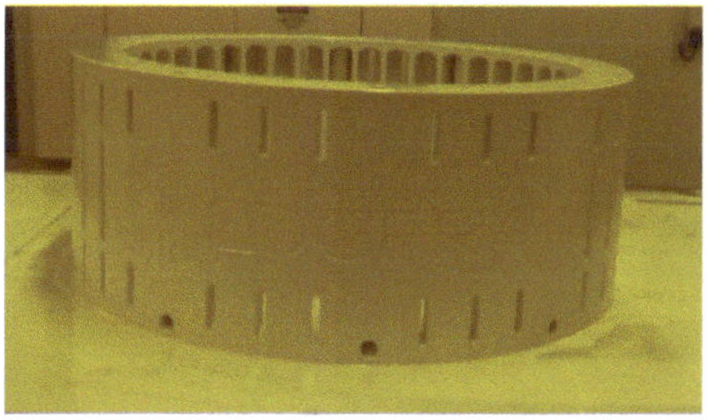

Figure 10.6 E-Chucker for semiconductor production.

10.3.2 Semiconductor Industry: From Prototyping to Industrial-Scale Production

The semiconductor sector demands high chemical, thermal, and electronic performance, as well as complex shapes for handling, processing, and deposition equipment. Traditional ceramic shaping is slow and generates excessive waste. The application of AM would result in the following advantages:

- **Technique versatility for critical components:** wafer transport trays, robot arms, radiant tubes, gas ports, electrostatic chucks, nozzles, rings, and heater elements, all benefit from the resilience of ceramics.
- **Process efficiency:** 3D printing eliminates multiple machining and joining steps, dramatically reducing lead times (e.g. Figure 10.6 shows a 550 mm diameter for this part dedicated to semiconductor fabrication, printed in a single run rather than assembled from many pieces).
- **Material flexibility:** Key ceramics include alumina, zirconia, cordierite, silicon nitride, aluminum nitride, and silicon carbide, tailored for each application's requirements.

10.4 Biomedical Applications

10.4.1 Personalized Healthcare: The Expanding Role of Ceramic 3D Printing in Biomedicines

The biomedical sector is experiencing a revolution in implant and device design, thanks to ceramic AM. The unique combination of biocompatibility, mechanical strength, and the potential for fully customized shapes has positioned technical ceramics at the forefront of orthopedics, craniofacial surgery, dental restoration, and surgical instrumentation. Over the past few years, SLA-based AM, in particular, has enabled a new paradigm – one where patient-specific solutions and industrial-scale efficiency coexist (Figure 10.7).

3DCERAM's SLA technology has been leveraged for more than a decade to produce custom-made bone substitutes and craniofacial implants. The workflow begins with data from CT scans, translated into 3D CAD files that capture both the anatomical defect and the surgical requirements.

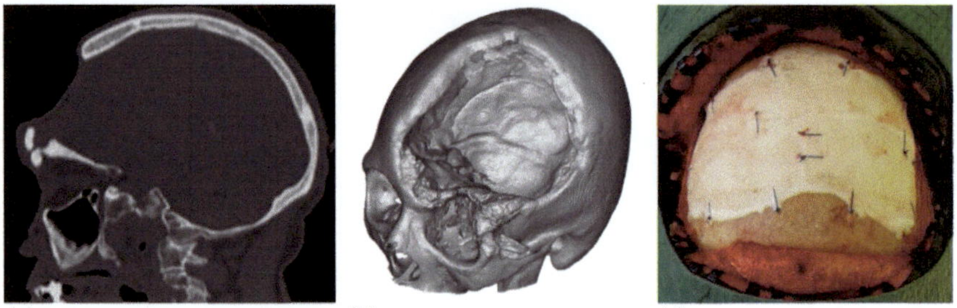

(a) (b) (c)

Figure 10.7 (a) Scanned images, (b) 3D-reconstructed model based on (a), and (c) during surgery.

Process highlights:

- **Precision modeling:** 3D CAD design allows for precise adaptation to individual patient anatomy, including the design of porous (osteointegrative) and dense (mechanically robust) regions.
- **Layer-by-layer fabrication:** The ceramic slurry is polymerized layer by layer with a UV laser, ensuring fine feature control.
- **Post-processing:** Debinding and sintering remove resin and densify the ceramic, achieving up to 98–99% densification and <20% shrinkage.
- **Quality control:** Implants are measured and assembled (if multipart).

Clinical results:

More than a dozen clinical cases (cranial, frontal, parietal, and facial reconstructions) have demonstrated excellent osseointegration, mechanical stability, and patient satisfaction. No encapsulation, inflammation, or implant rejection was observed in follow-up up to one year. Functional and aesthetic outcomes consistently met or exceeded expectations.

10.4.2 Orthopedic and Dental Solutions

Ceramic AM enables the manufacture of intervertebral cages, tibial wedges, and bone scaffolds with highly controlled porosity and pore interconnectivity (Figure 10.8). Compared to conventional foam-based or molded substitutes, AM components can achieve 3–5 times higher compressive strength while promoting bone in growth.

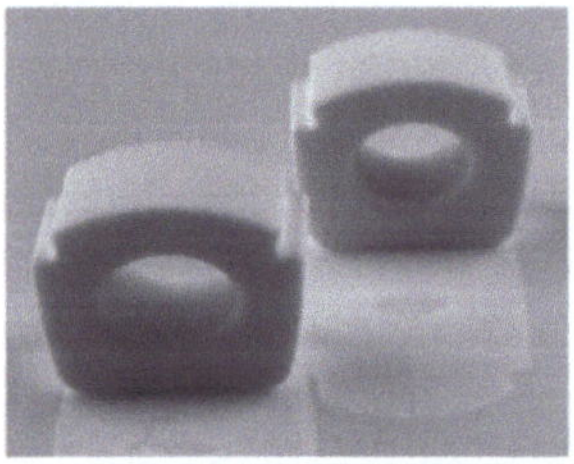

Figure 10.8 3D-printed spinal cage.

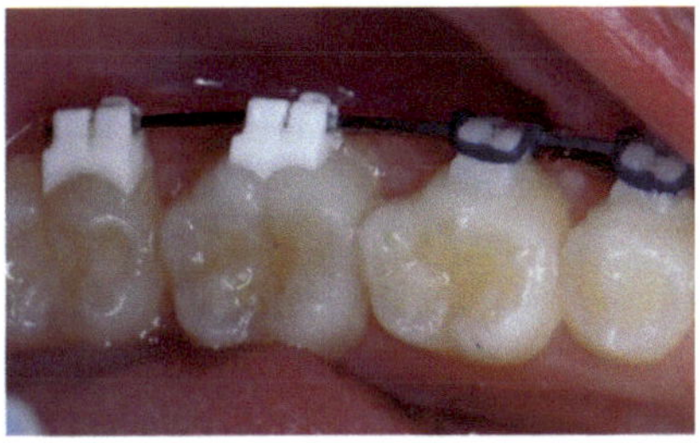

Figure 10.9 Custom-made brackets by 3D printing.

On the other hand, the dental market is transitioning to ceramic orthodontic brackets made from ultra-high purity alumina via AM (Figure 10.9). The benefits include:

- **Full customization:** Each bracket is individually tailored to the patient's dentition, improving fit and reducing treatment time.
- **Aesthetics and comfort:** Ceramic brackets are translucent, blending with natural teeth.
- **Industrial throughput:** Platforms such as the C1000 Flexmatic can produce over 2,100 brackets per run, with accuracy and repeatability on par with (or better than) traditional processes.

10.4.3 Next-Generation Surgical Tools and Devices

Injection molding, machining, and casting are established production methods for medical instruments. Plastic, metal, and ceramic AM technologies are increasingly addressing the challenges of conventional manufacturing. AM developments focus on automation, cost-efficiency, material diversity, and supply-chain integration, offering design freedom that exceeds the limitations of traditional techniques. Advances in various medical fields such as tissue engineering, patient-specific bone implants, or low-cost polymer instruments have been accelerated by AM. The medical instrument sector shows high potential for AM, given the diverse range of devices and the fact that each procedure often requires task-specific handling tools. Advances in AM now allow for the integration of sensors, channels, and composite structures into bioceramic devices. Multi-material systems support the co-printing of ceramics with metals or polymers, a crucial step toward "theranostic" (therapeutic + diagnostic) implants. AM facilitates the rapid prototyping and low-volume production of complex, miniaturized surgical tools, addressing evolving requirements in minimally invasive and image-guided surgery.

The main ceramics for biomedical AM include:

- **Hydroxyapatite (HAP):** Chemically similar to natural bone, excellent osseointegration for craniofacial and orthopedic use.
- **Alumina:** Exceptional hardness and wear resistance, suited for dental and orthopedic prostheses.
- **Zirconia:** High toughness and aesthetics, increasingly used in dental prosthetics.
- **TCP:** Resorbable, for temporary scaffolds in bone repair.

Ceramic AM is redefining what is possible in biomedical engineering – delivering fully customized, robust, and biocompatible implants and devices. As material portfolios broaden and multi-material solutions emerge, patient outcomes and healthcare efficiency stand to benefit even further. The next decade will likely see ceramic AM become a standard not only for complex implants but also for intelligent, multifunctional medical devices. The global market for 3D-printed medical ceramics is expanding rapidly: valued at $400 million in 2016, it is projected to reach $6 billion by 2027, with technical ceramics at the heart of this growth.

This section will illustrate the application of AM in the medical sector using a surgical gripper tip made from 3Y-stabilized zirconia via SLA as a representative case. The gripper was produced by the service provider Bosch Advanced Ceramics, Immenstadt, Germany. The ceramic component demonstrates how material selection, manufacturing technology, and design collectively determine the final product performance.

The efficiency of surgeries relies heavily on the design and performance of surgical instruments such as end-effectors that interact directly with tissue. Surgical grippers – used for manipulation, retraction, and grasping – are continuously refined to enhance precision and reduce tissue damage.

Common ceramic materials used in AM include zirconia, alumina, hydroxyapatite, and silicon nitride. The specific application dictates the choice of material. 3Y-stabilized zirconia offers a combination of high hardness, biocompatibility, chemical inertness, flexural strength, fracture toughness, and wear resistance, making it particularly suitable for surgical instruments subject to repeated sterilization and long-term use. The material also exhibits a thermal conductivity of approximately 2 W/m·K at 100 °C and a density exceeding 6.1 g/cm^3, supporting both heat dissipation and mechanical integrity.

The design shown in Figure 10.10 integrates multiple functionalities within a compact volume. The rounded tip ensures smooth interaction with tissue and minimizes tearing. SLA combined with thermal sintering eliminates the need for post-processing such as deburring, reducing hard machining efforts. An internal channel adds functionality by enabling vacuum application, fluid guidance, or sensor integration. A pressure sensor could monitor the applied force during delicate procedures. Electrical applications are also possible thanks to the insulating

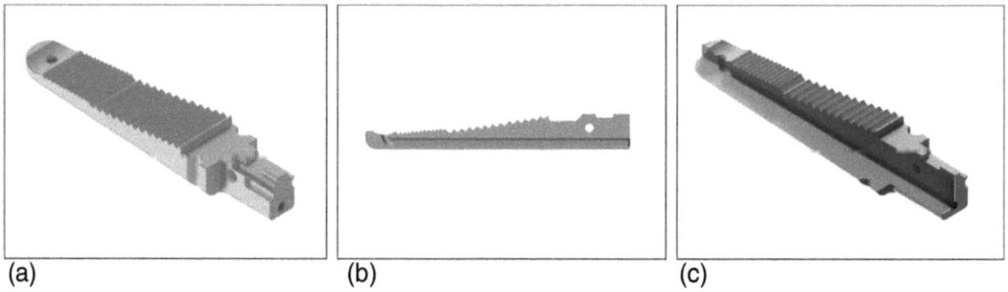

Figure 10.10 (a) CAD model of the tip of a surgical gripping instrument produced via additive manufacturing, (b–c) half-section views of the model showing the internal channel. Bosch Advanced Ceramics, Immenstadt, Germany.

properties of oxide ceramics like zirconia. This curved channel is not machined but directly integrated via AM, which would be extremely difficult to achieve with conventional methods.

The gripper tip's primary function is to contact and hold tissue. This is achieved by a sawtooth design with two different tooth sizes. Since the part is printed flat on the build platform, layer steps align horizontally along the teeth, increasing surface area and friction – an apparent disadvantage turned functional.

The attachment side shown in Figure 10.10 includes several features for mechanical connection (pins, clips, or plugs), with the channel connection at the rear. The part was printed with a 50 μm layer height in the green state, resulting in a 42 μm step height post-sintering. It weighs 0.57 g and measures 19.5 mm in length, 4 mm in width, and 2.5 mm in height, achieving a density greater than 6.1 g/cm^3 with no open porosity. Compared to alumina, the material also provides higher fracture toughness and bending strength, which is critical for long-term functionality and reliability in surgical settings.

The manufacturing process includes several critical thermal steps. After printing, the green body undergoes thermal debinding to remove residual organic binders, followed by high-temperature sintering. During sintering, the part densifies to near-theoretical density. Careful control of thermal profiles is required to avoid warping or cracking and to ensure uniform shrinkage.

Surface roughness is measured at $R_a = 0.6\,\mu m$ in the printing direction and $R_a = 0.4\,\mu m$ in the layer plane. While post-machining can improve these values, the process is time- and cost-intensive, as with all ceramic materials. Achievable dimensional tolerances depend on the technology. Microfabrication techniques like two-photon polymerization (TPP) can achieve ±1 μm. SLA technology, as used for this part, typically achieves tolerances of ±20–40 μm depending on geometry.

To ensure conformance with technical and clinical requirements, quality assurance procedures such as CT scanning, 3D metrology, and mechanical testing are integrated into the process chain. These steps validate the part's dimensional accuracy, surface integrity, and material properties before deployment in sensitive applications.

Although small in size, the gripper plays a key role in a range of surgical procedures, including micro-invasive and minimally invasive operations, where compactness, material inertness, and biocompatibility are crucial. Integrated channels can also allow multifunctionality within constrained instrument dimensions, such as fluid control or signal routing.

The main challenges of AM in medical applications include surface finish, dimensional accuracy, unit costs, and process validation. Nevertheless, AM supports fast prototyping and production of surgical tools, minimizing waste and reducing trauma by enabling precision. This is particularly relevant when using high-cost materials such as zirconia.

Ceramic AM allows for batch sizes ranging from one to several thousand parts, depending on geometry. Individual features or markings can be incorporated into each part. This capability supports rapid design iteration, customization, and efficient production throughout the product lifecycle. Unlike injection molding, there are no tooling costs. On-demand manufacturing further reduces inventory and increases supply-chain flexibility.

Compared to conventional metal or polymer-based instruments, ceramic AM grippers offer enhanced longevity, improved sterilization resistance, and functional integration. While metals allow for high toughness and polymers for cost-effective mass production, ceramics bridge the gap for demanding applications where mechanical, chemical, and thermal performance must be combined in a single part.

The key challenge in scaling AM in the medical field remains the unit cost, driven by build speed and build volume. As with all AM technologies, geometry, function, and material define the optimal solution. However, costs associated with equipment, process validation, regulatory compliance, and post-processing can limit broader adoption.

10.5 Detailed Case Studies of Ceramic AM Applications

Below, we are introducing two detailed case studies, showcasing the background, motivation, rationale, and resulting effect of ceramic AM in the energy and environment-related applications.

10.5.1 Solid Oxide Cells for High-Efficiency Hydrogen Production

The global energy scenario is currently undergoing a significant transformation, marked by a large-scale deployment of renewable energy sources [3]. Hydrogen emerges as a promising chemical energy storage vector, which is capable of enhancing the flexibility of the power systems by managing the generated surplus electricity while sustainably covering peaks of demand [4]. Green hydrogen, produced through water electrolysis powered by renewable energy sources, enables a sustainable connection between global gas and electrical networks (Figure 10.11) [5]. In these systems, the power-to-gas (P2G) process stores the surplus renewable electricity as green hydrogen via electrolysis, while the gas-to-power (G2P) route allows the reconversion of stored hydrogen into electrical energy by the application of fuel cells, offering a strategic advantage during periods of high electricity demand or low renewable energy generation.

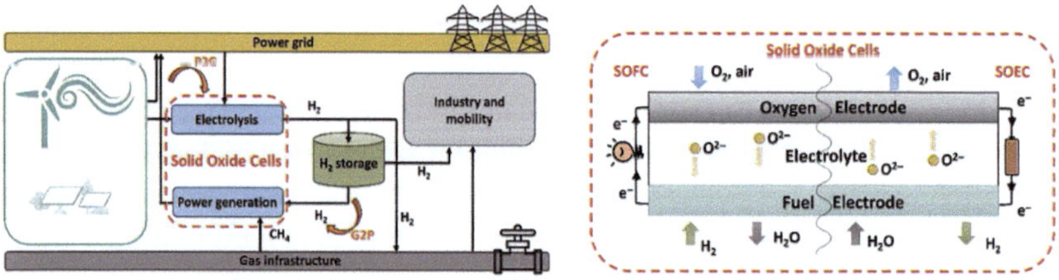

Figure 10.11 Schematic representation of the power-to-gas and gas-to-power routes using solid oxide cells.

High-temperature multilayer ceramic SOCs, which consist of three main functional components: gas-tight ceramic electrolyte membrane, and two porous electrodes supplied with different gas atmospheres [6], exhibit immense potential to be involved in P2G and G2P routes owing to their reversibility, allowing the operation in both fuel and electrolysis modes. Nevertheless, further advancements in SOC development remain constrained by the limitations of conventional ceramic manufacturing techniques, such as tape-casting, screen-printing, or extrusion, which restrict the shape of the cells to the relatively simple planar and tubular geometries [7]. In contrast, ceramic AM has emerged as a promising alternative, offering the capability to fabricate individual SOC functional layers with high geometric resolution along all axes [8]. More importantly, AM enables the transition to a new generation of SOC systems by drastically increasing the shape complexity of the fabricated devices, which is used to increase the active area of each cell, thereby enhancing the overall system performance [9]. Moreover, ceramic 3D-printing technologies address one of the key barriers to large-scale SOC commercialization: high manufacturing costs [10]. Compared to traditional ceramic fabrication methods, AM tends to be a more time- and cost-efficient approach with significant potential for digitalizing the production process [11]. Additionally, the concept of layer-by-layer fabrication realized by AM technologies minimizes the valuable material waste, which is crucial for industrial-scale SOC production.

In recent years, ceramic AM has been strategically employed to advance SOC production from multiple perspectives, for 3D structuration and surface engineering of electrodes and electrolytes, for reducing the use of expensive catalysts through automated infiltration techniques, and for production of monolithic and fully 3D-printed devices, based on the use of inkjet printing, VPP, and fused deposition modeling methods [12–16]. Moreover, the research efforts have been focused on developing reliable hybrid multi-material 3D-printing approaches using inkjet printing, SLA, and robocasting for the fabrication of complete SOC devices, laying the groundwork for the future digitalization and automation of SOC production pathways [17].

However, the most significant advancements in fabricating innovative SOCs with complex geometries have been achieved through VPP techniques. The application of VPP for the SOC electrolytes production has revolutionized the cell design by moving beyond conventional geometries to new concepts based on 3D structuration, as it could be clearly observed for 3D-printed zirconia-based SOC electrolytes.

The effectiveness of SLA in producing high-aspect-ratio electrolytes for SOCs was first demonstrated by Masciandaro et al. [18] by producing self-supported yttria-stabilized zirconia (3 mol.%, 3YSZ) flat electrolytes using a commercial Ceramaker C-900 3D printed (3DCERAM-SINTO, France). The fabricated cells exhibited electrochemical properties comparable to those produced by conventional methods (a peak power density of 100 mW/cm^2 at 900 °C), validating the suitability of the developed manufacturing approach. In the same study, the authors showed the transition to more advanced geometries by producing 260 μm-thick 3YSZ SLA-printed electrolytes supported by a honeycomb-like network consisting of 530 μm-thick and 220 μm-wide beams (Figure 10.12a), resulting in an increase in peak power density to 115 mW/cm^2 at 900 °C.

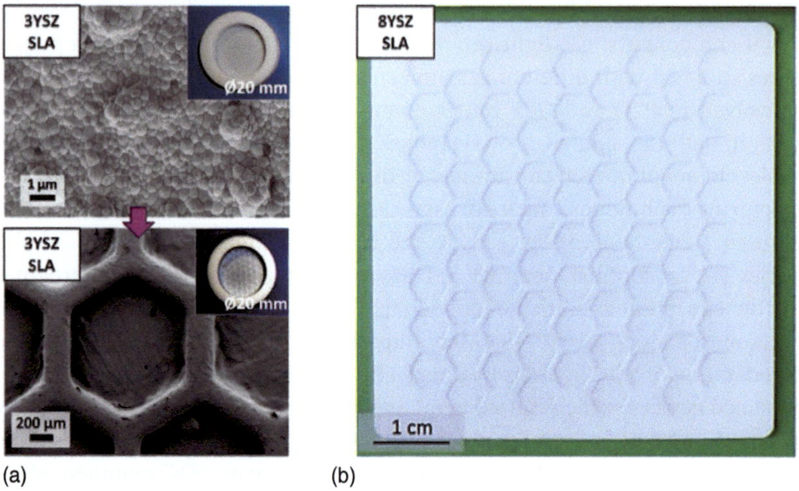

Figure 10.12 SLA 3D-printed SOC electrolytes with innovative geometries: (a) 3YSZ membrane supported by honeycomb beam network. [18] / with permission of Elsevier. (b) Large-area 8YSZ membrane supported by honeycomb beam network. [19] / with permission of Elsevier.

The scalability of the SLA-based manufacturing of SOC electrolytes with integrated honeycomb-like pattern was further demonstrated by Lira et al., who fabricated a SOC single repeating unit (SRU) with an active area of $16\,cm^2$ (Figure 10.12b) [19]. The corresponding device generated up to 3.5 W at 950 °C in SOFC mode and provided 7.3 W in SOEC mode at the same temperature, operating with the injecting current density of $340\,mA/cm^2$ at 1.3 V. Moreover, the stable operation in SOFC mode was demonstrated for over 1000 hours, while the functionality in the reverse mode was successfully validated during 100 hours of continuous operation.

Further implementation of SLA-based approaches for production of a new generation of SOC devices was demonstrated by Pesce et al. by fabricating 270 μm – thick YSZ (8 mol.%, 8YSZ) electrolytes with a radically new corrugated wavy-like geometry (Figure 10.13a) [20]. This 3D structuration increased the active area by approximately 60%, which directly translated into enhanced device performance, achieving a peak power density of $410\,mW/cm^2$ at 900 °C. Moreover, the functionality of the 3D-printed SOC devices was confirmed in co-electrolysis mode by converting steam and CO_2 into syngas, yielding a current density of $-0.6\,A/cm^2$ at 1.3 V at 900 °C, which is significantly higher compared to $-0.46\,A/cm^2$ for planar electrolytes.

Most recently, the technology has reached the stack level. Martos et al. reported successful testing of SOC substack (3 SRUs) and stack (18 SRUs) based on corrugated 3YSZ electrolytes fabricated using an industrial Ceramaker C-1000 3D-printing machine (3DCERAM-Sinto, France) [21]. The incorporation of the

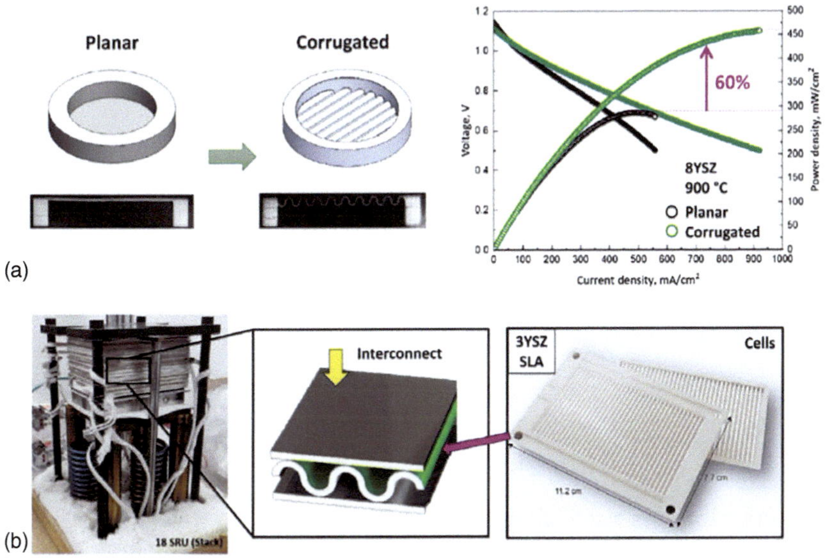

Figure 10.13 (a) 3D-printed 8YSZ corrugated cell with a total diameter of 20 mm and its electrochemical characteristics in SOFC mode at 900 °C compared to the planar cell. [20] / The Royal Society of Chemistry/ CC BY 3.0. (b) Stack of 18 SRUs based on 3D-printed 3YSZ corrugated electrolytes. [21] / Elsevier / CC BY 4.0.

innovative wavy-like pattern enabled by ceramic AM allowed to achieve an active area of ~72 cm^2 per cell (Figure 10.13b), contributing to excellent overall system performance. Specifically, the substack yielded a maximum power of 27 W (200 mW/cm^2) and a maximum injected current of −20 A at 4.2 V (−450 mA/cm^2) in SOFC and SOEC modes, respectively, while the complete stack reached −13 A at 25 V (equivalent to 330 W of power). Thus, the efficient operation and a remarkably small degradation rate (5%/kh for 500 hours) of the devices proved the durability of the developed manufacturing technology in the mid-term.

Moreover, SLA 3D printing offers flexibility in the feedstock material selection, as the range of printable functional ceramics continues to expand rapidly. This allows the application of materials with high ionic conductivities, such as ytterbium-doped scandia-stabilized zirconia, for advanced SOC electrolyte manufacturing (Figure 10.14), further enhancing the device performance [22].

At the current stage of development, innovative corrugated YSZ electrolytes with efficient active areas ranging from 3.2 to 125 cm^2 could be fabricated using SLA technology (Figure 10.15a), marking a significant step toward industrial-scale SOC production. In pursuit of this goal, the first pilot line for SOC device manufacturing via 3D printing was established at the Catalonia Institute for Energy Research (IREC, Spain), equipped with a Ceramaker C-2000 3D printer (3DCERAM-Sinto, France) capable of 24/7 continuous operation (Figure 10.15b).

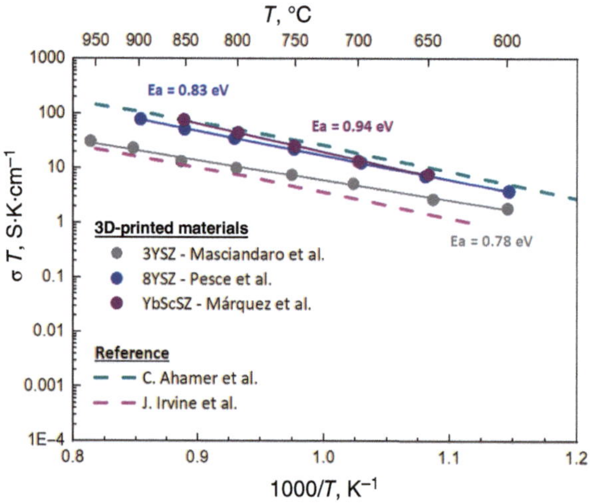

Figure 10.14 Arrhenius plot of the ionic conductivity of various SLA 3D-printed SOC electrolyte materials compared to the data reported in the literature.

(a)
(b)

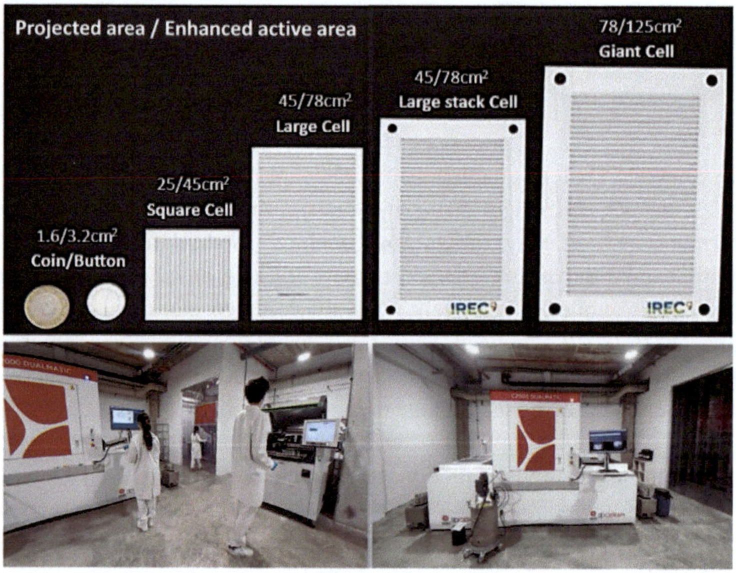

Figure 10.15 (a) SLA 3D-printed YSZ SOC electrolytes with the corrugated membranes; (b) the first SOC production pilot line based on SLA 3D printing at IREC.

Overall, ceramic AM offers unique capabilities to boost hydrogen production and energy generation by introducing new strategies for the fabrication of SOCs in both laboratory and industrial scales. The implementation of ceramic 3D printing in energy device manufacturing has the potential to overcome long-standing limitations in the conventional SOC production cycles by enhancing the specific performance, structural robustness, and operational flexibility of the systems.

10.5.2 Ecological Restoration for Environmental Protection

Ceramic AM can also have direct social and environmental impact. It can be used as an effective tool to help protect the environment. Here, we are showing an application case study with regard to 3D printing of ceramic substrates for sexually reproduced coral larvae settlement in the ecological restoration domain.

The Great Barrier Reef (GBR) makes up approximately 10% of the world's coral reef ecosystems and is referred to as being one of the best known and most complex natural systems on Earth [25]. It is an irreplaceable ecological and economic asset. Valued at $56 billion, it contributes $6.4 billion annually to Australia's economy and supports 64,000 jobs through tourism and fisheries [26]. Ecologically, its coral reefs provide coastal erosion protection, critical marine habitats, and carbon dioxide absorption to mitigate climate change. A $1 billion federal investment underscores its national significance, targeting conservation amid climate threats [27]. The GBR provides vital coastal erosion protection, marine habitats, and water purification while absorbing CO_2 to mitigate climate change. However, anthropogenic climate change drives severe threats: ocean warming forces coral bleaching by expelling symbiotic algae (zooxanthellae), acidification from absorbed CO_2 (30% since the 18th century) weakens skeletal growth, and deoxygenation stresses ecosystems [28]. Mass bleaching events (1998–2022) have severely impacted 80.7% of reefs, with recovery compromised by rising frequency/intensity. These cascading impacts degrade biodiversity, coastal resilience, and economic sectors, demanding urgent conservation action to prevent irreversible ecosystem collapse [29–31]. Active GBR management employs three approaches: sexual propagation (larval enhancement), asexual techniques (transplantation/gardening), and substratum enhancement (artificial reefs, stabilization) [32, 33]. These methods accelerate coral recovery by transplanting fragments to degraded reefs, yet require 10–100 million corals for net-positive impact. Critical success factors include substrate compatibility, environmental safety, social acceptance, and scalable low-cost manufacturing [34]. A key knowledge gap persists regarding how surface geometry affects larval settlement/survival. This might be addressed by testing alumina and hydroxyapatite substrates with five pore sizes (0–900 μm) across 1200 samples for Montipora, Mycedium, and Acropora larvae, as shown in Figure 10.16. 3D printing enables precise pore geometry control essential for optimizing restoration efficacy.

The designs were modeled for 3D printing using the computational design software nTop with pore sizes intended to range between 0 and 1000 μm. A total volume of 2 × 2 × 8 mm was selected to act as an extension to previous studies conducted by AIMS, with a dense tab with no pores initially selected to act as a base control to determine whether the larvae showed preferential signs to specific porosity, or just the material itself. Five increments were chosen through collaboration with AIMS, with these initially established to be 0, 400, 600, 800, and 1000 μm; 400 μm was selected to be the smallest pore size as this was the smallest size that could be manufactured and post-processed effectively. When modeling these designs, it was also found that the ratio of pores to flat surface had to be maintained across all pore sizes to ensure that the larvae were settling on a specific pore size because that

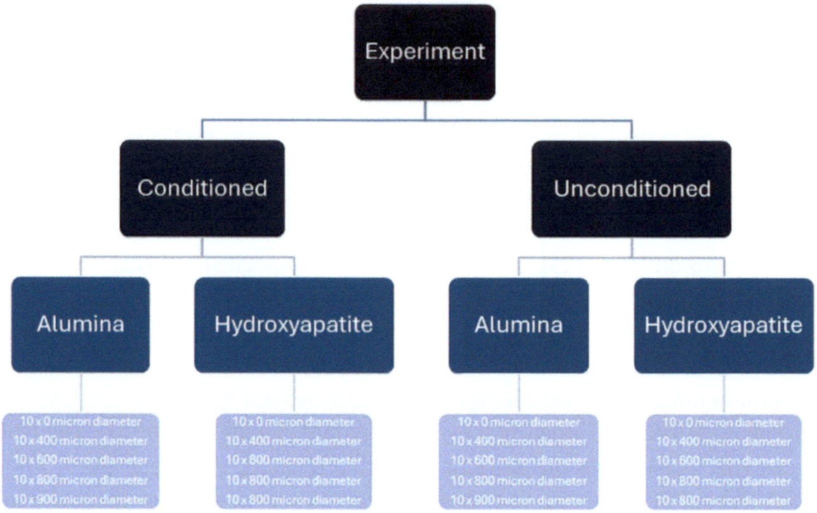

Figure 10.16 Hierarchy of settlement substrates for May/June spawning.

is what it prefers, rather than it simply being more abundant; this ratio was maintained between 0.779 and 0.8684. This value was not able to be completely consistent among pore sizes due to the parameters of the design requirements; pores had to be whole and could not be cut off by the 2 × 8 mm top surface restriction. As a result, 1000 μm could not be used, and this increment was lowered to 900 μm.

The alumina tabs were manufactured using Al-E03 slurry sourced from 3DCeram, while the hydroxyapatite tabs were manufactured using HAP-E01 slurry also sourced from 3DCeram. The alumina slurry can obtain the following microstructure properties: density of >3.9 g/cm^3 and a grain size after sintering of 2.2 μm. Its mechanical properties include a four-point bending strength of 397 MPa, a Vickers hardness of 16.4 GPa, and a fracture toughness of 4 MPa · m$^{1/2}$. Similarly, the HAP slurry has the following microstructure properties: densification rate of >96%, density of >1.5 g/cm^3, and a grain size after sintering of 2 μm. The mechanical properties include a four-point bending strength of 107 MPa, with general properties such as a crystallinity of >95% listed in addition [35]. All tabs were printed from 3DCeram CERAMAKER 100 printer (C100) utilizing SLA technology, which were then followed as per the post-processing requirements of binder removal through an inert environment debinding process followed by sintering as per 3DCeram's specified requirements to obtain the final mechanical properties as outlined above.

The purpose is to obtain an overall understanding of coral settlement patterns and either confirm or object to past literature presented. While not completely effective in maximizing the potential that ceramic tabs hold in relation to the fabrication of surface microstructure, the results have revealed many trends that can be further refined through future experimentation. Numerous inter- and intra-species trends were noted regarding the importance of conditioning, substrate material, and geometrical differences.

Regarding the effect of CCA conditioning, the results presented were unanimous; conditioned tabs were far more effective in promoting settlement compared to

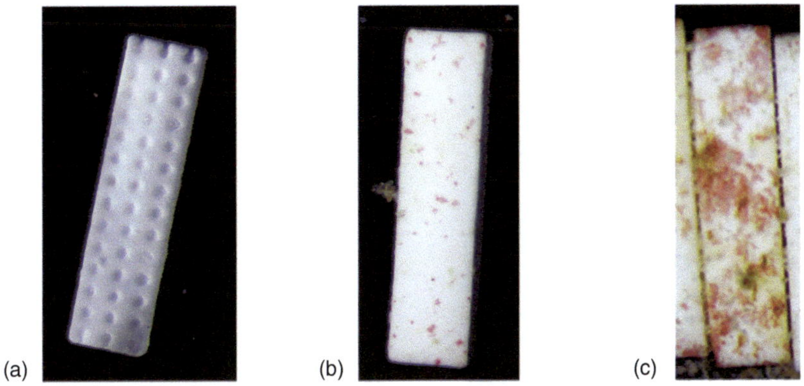

Figure 10.17 Conditioning coverage across varying treatment stages. (a) Unconditioned. (b) May conditioning. (c) June conditioning image included for comparison, despite not having been completed at this stage of recommended improvements.

unconditioned tabs (Figure 10.17). For the first experiment conducted on Maeq larvae within well plates, conditioned alumina exceeded the success of unconditioned alumina, increasing in percentage by 74.53% (6.63% to 81.16%), and conditioned HAP to unconditioned HAP by 39.86% (33.24–73.10%). These trends were replicated for Mele, albeit at a much lower magnitude, where unconditioned tabs were seen to have 0% settlement compared to conditioned alumina and HAP recorded to be 1.44% and 5.26%, respectively. Finally, at the same scale but during the June spawning, Alor results were found to increase by 110% from 17.32% to 36.46% for alumina when comparing unconditioned to conditioned, and 159% for HAP unconditioned to conditioned from 14.89% to 38.59% (Figure 10.18). At a larger scale, trends were again reproduced with Maeq having only 12 coral spats settle on unconditioned tabs, compared to 25 settling on conditioned tabs, indicating another increase greater than two times that of unconditioned, with similar results obtained for Alor larvae with 0 settlement experienced across all unconditioned tabs, compared to 310 coral spats settled on conditioned tabs. The overall trend observed aligns with past literature as CCA has been proven to be effective in enhancing settlement [36].

With respect to substrate material, two materials were analyzed for the purpose of this experiment: alumina and hydroxyapatite. For Maeq, HAP was preferred both at the small scale of the 10 ml well plates and at the larger scale of the 15 l tanks. An average settlement percentage of 53.17% was obtained for HAP within the well plates, compared to 43.89% for alumina, indicating only a slight preference, while the tank experiment showed a 540% increase from 5 coral spats settling on alumina, compared to 32 coral spats settling on HAP. For Mele, HAP was again the favored material, experiencing a 5.26% settlement on HAP tabs within the wells compared to only 1.44% on alumina. Unfortunately, due to limited tabs printed, this larval species could not be tested within a larger volume to validate the scalability of results. Finally, a slight deviation was seen when testing the Alor larvae within a small volume with the average settlement for HAP being 0.15% lower than that of alumina with settlement percentages of 26.74% and 26.89%, respectively.

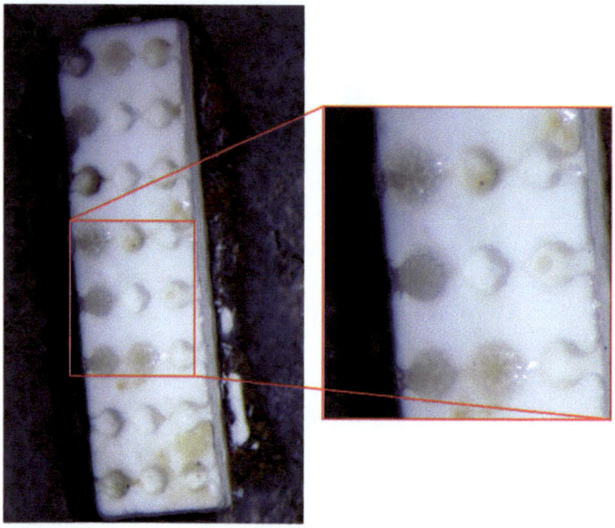

Figure 10.18 Alor larvae settled within 600 μm conditioned HAP pore.

However, the large-scale tank experiment confirmed the overall HAP proclivity with 201 coral spat settlements observed compared to only 109 across alumina. This finding presented new data as previously HAP had not been tested with alumina still undergoing testing to maximize its potential. Past literature commonly focused on mortars with cement-based mortars being ranked more effective than geopolymer-based mortars when considering the process holistically. Ceramics then began being included in discussions due to their chemical properties, with HAP now also investigated due to its similarities to the skeletal calcium carbonate found in reefs.

The final variable tested during this choice experiment was porosity through fabricated pore sizes. Maeq larvae had the highest settlement occurrences at 800 μm for both the small and large volume experiments at 18.15% and 6 coral spats, respectively, with Mele larvae also showing the highest proclivity to 800 μm at 4.70%. The Alor larvae differed slightly, with 900 μm having the highest settlement at 2.5% for the well plates; however, 800 μm was the second best at 0.8%. For the tank experiment, Alor larvae clearly preferred 800 μm, with 125 coral spats settling. The remaining pore sizes did not show as much of a distinct order of preference, with 900 μm and 600 μm often alternating for the second most favored, with 0 μm and 400 μm always showing the lowest preference for settlement.

In conclusion, a clear preference for settlement has been identified and is conducive across all three species tested. Conditioned tabs were drastically better than unconditioned with HAP presenting better results than alumina. The most effective pore size was almost unanimous at 800 μm, irrespective of larvae species and their individual diameters. Some of the trends observed were found to align with previous literature, some findings were new, and some were found to disagree with past research; thus, the final recommendation based on experimental findings was to use conditioned HAP tabs with a fabricated porosity of 800 μm to obtain the highest settlement percentage. Despite these findings, the process of producing the tabs must also be considered regarding the ease of fabrication and related expenses. The alumina was much easier to handle during manufacturing than HAP, resulting in fewer breakages and thus less wastage, decreasing the overall cost of production. Raw costs of the alumina slurry are also significantly lower than that of HAP, resulting in a compromise needing to be determined. In terms of obtaining ideal CCA coverage, the methodology should be improved by allowing for a longer lead time between microstructural design and implementation at spawning events. The June conditioning time of 28 days was almost enough but was found to still result in extremely variable coverage. This time should be increased to allow for consistent coverage between 10% and 20% across all tabs used. Finally, it is recommended to ensure healthy larvae are used for the experiment by completing competency tests prior to implementation, where a competency of 80% should be obtained. While difficult to control, this factor greatly affects the success of the experiment with standards needing to be maintained to ensure valid data. This may be improved by conducting experiments during their natural spawning events rather than simulated out-of-spawning during winter. In terms of design extensions, recommendations were also made based on the data obtained and trends processed. It has been observed that while the data has proved scalable between 10 ml and 15 l, the experiment should be extended to a larger volume before assuming its applicability at oceanic scale. A 50 l tank volume should be included in the method to ensure results can be replicated with the potential for additional species depending on spawning success to confirm the applicability of these results when implemented on mass.

10.6 Cross-Domain Value and Systemic Impact of Ceramic Additive Manufacturing

AM of technical ceramics transcends its role as a mere production tool, emerging as a catalyst for systemic transformation across industrial, economic, and environmental domains. This disruption stems from its capacity to decouple design complexity from manufacturing cost, enabling a fundamental shift from design-for-manufacturability to manufacturability-for-design. Where traditional ceramic processing – such as dry pressing or injection molding – imposes severe geometric constraints (e.g. uniform wall thickness and limited undercuts), AM empowers engineers to prioritize functional intent. This paradigm is exemplified by topology-optimized radiofrequency antennas achieving 40% weight reduction without sacrificing gain, or catalytic reactors with graded porosity enhancing mass transfer efficiency by 200%. Such functional supremacy redefines ceramic components from passive elements to active system enablers.

The ripple effects extend deeply into supply-chain logistics and manufacturing ecology. AM facilitates distributed, on-demand production, drastically reducing warehousing needs and mitigating risks associated with global supply-chain disruptions. A salient case lies in semiconductor manufacturing: replacement parts for wafer handling systems – previously requiring months-long lead times and costly mold fabrication – can now be produced locally within days, slashing inventory costs by up to 60%. This agility is complemented by remarkable material efficiency; near-net-shape fabrication minimizes machining waste, with ceramic utilization rates exceeding 95% versus <40% in subtractive processes. Such efficiency aligns with circular economy principles, particularly when coupled with closed-loop recycling of unused powder and support structures.

Economically, AM's impact manifests most profoundly in low-volume, high-complexity components prevalent in aerospace, energy, and biomedical sectors. While unit costs may remain higher than mass-production techniques, the total lifecycle cost – incorporating design iteration speed, tooling elimination, and performance gains – often reveals compelling advantages. For instance, the development of patient-specific bone implants leverages AM's customization capability to eliminate intraoperative adjustments, reducing surgery time by 30% and improving clinical outcomes. Moreover, the absence of dedicated tooling makes design iterations economically trivial, accelerating innovation cycles by orders of magnitude.

Underpinning these advances is the convergence of digital technologies. AM inherently generates vast datasets spanning process parameters, material behavior, and performance outcomes. Machine learning algorithms harness this information to predict optimal printing parameters, detect defects in real time, and autonomously adjust laser paths to mitigate thermal stress. Digital twin capabilities further amplify this synergy, enabling virtual validation of ceramic components under extreme thermomechanical loads before physical production. Such integration positions AM not as a standalone technology, but as the physical gateway to a fully digitalized industrial ecosystem.

In essence, ceramic AM's greatest legacy may lie not in the parts it produces, but in the ecosystems it enables: democratized manufacturing, accelerated innovation, and sustainable material use. Its true impact is measured not merely in technical metrics, but in its capacity to redefine how we conceive, produce, and value advanced ceramic components in a rapidly evolving technological landscape.

The recent progress in ceramic AM is not only a matter of materials, but above all, of process innovation. The shift to fully digital workflows – from design to industrial production – has enabled unprecedented levels of complexity, repeatability, and industrial scalability. Here, in particular, we show how digitalization and automation work in the current ceramic AM process.

The latest generation of SLA-based ceramic AM platforms now operate within integrated digital ecosystems. Each stage – from design (CAD), through process simulation and build optimization, to automated print monitoring and quality control – can be digitally orchestrated to minimize risk and maximize performance.

This digital chain allows for:

- real-time adjustment of parameters to suit the geometry and the chosen ceramic,
- advanced simulation and optimization of the build process,
- automated quality assurance and traceability, essential for mission-critical and regulated sectors.

Figure 10.19 CERIA interface.

A major leap in process reliability and productivity comes from the introduction of CERIA (Figure 10.19), 3DCERAM's proprietary software based on artificial intelligence. CERIA addresses some of the most persistent challenges in ceramic 3D printing:

- **Process repeatability:** By learning from thousands of builds, CERIA can predict potential failures, adjust process parameters in real time, and thus deliver consistent part quality across production runs.
- **Yield optimization:** AI-driven process control reduces scrap rates and increases machine uptime, which is crucial for cost-effective industrial production of high-value components.
- **Technology transfer:** CERIA enables rapid scaling from R&D prototypes to full industrialization – even for users new to ceramic AM – by guiding operators through best practices and automating complex tasks.

This integration of artificial intelligence is redefining the standard for reliability, scalability, and cost-efficiency in technical ceramic production. For industries where each component's performance and traceability are non-negotiable – such as aerospace, defense, and semiconductors – the impact is transformative.

10.7 Conclusion

Ceramic AM has emerged as a transformative force across high-performance industries, fundamentally redefining the design, production, and functional capabilities of advanced ceramic components. By transcending the geometric constraints of traditional manufacturing, AM enables unprecedented integration of complexity, functionality, and customization – from lightweight aerospace mirrors with embedded cooling channels to SOCs with corrugated electrolytes boosting active area by 60%. The technology's impact extends beyond performance enhancement to drive economic and operational efficiencies: supply-chain democratization through distributed production, material waste reduction via near-net-shape fabrication, and accelerated innovation cycles by eliminating tooling dependencies.

Cross-sector applications demonstrate AM's versatility: in energy, SOCs achieve >55% operational efficiency in reversible mode; in biomedicine, patient-specific implants improve surgical outcomes; in semiconductors, monolithic ceramic assemblies slash lead times and costs. Underpinning these advances are digital ecosystems integrating AI-driven process optimization, real-time monitoring, and predictive analytics, ensuring reproducibility at industrial scales.

Future trajectories will focus on multi-material printing, closed-loop sustainability, and hyper-scale production platforms. As AM evolves from prototyping to end-use manufacturing, it will continue to dismantle long-standing trade-offs between complexity, cost, and performance – solidifying its role as the cornerstone of next-generation ceramic technologies for a resilient, efficient, and sustainable industrial future.

References

1 Ertl, G., Knözinger, H., Schüth, F., and Weitkamp, J. (2008). *Handbook of Heterogeneous Catalysis*. Weinheim: Wiley-VCH.

2 Kollenberg, W. (2004). *Technische Keramik: Grundlagen-Werkstoffe-Verfahrenstechnik*. Essen: Vulkan-Verlag.

3 Mukelabai, M.D., Gillard, J.M., and Patchigolla, K. (2021). A novel integration of a green power-to-ammonia to power system: reversible solid oxide fuel cell for hydrogen and power production coupled with an ammonia synthesis unit. *Int. J. Hydrog. Energy* 46 (35): 18546–18556.

4 Kotowicz, J., Węcel, D., and Jurczyk, M. (2018). Analysis of component operation in power-to-gas-to-power installations. *Appl. Energy* 216: 45–59.

5 Guan, D., Wang, B., Zhang, J. et al. (2023). Hydrogen society: from present to future. *Energy Environ. Sci.* 16 (11): 4926–4943.

6 Zamudio-García, J., Caizán-Juanarena, L., Porras-Vázquez, J.M. et al. (2022). A review on recent advances and trends in symmetrical electrodes for solid oxide cells. *J. Power Sources* 520: 230852.

7 Singh, M., Zappa, D., and Comini, E. (2021). Solid oxide fuel cell: decade of progress, future perspectives and challenges. *Int. J. Hydrog. Energy* 46 (54): 27643–27674.

8 Farahani, R.D., Dubé, M., and Therriault, D. (2016). Three-dimensional printing of multifunctional nanocomposites: manufacturing techniques and applications. *Adv. Mater.* 28 (28): 5794–5821.

9 Tarancón, A., Esposito, V., Torrell, M. et al. (2022). Roadmap on 3D printing for energy. *J. Phys. Energy* 4 (1): 011501.

10 Mahmud, L.S., Muchtar, A., and Somalu, M.R. (2017). Challenges in fabricating planar solid oxide fuel cells: a review. *Renew. Sust. Energ. Rev.* 72: 105–116.

11 Ferreira, V.J., Wolff, D., Hornés, A. et al. (2021). 5 kW SOFC stack via 3D printing manufacturing: an evaluation of potential environmental benefits. *Appl. Energy* 291: 116803.

12 Tomov, R., Dudek, M., Hopkins, S. et al. (2013). Inkjet printing of direct carbon solid oxide fuel cell components. *ECS Trans.* 57 (1): 1359–1369.

13 Z. Chen, J. Ouyang, W. Liang, Z.-C. Yan, F. Stadler, C. Lao, Development and characterizations of novel aqueous-based LSCF suspensions for inkjet printing, *Ceram. Int.* 44(11) (2018) 13381–13388.

14 Jia, K., Zheng, L., Liu, W. et al. (2022). A new and simple way to prepare monolithic solid oxide fuel cell stack by stereolithography 3D printing technology using 8 mol% yttria stabilized zirconia photocurable slurry. *J. Eur. Ceram. Soc.* 42 (10): 4275–4285.

15 Peláez-Tirado, I.M., Marín-Rueda, J.R., Ramos-Fajardo, J.M. et al. (2024). Fused filament fabrication and characterisation of 3- and 8-YSZ-based SOFC electrolytes. *J. Eur. Ceram. Soc.* 44 (8): 5031–5040.

16 Zhu, Z., Gong, Z., Qu, P. et al. (2021). Additive manufacturing of thin electrolyte layers via inkjet printing of highly-stable ceramic inks. *J. Adv. Ceram.* 10: 279–290.

17 Anelli, S., Rosa, M., Baiutti, F. et al. (2022). Hybrid-3D printing of symmetric solid oxide cells by inkjet printing and robocasting. *Addit. Manuf.* 51: 102636.

18 Masciandaro, S., Torrell, M., Leone, P., and Tarancón, A. (2019). Three-dimensional printed yttria-stabilized zirconia self-supported electrolytes for solid oxide fuel cell applications. *J. Eur. Ceram. Soc.* 39 (1): 9–16.

19 Lira, M., Kostretsova, N., Babeli, I. et al. (2023). Large-area 3D printed electrolyte-supported reversible solid oxide cells. *Electrochim. Acta* 467: 143074.

20 Pesce, A., Hornés, A., Núñez, M. et al. (2020). 3D printing the next generation of enhanced solid oxide fuel and electrolysis cells. *J. Mater. Chem. A* 8 (33): 16926–16932.

21 Martos, A.M., Márquez, S., Pavlov, R.S. et al. (2024). 3D printing of reversible solid oxide cell stacks for efficient hydrogen production and power generation. *J. Power Sources* 609: 234704.

22 Márquez, S., Anelli, S., Nuñez, M. et al. (2024). 3D printed electrolyte-supported solid oxide cells based on Ytterbium-doped scandia-stabilized zirconia. *J. Phys. Energy* 6 (1): 015016.

23 Ahamer, C., Opitz, A.K., Rupp, G.M., and Fleig, J. (2017). Revisiting the temperature dependent ionic conductivity of yttria stabilized zirconia (YSZ). *J. Electrochem. Soc.* 164 (7): F790.

24 Ghatee, M., Hossain Shariat, M., and Irvine, J. (2008). Production of high conductivity composite zirconia solid oxide electrolytes with good mechanical strength through net-shape. *J. Mater. Chem.* 18 (43): 5237–5242.

25 Great Barrier Reef Marine Park Authority (2024). Reef facts. https://www2.gbrmpa.gov.au/learn/reef-facts. Accessed on 03/05/2024.

26 O'Mahony, J., Simes, R., Redhill, D. et al. (2015). At what price? The economic, social and icon value of the Great Barrier Reef.

27 David, P. and Bathgate, R. (2023). Great Barrier Reef. https://www.aph.gov.au/About_Parliament/Parliamentary_Departments/Parliamentary_Library/pubs/rp/BudgetReview202223/GreatBarrierReef. Accessed on 03/05/2024.

28 Great Barrier Reef Foundation (2024). Climate Change. https://www.barrierreef.org/the-reef/threats/climate-change. Accessed on 12/09/2024.

29 Australian Institute of Marine Science (n.d.). What is coral bleaching? https://www.aims.gov.au/research-topics/environmental-issues/coral-bleaching/what-coral-bleaching#top. Accessed on 03/05/2024.

30 Australian Institute of Marine Science (n.d.). Coral bleaching events. https://www.aims.gov.au/research-topics/environmental-issues/coral-bleaching/coral-bleaching-events. Accessed on 12/09/2024.

31 Hughes, T.P., Kerry, J.T., Connolly, S.R. et al. (2021). Emergent properties in the responses of tropical corals to recurrent climate extremes. *Curr. Biol.* 31 (23): 5393–5399.e3.

32 Strudwick, P., Seymour, J., Camp, E.F. et al. (2022). Impacts of nursery-based propagation and out-planting on coral-associated bacterial communities. *Coral Reefs* 41 (1): 95–112.

33 Boström-Einarsson, L.A.-O., Babcock, R.C., Bayraktarov, E. et al. (2020). Coral restoration – a systematic review of current methods, successes, failures and future directions, (1932-6203 (Electronic)).

34 Reef Restoration and Adaptation Program (n.d.). Coral aquaculture and deployment. https://gbrrestoration.org/program/coral-aquaculture-and-deployment/. Accessed on 12/09/2024.

35 3DCeram (2022). 3D mix, ceramics for 3D printing technical datasheet.

36 Lei, X., Jiang, L., Zhang, Y. et al. (2021). Coral larval settlement and post-settlement survival facilitated by crustose coralline algae with or without living tissue. *Mar. Biol.* 168 (8): 128.

11

New Developments and Perspectives

Fiona Spirrett and Soshu Kirihara

The University of Osaka, Joining and Welding Research Institute, Department of Lithographic Additive Manufacturing, 11-1 Mihogaoka Ibaraki City, Osaka, 567-0047, Japan

11.1 Introduction

Additive manufacturing (AM) of ceramics has undergone considerable evolution in recent years, offering alternative fabrication methods for ceramic components without the rigid constraints of traditional shaping methods. As outlined in this book, advances in vat photopolymerization (VPP), material jetting (MJT), material extrusion (MEX), powder bed fusion (PBF), and directed energy deposition (DED) have gained popularity for producing geometrically complex and functionally optimized ceramic components. Despite these advancements, it could be argued that current techniques remain limited in scope when considering sustainability, adaptability, and real-time responsiveness of parts and processes.

In this chapter, we consider emerging developments that aim to overcome these limitations and redefine the boundaries of ceramic AM. Concepts such as green manufacturing, digital integration, and functionally responsive printing, broadly referred to as smart design, are becoming increasingly relevant to modern industry. At the same time, innovations such as 4D and 5D printing offer new opportunities for printing either adaptive structures or on nonplanar surfaces, enabling more intricate geometrical control and the potential for broader application.

The shift toward these novel developments is not purely technical but is influenced by a growing awareness of environmental responsibility, resource limitations, and the increasing need for adaptive materials in high-demand sectors such as biomedical, energy, and aerospace. The Sustainable Development Goals, adopted in 2015 by all United Nations members, give clear targets for realizing a sustainable society. With goals such as 9. Industry, Innovation, and Infrastructure, 11. Sustainable Cities and Communities, and 12. Responsible Consumption and Production, the global AM research communities must respond with innovation that aligns with these efforts [1]. As a result, innovation is no longer confined to just technical development but is increasingly shaped by the demand for creative, interdisciplinary solutions that integrate digital fabrication, circular economy principles, and advanced materials science. These efforts are not only accelerating the development of greener manufacturing processes but also expanding the potential of ceramics in applications where sustainability, adaptability, and resilience are a priority. This new trend toward digitalization in AM, working toward sustainable manufacturing and interdisciplinary material design, will hopefully lead to a future in which AM ceramic components are not only geometrically complex but also multifunctional and sustainably produced.

This chapter will first explore the concepts of green and smart design in ceramic AM before discussing 4D and 5D printing. Furthermore, we will discuss the broader potential and challenges of these technologies and present potential future developments and perspectives that may transform ceramic AM beyond its current capabilities.

11.2 Green/Smart Design and Manufacturing

Green manufacturing, also known as sustainable manufacturing, or green production, broadly refers to manufacturing strategies that prioritize environmental sustainability. As outlined in the Brundtland Commission in 1987, sustainable development means "meeting the needs of today's generations, without compromising the ability of future generations to meet their needs" [2]. Generally speaking, the key to green manufacturing practices is reducing the overall carbon footprint of manufacturing processes and minimizing their negative environmental impacts [3]. This may be realized by considering efficient use of resources, minimizing waste, sourcing sustainable raw materials, using renewable energy sources, and reducing harmful emissions. These practices enhance the sustainability of manufacturing processes while also offering significant advantages for businesses, including cost reductions, adherence to regulatory standards, improved reputation, and increased safety for consumers and workers. In the context of ceramic AM, green strategies must consider the full life cycle of ceramic parts, from the origin of raw materials to the energy consumed during post-processing.

Smart design and manufacturing is also an important direction for industrial development. How materials and products are conceptualized, produced, and utilized, particularly in the era of Industry 4.0, is a significant consideration for manufacturing. In modern industry, smart manufacturing systems are utilizing technological advances such as cloud computing, the Internet of Things (IoT), big data analytics, and artificial intelligence (AI) to improve their efficiency, flexibility, and responsiveness across production [4]. In AM, the adoption of smart technologies is a promising direction for developing sustainable manufacturing. By using smart design tools such as computational modeling and simulations, and integrated sensors for process monitoring, AM processing can be optimized from part design through to production, allowing greater control over material usage, energy efficiency, quality assurance, and customization to meet specific application requirements. Computational modeling can be used to reduce the trial-and-error of experimental iterations typically associated with process optimization, which is beneficial for energy, time and cost savings. Many researchers now utilize these tools to explore intelligent design frameworks that incorporate real-time data feedback, digital twins, and predictive modeling to adaptively guide the AM process. Such approaches not only improve the precision and performance of printed components but also enable the creation of more complex, functionally graded, and application-specific structures that would be difficult to design using conventional methods. Intelligent AM ultimately supports the shift toward closed-loop, data-driven systems that foster efficiency, sustainability, and innovation throughout the product life cycle [5].

11.2.1 Sustainable Feedstocks and Material Circularity

One of the advantages often advertised by the AM industry is the ability to use recycled materials as feedstocks and reuse materials that are not joined during processing, limiting material waste associated with manufacturing. For metals and plastics, recycling loops can already be partially closed: excess powders can often be sieved and reused in subsequent builds, and post-consumer waste can be mechanically or chemically processed into feedstock forms like filaments or powders. In metal AM, for example, used titanium powder that meets morphological and purity standards can be reintroduced into the production cycle with minimal loss in performance [6]. Similarly, polymer AM systems are increasingly compatible with recycled thermoplastics, and new initiatives are working to improve the recyclability and traceability of polymer waste streams [7].

However, the path toward material circularity in ceramic AM is perhaps more complex. Ceramics pose specific challenges due to their brittleness, high melting points, and the irreversibility of sintering, which often renders post-processed parts nonrecyclable in the traditional sense. Nonetheless, recent studies and emerging practices are beginning to explore recycling routes for ceramic AM. In the layer-wise slurry deposition (LSD) method, for example, water-based ceramic slurries are prepared and deposited using a method similar to tape-casting and slip-casting [8]. Subsequent layers are dried and then processed by laser or binder jetting methods. With this technique, large, complex ceramic components can be fabricated and supported by dried slurry layers. After processing, the slurry layers, akin to a green body in terms of solidity, are washed in water, redispersing the ceramic powder, which can then be reused (Figure 11.1). Recycling unfired ceramic waste, such as green bodies or excess slurry/paste from lithography-based ceramic AM, in this way, is an important direction for developing sustainable practices in ceramic AM.

Sustainable ceramic AM can also be advanced through the strategic sourcing of recycled or waste-derived ceramic feedstocks. Recent research highlights the incorporation of secondary raw materials, such as construction and demolition waste, industrial by-products, or post-consumer glass, into printable ceramic feedstocks [9]. These alternative sources not only help divert substantial volumes of waste from landfills but also reduce the environmental impact of the extraction and processing of virgin raw materials. Materials such as waste-derived glass powders, fly ash, metakaolin, and ceramic sludge from industrial processes can

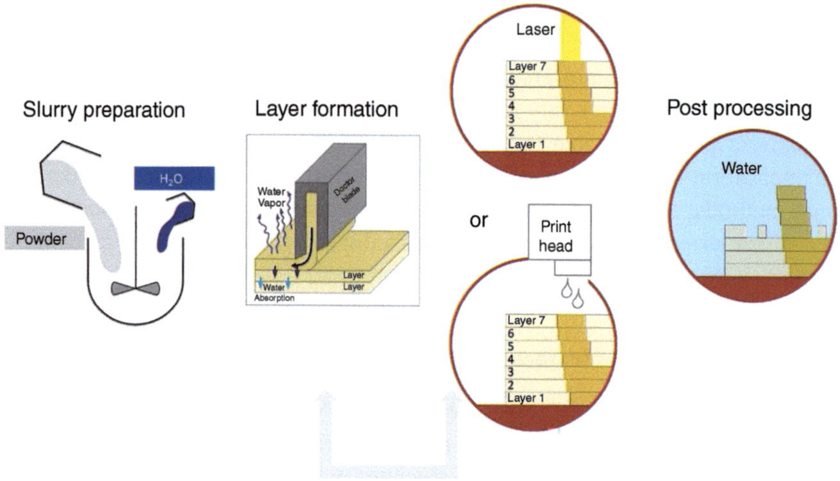

Figure 11.1 Schematic of the LSD process showing water-based slurry preparation, layer formation, and washing process. [8] / with permission of Elsevier.

be reprocessed and formed into ceramic pastes or powders for AM techniques like robocasting, binder jetting, and stereolithography [10–12]. For silica-rich feedstocks made from finely milled waste glass, sintering can be achieved at relatively lower temperatures compared to traditional ceramic compositions. Construction and building applications will further benefit from utilizing waste glass as aggregates for 3D-printed concrete and cement [13, 14]. While there are many studies that highlight the inherent recyclability of glass materials in AM, recycled ceramic feedstocks remain largely unexplored.

For sintered ceramics, the recycling route is considered more challenging. Sintered ceramics must be ground into powders, which may then be used as inert fillers or even as primary feedstocks for composite materials. Although the mechanical and structural properties of these secondary ceramics may be compromised, this approach can be suitable for nonstructural applications, decorative components, or lightweight parts where full density or strength is not essential.

Integrating recycled materials into ceramic AM systems presents challenges in ensuring batch-to-batch consistency, controlling particle morphology and size, and compatibility with binders or resins used in various AM processes. High levels of impurities or variable chemical compositions in waste-derived materials can affect rheology, shrinkage during sintering, and final part performance. For this reason, the importance of developing standardized protocols for characterization, pre-treatment, and blending of recycled feedstocks to ensure reproducibility and printability has been highlighted [9]. Techniques such as sieving, calcination, or surface functionalization may be required to adapt secondary materials for specific AM systems.

Despite these challenges, the use of recycled and industrial by-product-based materials represents an important step toward realizing a circular economy in ceramic AM. These practices not only reduce material costs and energy consumption but also promote more sustainable supply chains, particularly in applications where ceramics are used in high volumes or must withstand demanding environmental conditions. The role of responsible mineral sourcing and reuse should also not be understated as innovation continues, and frameworks for integrating ceramic AM into sustainable production systems must be developed [15, 16]. As material circularity becomes increasingly important in global manufacturing strategies, integrating responsibly sourced ceramic feedstocks into AM workflows will be essential for aligning innovation with environmental responsibility. If supported by intelligent design, process monitoring, and responsible end-of-life strategies, efforts to improve the material circularity of ceramic AM will help to reduce raw material dependency and overall environmental impact.

11.2.2 Environmental Impact of Ceramic Sintering

Ceramic AM processes, particularly those involving sintering, are notably energy intensive. Sintering often requires long exposure to high temperatures over 1200°C, especially for materials like alumina and zirconia. This high energy demand not only increases operational costs but also contributes significantly to greenhouse gas emissions, especially when fossil fuels are used as the primary energy source. Therefore, the environmental footprint of ceramic AM is a pressing concern that requires comprehensive strategies for energy reduction and emission control.

In recent years, efforts to improve the energy efficiency of sintering have been a focus of research. Various methods such as incorporating additives to lower sintering temperatures [17], microwave-assisted sintering (MAS) [18], and two-step sintering strategies have been explored to reduce the thermal load required for ceramic densification. MAS, for example, enables volumetric heating of the ceramic material by transferring energy at the molecular level, significantly reducing sintering times and peak temperatures compared to conventional furnaces. Two-step sintering, which involves an initial high-temperature stage followed by a prolonged dwell at a lower temperature, has been shown to minimize grain growth while achieving high density, thereby lowering overall energy consumption [19]. These methods not only reduce energy demand but can also enhance microstructural control and mechanical performance of the final components. In combination with energy-efficient furnace technologies and optimized sintering profiles, such innovations offer promising routes to reduce the carbon footprint of ceramic AM.

Optimization of Yttria-Stabilized Zirconia (YSZ) pastes for ceramic stereolithography has been shown to impact sintering requirements [20]. The discrete element method, a numerical method for computing the motion and effect of a large number of small particles, was used to make particle models of hypothetical YSZ paste formulations (Figure 11.2). The researchers simulated various combinations of YSZ particle size distributions and calculated the number of particle contact points per weight in the container. Three-dimensional surface graphs were presented to show the effect of particle size and standard deviation on particle contacts. Several formulations of the YSZ pastes were then replicated experimentally and processed by ceramic stereolithography. The green components were sintered, and results showed

Material	Median diameter (μm)	Standard deviation (μm)	Particle size distribution (μm)
Paste 1	0.48	0.19	0.2–1.0
Paste 2	0.48	0.43	0.27–3.9
Paste 3	0.47	0.55	0.08–3.3
Paste 4	0.62	0.67	0.27–3.9
Paste 5	0.72	0.79	0.06–3.9
Paste 6	0.87	0.69	0.27–5.0

Figure 11.2 DEM model of YSZ particles for optimization of paste composition for sintering and physical properties of simulated paste compositions. Composition that achieved reduced sinter time is highlighted.

that a faster heating rate could be tolerated in samples that were made with particles with a large median diameter and standard deviation, and a low number of particle contact points. From this analysis, the sintering time was reduced to 1/10th of the conventional heat treatment time for YSZ. Efforts such as this, utilizing computational methods to reduce experimental iterations and optimize sintering schedules, are expected to reduce the environmental impact of ceramic AM.

11.2.3 Design for Sustainability – Computational Modeling

The geometric freedom of AM techniques is enabling the redesign of functional ceramic components that have previously been limited by traditional forming methods. Technical components can now be optimized for various requirements, including making existing components more sustainable through lightweighting and reducing material usage, or improving the performance. Computational modeling, such as finite element analysis (FEA), plays an important role in smart design and optimization, allowing for the simulation and analysis of complex geometries and material behaviors before component fabrication.

For the ceramic stereolithography method, computational modeling has been utilized in various ways for smart design of components. A YSZ electrode was designed using lattice cell structures to enable aluminum smelting with reduced carbon emissions [21]. Traditional carbon electrodes produce CO_2 gas during the electrolysis of molten cryolite. Thanks to the high oxygen ion conductivity of YSZ ceramics, these concept electrodes promise a greener electrolysis process, only producing O_2 gas. Lattice cells of varying coordination number and aspect ratio were designed for the concept porous electrodes with maximized specific surface area, and the fluid flow was visualized by continuous fluid dynamics (CFD). These fluid simulations were used to identify the lattice cell with well-distributed streamlines with consistent fluid velocities to enable efficient ion transport. Other FEA methods were utilized to visualize the pressure and temperature distribution to ensure that the concept electrodes would be able to function in the environmental conditions of the aluminum smelting process.

Computational methods were also utilized for the smart design and fabrication of alumina components by ceramic stereolithography [22]. A hexagonal honeycomb thermoacoustic stack element was designed with gradient flow path diameters from 1 to 1.5 mm. This concept was designed to improve the thermoacoustic conversion efficiency, and utilized alumina ceramic due to its suitable thermal conductivity and specific heat. Furthermore, the use of discrete element modeling and ray tracing computational simulations was demonstrated to assist in material and process optimization. DEM was used to create particle models using the physical properties of the alumina feedstock, which was used to predict the maximum solid loading for the photosensitive paste. The prediction was close to the experimental value, allowing efficient optimization of materials. The DEM model was further analyzed by ray tracing, a simulation method for analyzing light and optical effects. Using this computational method, the UV laser equipped in the stereolithography device was simulated, and the interaction of the laser light and the particle model was calculated. The resulting predictions of curing depth based on laser parameters allowed efficient optimization of processing parameters, reducing the need for many experimental iterations and therefore reducing the time and cost of the optimization timeline.

Pelanconi et al. [23] presented a comprehensive overview of how computational modeling and digital design have transformed the development of functional ceramic components. In the context of functional optimization, this work shows how simulations can aid the design of ceramic structures with enhanced mechanical performance, tailored porosity, and application-specific geometries. Finite element modeling (FEM) and other numerical tools have allowed researchers to predict the mechanical behavior of cellular ceramic architectures before fabrication, such as their strength, stiffness, and fracture response. These predictions are particularly important for ceramics, where brittleness and processing constraints can limit traditional prototyping methods. The review highlights how AM enables fabrication of these digitally optimized designs, including both periodic and stochastic cellular networks, which can be tuned for targeted energy absorption, thermal insulation, or fluid permeability.

Furthermore, the study underlines how digital tools now facilitate not only structural analysis but also the generation of increasingly complex design motifs, including functionally graded lattices and hierarchical cellular structures. For example, the use of boundary-anchored digital connections between struts can enhance the structural integrity of the components, while simulation-based optimization ensures material is joined precisely. The paper presents design strategies inspired by atomic grain boundary configurations, which allow smooth transitions between different cellular patterns within a single part. This has promising implications for developing multifunctional ceramic components in aerospace, energy, and biomedical applications, where regions of differing mechanical or thermal performance are required within the same part. The integration of these computational modeling methods into ceramic AM workflows could be essential not just for efficiency and resource reduction, but for unlocking entirely new capabilities in ceramic part performance as well.

An interesting direction for functional component design in ceramic AM is toward utilizing fractal patterns. These self-similar geometries that repeat at multiple length scales in two dimensions and three dimensions are generated through recursion

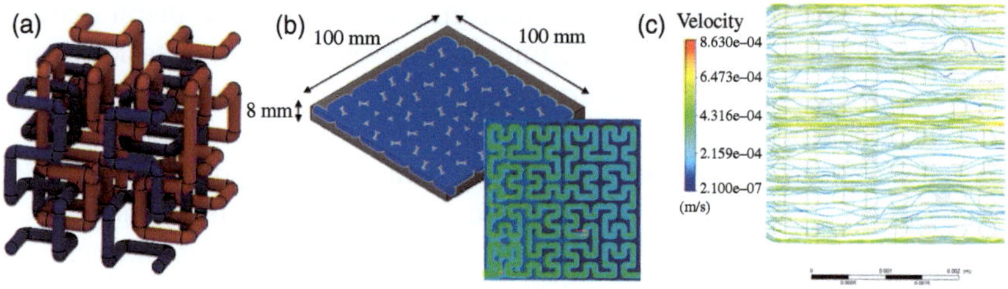

Figure 11.3 Examples of fractal pattern designs and computational simulations for (a) environmental control in structural applications, (b) thermal transfer, and (c) biological fluid flow. [24] / with permission of Elsevier.

and offer interesting design opportunities. Computational modeling of Hilbert curve structures was carried out to explore functions for various applications [24]. FEA was used to evaluate the mechanical performance, thermal conductivity, and fluid dynamics of planar and 3D Hilbert curve structures, highlighting their potential for advanced applications such as structures for environmental control in construction, bone scaffolds in biomedical engineering, and efficient circuit boards (Figure 11.3). The computational simulations were used to demonstrate the performance of such complex components before physical fabrication. Samples were subsequently fabricated by the ceramic stereolithography method using a photosensitive paste made of soda lime silica glass and zirconia. Using a mix of glass and ceramic enabled sintering at a relatively low temperature of 1100 °C. Exploiting the geometric freedom of AM to explore fabrication of ceramic fractal pattern geometries is sure to reveal further opportunities in smart design for novel, functionally optimized components with enhanced performance and sustainability in mind.

11.2.4 Challenges and Barriers to Green Design and Manufacturing

Despite these promising developments, the implementation of green/smart manufacturing strategies for ceramic AM faces several barriers. The integration of recycled or bio-based materials often involves trade-offs in mechanical strength or printability. Smart monitoring systems can be costly and are still limited in spatial or temporal resolution, especially in high-temperature environments. There is also a lack of standardized metrics for assessing the environmental performance of AM processes, making benchmarking and improvement tracking difficult.

However, as regulatory pressures increase and supply chains become more volatile, the case for sustainable, intelligent ceramic AM becomes stronger. Future work is expected to focus on developing holistic frameworks that combine green materials, energy-efficient processes, and AI-driven optimization for truly next-generation manufacturing of ceramic components.

11.3 4D/5D Printing

4D and 5D printing represent emerging fields in AM that extend the capabilities of traditional 3D printing. While 3D printing involves the layer-by-layer fabrication of parts in three spatial dimensions, 4D printing introduces time as a fourth dimension. This refers to fabricated structures that undergo controlled changes in shape, property, or function in response to an external stimulus such as temperature, humidity, or light. In ceramics, 4D printing is an especially novel concept, as it requires integrating smart or responsive material systems into a rigid and typically brittle medium. On the other hand, 5D printing refers not to the addition of time but to an advanced form of 3D printing that employs five-axis movement. In this approach, both the print head and the build platform may have the ability to move simultaneously across multiple axes, allowing the deposition of material along curved paths and complex orientations. This enables stronger, more resilient components due to optimized internal structures and anisotropic control, which is particularly valuable for the mechanical performance of ceramics. While both approaches are still in early stages of development for ceramic processing, they hold promise for creating smarter, stronger, and more efficient components beyond what is currently achievable with conventional 3D-printing methods.

11.3.1 4D Ceramic Printing

The concept of 4D printing was introduced by Skylar Tibbets in 2013, presenting 3D-printed components that could transform shape, proposing an innovative method to fabricate components that could perform self-assembly (Figure 11.4) [25]. For soft materials, 4D printing has shown great promise, allowing easy shape-changing properties based on external stimuli such as heat, light, pH, moisture, electricity, or magnetic field [26]. 4D printing of various polymers has shown impressive results such as self-expandable scaffolds for biomedical applications [27] and light-responsive membranes for tailored and controlled CO_2 transfer [28]. For ceramic materials, typical 3D printing with post-process sintering results in rigid, dense, high-temperature components that makes the concept of 4D-printing challenging. 4D-printed ceramic components must be designed to change in response to an external stimulus. Despite the brittle and inert nature of ceramics, the emergence of new design strategies, material systems, and composite processing techniques has promised opportunities for integrating responsive behavior into ceramic-based components. This integration requires a shift from simply fabricating stable, dense parts to engineering material systems that can undergo predictable and potentially reversible changes post-fabrication.

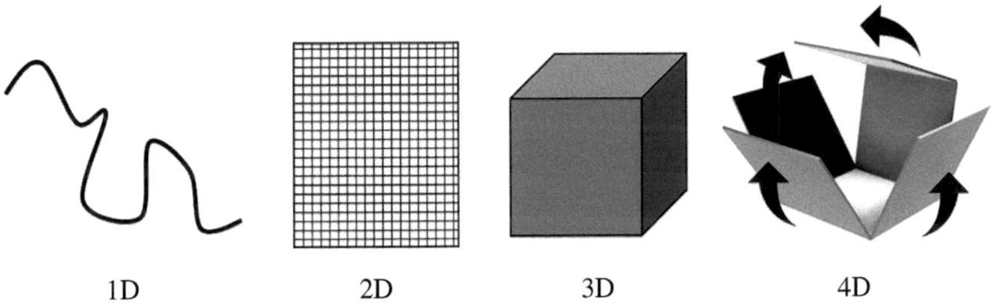

Figure 11.4 Schematic illustration of 1D–4D fabrication.

Stimuli-responsive behavior in ceramics is particularly relevant for applications involving extreme environments where traditional responsive materials like polymers degrade. For example, transformations triggered by temperature can be induced using thermally activated phase changes or differential sinter shrinkage. Electric-field-induced changes in piezoelectric ceramics could be used to create actuating or sensing systems, while humidity-sensitive ceramic membranes could modulate fluid or gas transport in filtration and environmental control applications. One unique advantage of using ceramics in 4D printing is the potential for stability and reliability over prolonged use in harsh conditions, such as high temperatures, corrosive substances, or vacuum environments, making them particularly suitable for aerospace, energy, and industrial automation applications. Computational tools are increasingly used to simulate and optimize these stimuli-responses, integrating multi-physics models to account for mechanical, thermal, and electrical coupling in design.

An example of 4D printing of ceramics was presented by Wang et al. [29]. This innovative work used a material-centric approach to exploit the behavior of ceramic materials during processing. Specifically, the material property of differential shrinkage during sintering was used to control post-sintering shape transformation of zirconia parts fabricated by direct ink writing (DIW). This approach moves away from the reliance on externally triggered smart materials and instead utilizes the natural physical changes of ceramics during thermal post-processing. In the study, bilayer zirconia (ZrO_2) structures were fabricated using DIW, with the two layers differing in solid loading. When sintering, the mismatch in shrinkage between the high and low solid content layers induced controlled bending and curling of the printed parts, enabling the transformation of initially flat geometries into complex three-dimensional shapes such as twisted flower structures (Figure 11.5). The extent and direction of deformation were controlled by adjusting the ink formulation and print path, allowing a considerable degree of design flexibility within the system. This method was also demonstrated for processing alumina ceramics into twisting structures that deformed after sintering [30].

Figure 11.5 Example of 4D printing of ceramic UV inks to create a blooming flower structure by modifying the solid content of the ink for different layers and stacking them. [29] / with permission of Elsevier.

These examples of 4D-printing work toward addressing the key challenge of the limited responsiveness of ceramics to common external stimuli due to their brittle and inert properties. By designing the shape change trigger mechanism into the sintering process itself, these studies effectively expand the design space for ceramic AM and create opportunities for fabricating shape-changing ceramic components without requiring smart material additives or multi-material systems. However, despite its potential, the technique remains limited by a reliance on fine control of material formulation and an understanding of material shrinkage behavior, which may be sensitive to environmental factors and process variability. Additionally, the transformation is a single effect bound to the irreversible sintering process, limiting its application in scenarios requiring reversibility or multiple response cycles. Nonetheless, this method introduces a practical and scalable strategy for incorporating fourth-dimensional characteristics into ceramic structures, particularly for applications in biomimetic systems, microfluidics, and lightweight architectural elements where controlled curvature and form adaptation can be functionally advantageous.

Wang et al. developed a bilayer system composed of a photocurable ceramic elastomer and a hydrogel to create shape-changing ceramic structures [31]. Using a multi-material digital light processing (DLP) technique, the researchers printed flat composite laminates using the hydrogel layer to induce programmed bending through dehydration. This bending was preserved during subsequent thermal treatment, where the hydrogel was removed, and the ceramic was sintered, producing a final monolithic ceramic structure with a complex 3D geometry. The authors used the method to create various structures such as cubes, Miura-origami folds and cranes, a twisted fan blade, and a miniature scorpion figure (Figure 11.6). They also introduced a theoretical model to predict shape changes based on bilayer thickness, modulus, and shrinkage parameters, enabling rational design of the shape-change behavior.

The results represent a significant advance in the field by utilizing the capabilities of polymer-based 4D printing and the structural integrity and functional properties of ceramics. By incorporating a sacrificial hydrogel layer as an active shaping component, the process overcomes the challenge of ceramic materials' limited intrinsic responsiveness and promises programmable deformation for future applications. Unlike systems relying solely on sintering mismatch, this approach introduces a broader design strategy by utilizing multi-material interactions and time-controlled actuation in the green body stage. There are still certain limitations to this 4D-printing method, such as material and processing challenges due to the need for precise co-printing of hydrogel and ceramic slurries, and the final shapes are still determined by irreversible shape changes, limiting applicability in reversible systems. Moreover, the thermal removal of hydrogel may introduce porosity or defects at the ceramic interface, and scale-up would require optimization. Even so, this method offers a creative solution to integrating time-dependent shape change into ceramics and highlights the potential of hybrid material systems in future 4D ceramic AM applications.

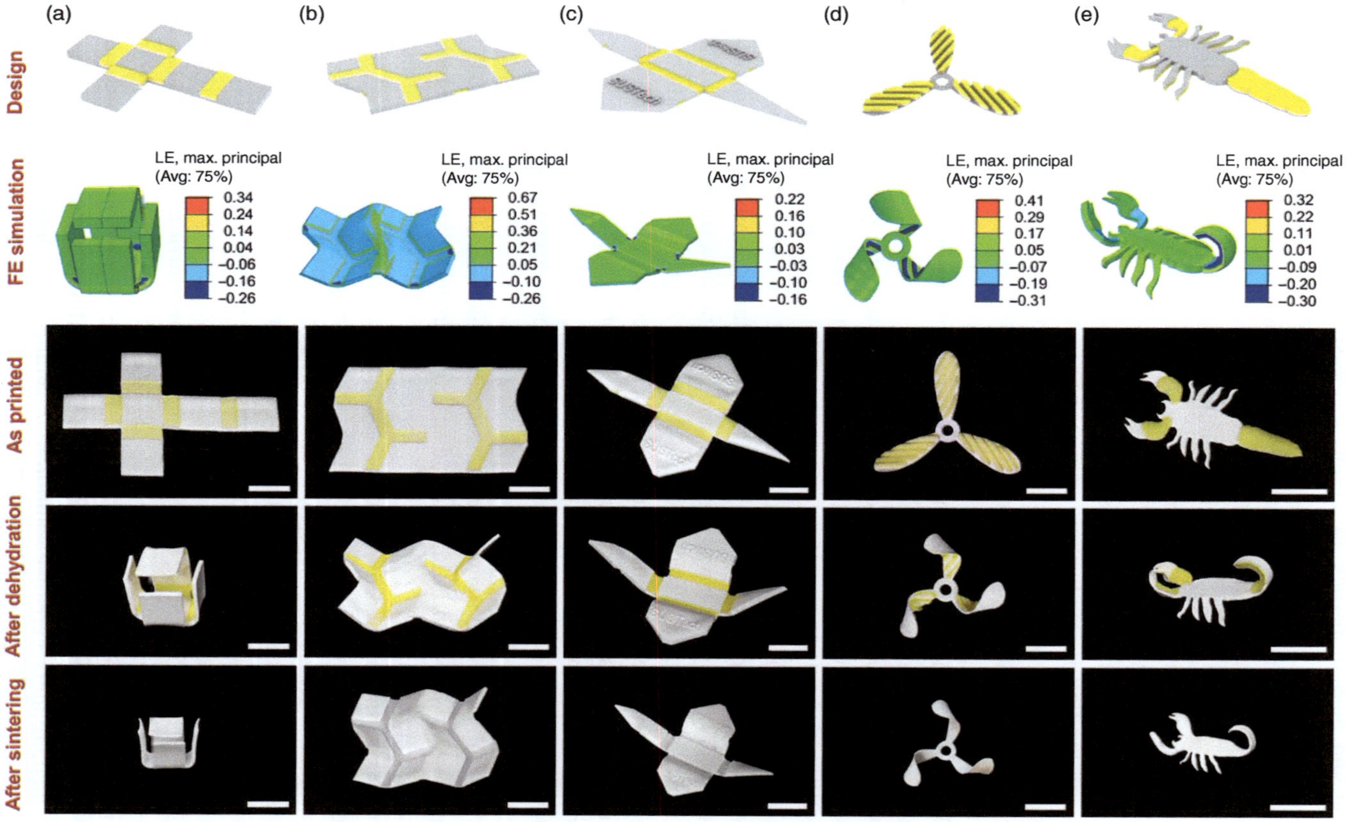

Figure 11.6 Examples of 4D printing of ceramic structures using hydrogel and ceramic materials to induce shape-change characteristics after heat treatment. [31] / Springer Nature / CC BY 4.0.

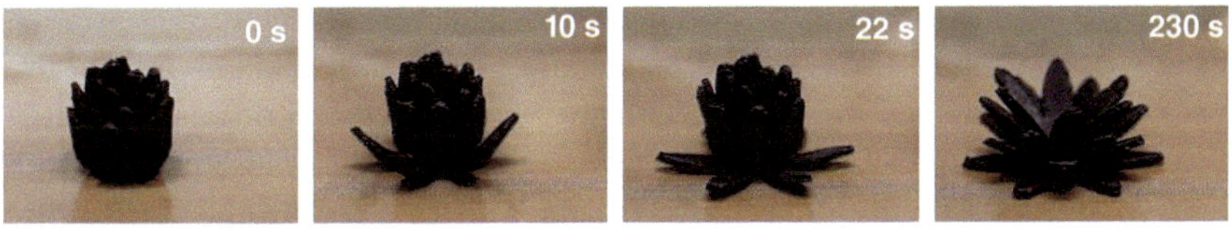

Figure 11.7 4D-printed flower shape when heated on a hot plate, demonstrating the shape recovery process from the programmed temporary shape to the recovered permanent shape. [32] / with permission of Elsevier.

4D printing of ceramic components with reversible shape memory characteristics has also been demonstrated in recent years. Fabrication of complex ceramic structures, such as a blooming flower shape, was carried out using shape memory polymer-derived ceramics (PDCs) (Figure 11.7) [32]. The researchers developed a multifunctional preceramic polymer composite that shows both reconfigurability and shape memory effects. By optimizing the material compositions, they created precursor inks suitable for 4D printing, which transformed into lightweight, self-shaping ceramics by pyrolysis. The preceramic polymers underwent a stepwise partial-to-full crosslinking process to form interpenetrating polymer networks, allowing the printed parts to be transformed into more complex shapes. Unlike traditional precursors with static shapes, the precursor used had shape memory capability ($Tg \sim 95\,°C$). During the sintering process, the 3D-printed precursor successively underwent programmable active shape change with a high recovery rate and was converted into lightweight ceramic by pyrolysis. Ceramic temperature indicators with complex geometry to warn about overheating were proposed as an application. The samples showed a wide temperature detection range (25–750 °C) and could be tailored to various surface shapes.

Applications of 4D printing in ceramics are still in their early stages compared to polymers and metals, but the potential is significant. These efforts to apply 4D printing to ceramic materials show great promise for creating complex structures with shape-changing capabilities. Despite the challenges that ceramic materials present in this field, researchers are finding creative solutions to utilize this technology. While there are still limitations, particularly regarding reversible shape changes and process control, the field is likely to continue expanding due to the many potential applications of 4D-printed ceramics such as intelligent sensing, adaptive structures, thermal warning systems, and biomimetic devices.

11.3.2 5D Ceramic Printing

While most AM technologies operate within the constraints of three linear axes (hence the term "3D printing"), emerging developments have given rise to 5D printing, which expands the degrees of freedom by incorporating multi-axis movement during processing. By incorporating rotational and tilting movement to the print head and/or the build platform, curved layer deposition can be achieved in this 5D-printing method [33]. Compared to traditional 3D printing, where most processes are limited to material deposition in flat, planar slicing, 5D printing presents opportunities for even greater geometric freedom. This approach also allows for the creation of components with greater mechanical performance, particularly in curved or stress-concentrated regions, by aligning material deposition with stress trajectories. First introduced by William Yerazunis of Mitsubishi Electrical Research Lab, the 5D-printing method was used to create a curved pressure cap. The resulting mechanical strength of the 5D-printed cap was three to five times stronger than the same cap printed by the 3D method [34]. For ceramics, which are typically brittle and prone to cracking under mechanical loads, 5D printing could achieve improved structural integrity through more intelligent deposition paths. The technique also enables the fabrication of geometrically complex or nonplanar parts without extensive support structures, which is particularly beneficial in ceramic processing, where support removal and warping during sintering can be problematic. Although still in its early stages, 5D printing of ceramics has begun to attract attention as a promising innovation, offering new possibilities for advanced structural components, biomedical implants, and functionally graded materials that benefit from curvature-conforming fabrication strategies.

The use of robotic arms is becoming more popular for 5D-printing applications. Five-axis tooling was developed for use with an industrial robotic arm for multi-axis extrusion of ceramic materials [35, 36]. Clay material was deposited to make complex brick shapes with various architectural features such as folding, coiling, and the meandering of viscous threads, along with variations in tooling structures between consecutively deposited layers. The potential for designing complex internal structures for tailored thermal exchange within architectures was presented. Furthermore, this method could achieve high print speeds and cantilevers of 40° with further development. Concept architectural designs were developed, showing promising geometric freedom and creativity for future applications of five-axis ceramic extrusion in construction.

While five-axis printing promises benefits such as allowing parts to be made without support structures, faster fabrication, and improved mechanical properties thanks to optimized orientation of layer lines, these additional degrees of freedom introduce challenges with slicing software. Because each layer can extend through multiple dimensions, optimization of toolhead movement is more complex than with 3D-printing methods [33]. As a result, software must consider a greater number of variables when generating toolpaths, which are no longer strictly linear or layer-based, leading to increased complexity. This poses a significant challenge for 5D systems for commercialization as most commercial computer-aided design software is intended for 3D design [37]. One particular challenge in toolpath planning for 5D printing is in complex surfaces that require many axis reversals. These reversals will usually result in tools slowing down to change direction, causing overbuilding or defects in parts as the deposition rate remains constant. To limit excess motion and path reversals, graphs and dynamic programming can be used to smooth surfaces [38, 39]. Furthermore, incorporating advanced control systems and in situ monitoring is expected to help mitigate overbuilding and defects in these AM systems [40].

An important note to mention is that there are multiple interpretations of what "5D printing" actually means, and the term is not yet strictly standardized across the field of AM. While the above-described multi-axis printing is possibly the most commonly accepted definition of "5D printing," there are other interpretations that should be considered. One alternative characterization of 5D printing is the combination of printing technologies in 3D with 2D surface modifications (3D + 2D = 5D). Surface topographies are known to have significant effects on various component properties, such as wear-resistance [41], wettability [42], biocompatibility [43], appearance, and much more. Combining these surface modifications with 3D-printing methods could achieve the fabrication of complex ceramic geometries with engineered surfaces at the micro- and nanoscale, further tailoring their functional properties.

One example of these surface modifications was presented by Kirihara et al. [44]. A ceramic paste was created by mixing nano-sized titania particles with a photosensitive acrylic resin. Micro-stereolithography and aqueous solution techniques were used to create titania micro-patterns. These titania artificial interfaces had precise hexagonal holes of 120 μm length. The planar patterning for photonic crystal structures was proposed for applications in microreactors and artificial photosynthesis devices. It is possible to imagine combining this technique with the fabrication of larger 3D architectures for further functional applications in ceramics.

Micro-textured ceramic components have been successfully fabricated by the DLP method [45]. Two biodevices were designed with micron-scale surface textures for fabrication by DLP lithography of alumina slurries (Figure 11.8). One device demonstrated cell culture platforms with controlled surface topographies with spiked features. The other device was designed for studying cell motility and had micro-textured channels. With particular promise in biomedical applications, 3D-printed structures with controlled surface textures such as these are expected to enhance component performance. Optimization of "2D" surface patterning on customized, complex medical devices could significantly affect cell behavior and tissue integration. As it is challenging to impart surface patterning on sintered ceramics using conventional micromachining techniques, being able to integrate topography modifications into the 3D-printing process has great potential. Furthermore, it is possible to obtain textured ceramics in AM by manipulating factors such

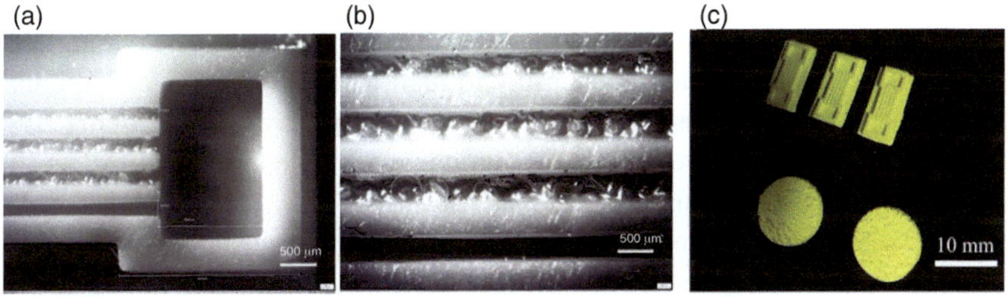

Figure 11.8 Micro-scale surface texturing by DLP processing of a cell motility testing device (a–c) and cell culture platforms with spiked features (c). [45] / Springer Nature / CC BY 4.0.

as the rheological properties of the feedstock, printhead geometry, and printing speed [46]. This control of microstructure and texture in printed ceramics could be used to tailor the mechanical and functional properties of ceramic components for applications in bio-, structural-, and electro-ceramics.

11.4 Potentials and Challenges

Ceramic AM is an established manufacturing method that will continue to develop as the demands of industry and modern society evolve. With the increasing pressure to prioritize sustainable fabrication, green and smart manufacturing will play a significant role in the future of ceramic AM. As researchers continue to develop fabrication methods for complex, high-performance ceramic components, the issue of sustainable manufacturing, responsible material sourcing, and recycling will need to be considered. Recent advances in material science, computational modeling, and multi-material printing are expanding the design space and functional capabilities of AM ceramic components, and various industries are already benefitting from these advances, utilizing the design freedom of AM and the desirable properties of ceramic materials to create components without the limitations of traditional manufacturing. However, despite these developments, there are still significant challenges to overcome, particularly in scalability, material limitations, process standardization, and reliability.

As AI becomes more established in industrial applications, ceramic AM is likely to benefit from enhanced process control, real-time process parameter optimization, and more robust computational simulations and predictions. Digital twin technologies offer the potential to virtually replicate AM processes, enabling continuous monitoring and accurate process simulations. In situ monitoring is another powerful tool, allowing real-time observation and control of AM processes. With appropriate monitoring, such as optical imaging, thermal cameras, and laser profilometry, defects during fabrication can be detected early, such as layer delamination, cracking, or uneven deposition. Together, these tools can improve manufacturing efficiency, reduce defects, and accelerate design iterations. However, challenges remain, particularly in collecting high-quality, domain-specific data for training AI models, and in accurately modeling the complex sintering behavior and shrinkage dynamics of ceramic materials within digital twin frameworks. Implementation of in situ monitoring may be limited by the typical environments in ceramic AM systems, such as exposure to high temperature or UV. Combined with feedback control loops and AI, in situ monitoring may help to improve AM ceramic components. Overcoming these limitations will help to fully realize the benefits of smart manufacturing in ceramic AM.

Expanding the catalogue of processable ceramic materials and improving scalability will help the industry realize the full potential of ceramic AM. Recent advances have enabled the fabrication of a wider range of technical ceramics, such as zirconia, alumina, and silicon nitride, yet there are a range of other ceramics that have not yet been established in AM. Ultrahigh temperature ceramics and complex composites remain difficult to process due to challenges such as high sintering temperatures, powder rheology, optical properties, or binder compatibility. Additionally, while the most established use-case for AM is in high-value, low-production volume manufacturing, industrial production presents challenges in scalability and repeatability. Ceramic AM technologies are still relatively slow and expensive compared to traditional mass-manufacturing methods like pressing and injection molding. While AM excels in customization and prototyping, scaling up for mass production remains constrained by equipment throughput, material costs, and labor-intensive post-processing. For ceramic AM to be widely adopted in industry, improvements must be made in consistency, cost effectiveness, and maintaining uniform material properties. Variability in feedstock quality, shrinkage during sintering, and long process times further limit applicability to industrial-scale production. Overcoming these limitations would require standardized processing protocols and scalable post-processing workflows, as well as material developments. As a result, many AM ceramic parts are still confined to research and development rather than mainstream commercial production. While unlikely to be changed in the very near future, successfully expanding the material catalogue and demonstrating scalability would improve the versatility and relevance of ceramic AM to industry.

4D and 5D printing, as introduced in this chapter, are starting to be explored for ceramic processing. The current research is limited to a few ceramic compositions and geometries but shows great potential for fabricating components with greater geometric complexity or functionality. While there are various challenges to overcome, advances in 4D and 5D printing of ceramics could allow for high-performance ceramic parts with potential in adaptive biomedical devices, aerospace components, and responsive architectural materials. Developments in hybrid AM, a combination of additive and subtractive post-processing in a single system, are expected to achieve higher precision, better surface finish, and integration of components such as electrodes or sensors directly into ceramic components. Developments in hybrid AM promise multifunctional ceramic parts across a wide range of applications, such as electronics, aerospace, and energy. As with any ceramic processing, the brittle nature of ceramics and their sensitivity to thermal and mechanical stresses make them difficult to machine or process in green or sintered states without inducing defects or cracks. Coordinating additive and subtractive steps, especially in terms of toolpath planning, fixturing, and thermal management, requires advanced control systems and precise alignment, which are still underdeveloped for ceramics compared to metals or polymers. Furthermore, the range of ceramic compositions means that hybrid processing must often be tailored on a case-by-case basis, limiting standardization and widespread adoption. Material compatibility between AM feedstocks and other hybrid process materials, like binders or coatings, also remains a significant limitation. Future advances in these technologies may enable ceramic processing beyond the current constraints of conventional and even 3D-printing techniques.

Ceramic AM is an advancing field with great potential to disrupt traditional manufacturing. The ability to fabricate geometrically complex, functionally graded, and application-specific parts introduces new possibilities across fields from biomedical engineering to architecture. Innovations such as 4D printing, hybrid multi-material approaches, and AI-driven design are expected to drive further developments in the field. However, the technology still faces considerable barriers to widespread adoption, particularly in materials processing, cost-effectiveness, and reproducibility. Addressing these challenges, as well as meeting sustainability goals, will require concerted efforts across multiple disciplines to fully realize the capabilities of ceramic AM.

References

1 Faheem, A., Hasan, M.A., Badhotiya, G.K., and Ali, S.M. (2024). A systematic literature review to examine the role of additive manufacturing in achieving sustainable development goals. *Sustain. Oper. Comput.* 5: 216–226.

2 World Commission on Environment and Development (1987). Brundtland Report 'Our Common Future'.

3 Deif, A.M. (2011). A system model for green manufacturing. *J. Clean. Prod.* 19: 1553–1559.

4 Zheng, P., Wang, H., Sang, Z. et al. (2018). Smart manufacturing systems for Industry 4.0: conceptual framework, scenarios, and future perspectives. *Front. Mech. Eng.* 13: 137–150.

5 Xiong, Y., Tang, Y., Zhou, Q. et al. (2022). Intelligent additive manufacturing and design: state of the art and future perspectives. *Addit. Manuf.* 59: 103139.

6 Denti, L., Sola, A., Defanti, S. et al. (2019). Effect of powder recycling in laser-based powder bed fusion of Ti-6Al-4V. *Manuf. Technol.* 19: 190–196.

7 Mishra, V., Negi, S., and Kar, S. (2023). FDM-based additive manufacturing of recycled thermoplastics and associated composites. *J. Mater. Cycles Waste Manag.* 25: 758–784.

8 Zocca, A. and Günster, J. (2024). Towards a debinding-free additive manufacturing of ceramics: a development perspective of water-based LSD and LIS technologies. *Open Ceram.* 19: 100632.

9 Villa, A., Gianchandani, P., and Baino, F. (2024). Sustainable approaches for the additive manufacturing of ceramic materials. *Ceramics* 7: 291–309.

10 Villaquirán-Caicedo, M.A., Fernández-González, A., Fernández-García, D.A., and Mejía de Gutiérrez, R. (2023). Valorization of a low-quality coal ash, in the preparation of alkali activated inks for applications in 3D additive manufacturing. *Constr. Build. Mater.* 399: 132598.

11 Deng, Q., Zou, S., Xi, Y., and Singh, A. (2023). Development and characteristic of 3D-printable mortar with waste glass powder. *Buildings* 13: 1476.

12 Marchelli, G., Prabhakar, R., Storti, D., and Ganter, M. (2011). The guide to glass 3D printing: developments, methods, diagnostics and results. *Rapid Prototyp. J.* 17: 187–194.

13 Annapareddy A, Li MT, Tan MJ, et al. (2018). Effect of recycled glass gradation in 3D cementitious material printing. *Proceedings of the 3rd International Conference on Progress in Additive Manufacturing (Pro-AM 2018).* 14-17 May 2018, Singapore.

14 Cuevas, K., Chougan, M., Martin, F. et al. (2021). 3D printable lightweight cementitious composites with incorporated waste glass aggregates and expanded microspheres – rheological, thermal and mechanical properties. *J. Build. Eng.* 44: 102718.

15 Giurco, D., Littleboy, A., Boyle, T. et al. (2014). Circular economy: questions for responsible minerals, additive manufacturing and recycling of metals. *Resources* 3: 432–453.

16 Tavares, T.M., Ganga, G.M.D., Godinho Filho, M., and Rodrigues, V.P. (2023). The benefits and barriers of additive manufacturing for circular economy: a framework proposal. *Sustainable Prod. Consumption* 37: 369–388.

17 Xiao, Y., Cheng, D., Li, G. et al. Preparation of MgO ceramics by low temperature sintering with MgF_2 and Al_2O_3 as sintering additives. *J. Electroceram.* 53: 281–292.

18 Ćurković, L., Veseli, R., Gabelica, I. et al. (2021). A review of microwave-assisted sintering technique. *Trans. FAMENA* 45: 1–16.

19 Lóh, N.J., Simão, L., Faller, C.A. et al. (2016). A review of two-step sintering for ceramics. *Ceram. Int.* 42: 12556–12572.

20 Takahashi, M., Spirrett, F., and Kirihara, S. (2022). Reduction of dewaxing and sintering time by controlling the particle size of YSZ particles for stereolithography. *Ceramics* 5: 814–820.

21 Takahashi, M. and Kirihara, S. (2021). Stereolithographic additive manufacturing of zirconia electrodes with dendritic patterns for aluminum smelting. *Appl. Sci.* 11: 8168.

22 Spirrett, F., Ito, T., and Kirihara, S. (2022). High-speed alumina stereolithography. *Appl. Sci.* 12 (19): 9760.

23 Pelanconi, M., Rezaei, E., and Ortona, A. (2020). Cellular ceramic architectures produced by hybrid additive manufacturing: a review on the evolution of their design. *J. Ceram. Soc. Jpn.* 128: 595–604.

24 Spirrett, F., Oi, A., and Kirihara, S. (2024). Additive manufacturing of composite glass/ceramic structures with self-similar geometries. *Open Ceram.* 17: 100509.

25 Tibbets, S. (2013). The emergence of '4D printing'. http://www.ted.com/talks/skylar_tibbits_the_emergence_of_4d _printing. 11th April 2025.

26 Fu, P., Li, H., Gong, J. et al. (2022). 4D printing of polymers: techniques, materials, and prospects. *Prog. Polym. Sci.* 126: 101506.

27 Wan, X., Wei, H., Zhang, F. et al. (2019). 3D printing of shape memory poly(D,L-lactide-co-trimethylene carbonate) by direct ink writing for shape-changing structures. *J. Appl. Polym. Sci.* 136: 48177.

28 Gillono, M., Roppolo, I., Frascella, F. et al. (2020). CO_2 permeability control in 3D printed light responsive structures. *Appl. Mater. Today* 18: 100470.

29 Wang, F., Liu, C., Yang, H. et al. (2023). 4D printing of ceramic structures. *Addit. Manuf.* 63: 103411.

30 Wan, L., Mao, Z., Liu, H. et al. (2023). Direct 4D printing of gradient structure of ceramics. *Chem. Eng. J.* 465: 142804.

31 Wang, R., Yuan, C., Cheng, J. et al. (2024). Direct 4D printing of ceramics driven by hydrogel dehydration. *Nat. Commun.* 15: 758.

32 Chen, S., Li, J., Shi, H. et al. (2023). Lightweight and geometrically complex ceramics derived from 4D printed shape memory precursor with reconfigurability and programmability for sensing and actuation applications. *Chem. Eng. J.* 455: 140655.

33 Duarte, J., Espírito Santo, I., Monteiro, M.T. et al. (2022). Curved layer path planning on a 5-axis 3D printer. *Rapid Prototyp. J.* 28: 629–636.

34 Kumaresan, R., Kulandaivel, S., Samykano, M. et al. Additive manufacturing: an overview of printing technologies. *J. Adv. Res. Appl. Sci. Eng. Technol.* 59.

35 Hinz, K. (2025). Brick geometries: 5-axis additive manufacturing for architecture. https://research.gsd.harvard.edu/maps/2017/03/24/brick-geometries-5-axis-additive-manufacturing-for-architecture/ (accessed 9 May 2025).

36 Hergel, J., Hinz, K., Lefebvre, S., and Thomaszewski, B. (2019). Extrusion-based ceramics printing with strictly-continuous deposition. *ACM Trans. Graph.* 38: 1–11.

37 Feldhausen, T., Barnes, A., Roschli, A. et al. (2024). Toolpath considerations for 5-axis systems. In: *Motion and Path Planning for Additive Manufacturing* (ed. A.C. Roschli, M.C. Borish, A.K. Barnes, et al.). Elsevier.

38 Plakhotnik, D. and Lauwers, B. (2014). Graph-based optimization of five-axis machine tool movements by varying tool orientation. *Int. J. Adv. Manuf. Technol.* 74: 307–318.

39 Sun, Y., Xu, J., Jin, C., and Guo, D. (2016). Smooth tool path generation for 5-axis machining of triangular mesh surface with nonzero genus. *Comput. Aided Des.* 79: 60–74.

40 Gibson, B.T., Richardson, B.S., Sundermann, T.W., and Love, L.J. (2019). Beyond the toolpath: site-specific melt pool size control enables printing of extra-toolpath geometry in laser wire-based directed energy deposition. *Appl. Sci.* 9: 4355.

41 Bhushan, B., Israelachvili, J.N., and Landman, U. (1995). Nanotribology: friction, wear and lubrication at the atomic scale. *Nature* 374: 607–616.

42 Zhang, P., Wang, S., Wang, S., and Jiang, L. (2015). Superwetting surfaces under different media: effects of surface topography on wettability. *Small* 11: 1939–1946.

43 Buxboim, A. and Discher, D.E. (2010). Stem cells feel the difference. *Nat. Methods* 7: 695–697.

44 Kirihara, S., Itakura, Y., and Tasaki, S. (2013). Creation of titania artificial interfaces with geometric patterns by using microstereolithography and aqueous solution techniques. *Int. J. Appl. Ceram. Technol.* 10: 468–473.

45 de Blas, R.A., Pfaffinger, M., Mitteramskogler, G. et al. (2017). Lithography-based additive manufacture of ceramic biodevices with design-controlled surface topographics. *Int. J. Adv. Manuf. Technol.* 88: 1547–1555.

46 Walton, R.L., Kupp, E.R., and Messing, G.L. (2021). Additive manufacturing of textured ceramics: a review. *J. Mater. Res.* 36: 3591–3606.

Index

μXCT 246
1 + 1 = 3 effect 211
3D Model Maker 86, 87
3D pore morphology 259
3D printing company 70
3D printing configuration 225
4D/5D printing
  1D-4D fabrication 329
  advantage 330
  blooming flower shape 330, 333
  example of 330–332
  polymer-based 331
5-phase 68
5D ceramic printing 333–336

a

acoustic DOD inkjet printing (AIJ) 59
additive manufacturing (AM) of ceramics
  binder jetting (BJT) 11, 13, 16, 62–84
    definition and working principle 62
    historical development and applications 63
  challenges and the need 9–10
  directed energy deposition (DED) 12, 13, 16, 195–207
  features and characteristics of 14
  history and development of 12, 13
  inkjet printing
    drop-on-demand (DOD) 52–59
    theory of ink printability 59–62
  inkjet printing (IJP)
    continuous 52–54
  material extrusion (MEX) 10, 13, 15, 109–139
  material jetting (MJ) 11, 13, 15
    applications 94–96
    challenges in 90–94
    definition and classification of 84–86
    historical development 86–88
    post-processes 98–100
    potential of 96–98
    process variations 88–90
  powder bed fusion (PBF) 11, 13, 16, 153–187
  sheet lamination (SHL) 11–13, 16
  Vat photopolymerization (VPP) 10, 11, 13, 15, 25–46, 295, 298, 300, 307
advanced ceramics 1–5, 7–8, 14, 18, 70–73, 75, 76, 96, 98–100, 112, 153, 161, 197, 207, 275, 300, 304, 317
aerosol jet printing (AJP) 226
alumina 3–5, 77, 88, 136, 157, 163–165, 175, 196, 201, 202, 223–225, 300, 303, 313–315
alumina (A) surface layers 279–280, 282
Alumina slurry 166, 282, 312, 315, 335
alumina toughened zirconia (ATZ) 136, 166, 169, 170, 279–282
angle of response (AoR) method 249
anisotropic shrinkage 271, 273
aqueous solution technique 335
Archimedes' method 258
artificial intelligence (AI) 295, 298, 317, 318, 322, 328, 336

b

B3B test 280, 284
batch-level shrinkage 257
beam fusion 154
beam speed (BS) 161
BET method 258, 259
biaxial bending strength 284
binder bleeding 83
binder jetting (BJT) 11, 13, 16
  definition and working principle 62–63
  historical development and applications
    AM techniques 63
    concrete and construction materials 65–70
    of gypsum plaster models 64
    sand casting molds and cores 64–65
    technical applications 70–77
  post-processes
    cleaning 98
    debinding 98
    drying and/or curing 98
    sintering 99–100
  powder feedstock properties 77–80
    layer recoating 80–82
    process parameters and binder saturation 82–84
bioactive ceramics 72
bioceramics 1, 5, 72, 76
bioinert ceramics 72
biomedical engineering 273–274, 304, 328
BJ/MEX 212, 214
boron carbide (B_4C) 4, 112, 132, 160, 185–186, 224
brittility factor (Bt) 115
bubble jet 57
bulk density 258
bulk-solid based AM 251–252
buoyancy measurements 258

c

calipers 257
cathode current (BC) 161
ceramic AM
  bulk-solid based AM 251–252
  composition and microstructure characterization 255–256
  density and porosity 257–259
  dimensional and geometrical stability 256–257
  liquid/slurry based AM 250–251
  mechanical properties 252–254
  model design 243–244
  and numerical simulation 244–245
  particle size distribution (PSD) 246–248
  powder-based AM 245–250
  powder flow and spreadability 249–250
  powder morphology 246
  powder packing density 248
  surface roughness 256
  topological optimization 244–245
ceramic coatings 202, 203, 206
ceramic components 7–10, 12, 14, 17, 18, 37, 45, 51, 76, 88, 97, 98, 111, 124, 131, 164, 169, 182, 184, 186, 195, 196, 203, 206, 212, 215, 217–221, 227, 229–231, 271, 272, 275, 277, 279, 282, 286, 289, 295, 304, 315–317, 321–323, 326–329, 331, 333, 335–337
ceramic green parts 218
ceramic heating elements 230
ceramics
  additive manufacturing 9–14
    aerospace applications 296–297
    binder jetting 11
    challenges and the need 9–10
    chemical applications 298–299
    cross-domain value and systemic impact of 315–317
    defense sectors 297–298
    directed energy deposition (DED) 12
    downholders for welding 299–300
    environmental protection, ecological restoration for 311–315

features and characteristics of 14, 15
history and development of 12, 13
lightweight and disruptive designs 296
material extrusion (MEX) 10
material jetting (MJT) 11
next-generation surgical tools and devices 303–306
orthopedic and dental solutions 302–303
personalized healthcare 301–302
powder bed fusion (PBF) 11
semiconductor industry 301
sheet lamination (SHL) 12
solid oxide cells 306–310
space communication 297
Vat photopolymerization (VPP) 10
advanced 1, 4
biocompatiblity 3
chemical stability 2
cutting-edge applications 4
defined 153
domains and relevant properties 5
electrical insulation 2
hardness 3
manufacturing techniques 8–9
mechanical strength 2–3
optical properties 3
processing routes for the fabrication
drying and binder removal 7
feedstock processing 7
green body formation 7
post-sintering treatments 8
powder synthesis 7
sintering 8
properties of 170
structural 3
thermal stability 2
traditional 1
ceramic 3D printing 38, 45, 301–302
chemical and process engineering 274
chemical vapor infiltration (CVI) 100
chemically induced binding 156
Chevron-Notch Beam (CNB) method 254
composite extrusion modeling (CEM) 126
Computer Aided Design software 334
constant load 251
Continuous Fluid Dynamics (CFD) 326
continuous inkjet printing 52, 54
co-sintering 17, 215, 218–220, 228, 229, 232, 286
Coulomb repulsion 161
crack path visualization 254
creep tests 251
crystallization temperature (Tc) 157

d

D3Vero 67, 68
debinding 7, 12, 17, 18, 25, 27, 31, 32, 34, 70, 98, 127, 129, 131–135, 214, 220, 222, 271–275
debinding process 70, 127, 129, 132–135, 220, 312
debinding step 76, 88, 90, 127, 132–135, 157, 220
DEM 81, 327
density, and microstructure 272
density, and porosity 257–259
Desamanera Italian Company 67
design complexity 155, 315
Digital Grotesque 66
digital light processing (DLP) 27–28, 33, 215, 276, 331
digital slicing 154
dilatometry 256
dimensional accuracy and precision 272
direct ink writing (DIW) 10, 212, 214, 224
feedstock characterization
LAOS analyses 117–118
printability quantification 121
rheological properties 115
rheology and printability 120
steady shear experiments 115–116
strain-controlled oscillatory shear experiments 117

direct ink writing (DIW) (*contd.*)
structure-rheology perspective, printability 118–119
YSF 114
feedstock preparation 111–113
colloidal gels or polymeric systems 113–114
colloidal sol-gel systems 113
variants 114
fiber-reinforced composites 135–137
fundamental working principle 110
heat treatment and post-processing 132
non-Newtonian fluids mechanics 110
printing parameters 121–124
robocasting or robotic assisted deposition 109
directed energy deposition (DED) 12
advantages of 197–99
energy source 195
evolution of 195
limitations of 197–199
metal matrix composites (MMC) 203–205
monolithic ceramics 201–203
powder characteristics 199–200
processing 200–201
Direct-Powder Bed Selective Laser Processing (PBSLP) 161
Discrete Element Modelling and Ray Tracing 327
dithering 84
downholders for welding 299–300
drawability 119, 120
drop-on-demand (DOD)
AIJ 59
EHD 58
EIJ 58
PIJ 54–55
TIJ 57–58
VIJ 58–59
D-Shape technology 67
Dutch company CONCR3DE 66
dynamic fracture resistance 254
dynamic image analysis (DIA) 246
dynamic light scattering (DLS) 247
Dynamic Mechanical Analysis (DMA) 254

e

electrohydrodynamic DOD inkjet printing (EHD) 58
electrohydrodynamic inkjet printing (EHDP) 226
electron beam melting (EBM) 17, 153, 161–164
electronics, and micro-systems 274
Electrostatic DOD inkjet printing (EIJ) 58
energy and aerospace engineering 274
energy dispersive spectroscopy (EDS) 255
energy dispersive x-ray spectroscopy (EDS) 204, 255
environmental protection, ecological restoration for 311–315
extrusion deposition additive manufacturing (EDAM) 126

f

feedstock preparation 111–114, 126–127
fiber-reinforced composites
DIW printing 135–137
fused filament fabrication (FFF)
carbon/carbon composites 138
chemical vapor impregnation 138
fiber oxide ceramic matrix composites 138–139
liquid silicon impregnation 138
preceramic polymer impregnation 138
finite element modelling (FEM) 327
first approach, LCM multi-material printing 279
flowability 69–72, 78–80, 97, 119–120, 166, 172–173, 196, 199, 249
focus offset (FO) 161
fracture toughness 4, 129, 136, 252, 253, 304, 305, 312
full melting 17, 156, 182

functional ceramics 1, 5, 229, 309, 326, 327
functional integration 17, 18, 296
functionally graded material (FGM) 196, 219, 279
fused deposition modeling (FDM) 10, 124, 221, 251
fused "deposition of ceramics" (FDC) 126
fused feedstock deposition (FFD) 126
fused filament fabrication (FFF) 124
  feedstock characterization 128–129
  feedstock preparation 126–127
  fiber-reinforced composites
    carbon/carbon composites 138
    chemical vapor impregnation 138
    fiber oxide ceramic matrix composites 138–139
    liquid silicon impregnation 138
    preceramic polymer impregnation 138
  heat treatment and post-processing 132–135
  printhead systems 124–125
  printing parameters 129–131
fused granular fabrication (FGF) 126

g

gas adsorption techniques 259
gas pycnometry 258, 259
Gas-to-Power (G2P) route 306
Generalized Newtonian fluids (GNFs) models 115
geopolymers 68, 69
granulation 72, 126, 167
graphene–vanadium nitride quantum dots (G-VNQDs) 229
Great Barrier Reef (GBR) 311
green/smart design and manufacturing
  challenges and barriers 328
  environmental impact of ceramic sintering 325–326
  sustainability design 326–328
Group's IntrinSiC® product 71

h

hatching distance 177
hatch space 171, 177
Hausner ratio 173, 249
heat treatment and post-processing 131–135, 182–186
Herschel-Buckley model (HB) 115
high-permittivity ceramics 297
hybrid additive manufacturing
  BJ/MEX 214
  laser-assisted MEX 214
  MEX/machining 215–217
  multi-material ceramic components 217–219
  UV-assisted MEX 212–213
  VP/2PP 215–216
  VP/MEX 214–215
hybridization 14, 39–41, 211, 229
hydroxyapatite (HAP) 39, 167, 169, 202, 213, 221, 273, 303, 304, 311–315

i

Indentation Fracture Toughness (IFT) method 254
Indirect-Powder Bed Selective Laser Processing (PBSLP) 157–161, 183, 184
industrialization 187, 297, 317
infiltration 8, 64, 67, 70, 100, 183–187, 288, 307
ink transmittance 215
inkjet printing 51–52, 91, 226, 275, 307
  continuous 52–53
  drop-on-demand (DOD) 53–59
  theory of ink printability 59–62
internal flaw analysis 254
Internal shrinkage patterns 257
Ishikawa (fishbone) diagram 82
isostatic pressing 8, 183, 187
iterative building 154

j

Jenike shear cell 250

k

Kamani-Donley-Rogers (KDR) 115

l

Large Area Maskless Photopolymerization (LAMP) 36
laser-assisted MEX 214, 217
laser beam, multiple reflections of 178
layer-by-layer approach 46, 245, 277, 280
layer-by-layer fabrication 41, 245, 253, 302, 307, 328
layer-by-layer porosity 259
laser diffraction 246, 247
laser energy density 177, 179
  function of 178
laser fusion 157
laser power 158, 168, 170, 175–178, 180, 181
laser tools 154, 164
laser wavelengths 166, 168, 170, 174, 175, 187
layerwise slurry deposition (LSD) 74, 214, 323, 324
layer thickness 37, 62, 67, 68, 74, 76, 82, 83, 90, 130, 154, 166, 170, 172, 177, 253, 259, 277, 280
light microscopy (LM) 216, 246
liquid-phase sintering and partial melting 156
liquid/slurry based AM 250–251
lithography-based ceramic manufacturing (LCM) 275–277
  multi-material printing 277–279
  printing applications 279–286
LSD-print process 74

m

magnesium oxychloride cement (MOC) 67
magnesium phosphate cement (MPC) 67, 68
mask-based SL 27
material extrusion (MEX) 10
  direct ink writing (DIW)
    feedstock characterization 114–121
    feedstock preparation 111–114
    fiber-reinforced composites 135–137
    fundamental working principle 110–111
    heat treatment and post-processing 131–135
    non-Newtonian fluids mechanics 110
    robocasting or robotic assisted deposition 109
  fused filament fabrication (FFF)
    feedstock characterization 128–129
    feedstock preparation 126–127
    fiber-reinforced composites 138–139
    heat treatment and post-processing 131–135
    print head systems 124–125
    printing parameters 129–131
material intelligence 298
material jetting (MJ)
  applications 94–96
  challenges in
    curing and crack formation 93–94
    droplet ejection, impingement and fusion 92–93
    leveling 93
    nozzles failures 90–91
  definition and classification of 84–86
  historical development 86
    Micro-Particle Jetting (MPJ) 88
    Multi-Material Jetting (MMJ) 87–88
    Nano-Particle Jetting (NPJ)TM 87
  post-processes
    cleaning 98
    debinding 98
    drying and/or curing 98
    sintering 99–100
  potential of 96–98
  process variations 88–90
material strength 119, 120, 139
mechanical integrity 252–253, 256, 274, 304

mechanical properties
  and fractography 272
  anisotropy in 253
mechanical reliability 256
mechanical testing 253, 305
melting temperature (Tm) 135, 153, 157, 159, 166, 178
Mercury Intrusion Porosimetry (MIP) 259
mesoporosity 259
metal matrix composites (MMC) 203–206
MEX/machining 215–217
Micro Particle Jetting (MPJ) 87–91, 94, 96, 97, 100
microcracks 185, 256
microhardness 204, 205, 252, 253, 256
microhardness testing 253
micrometers 90, 159, 257
microporosity 259
micro-stereolithography technique 335
microstructure, density and 272
micro-systems, electronics and 274
micro-textured ceramic components 335
microwave assisted sintering (MAS) 325
microwave (MW) dielectric ceramics 228
micro X-ray computed tomography (μ XCT) 246
mini extruder deposition (MED) 126
minimal material waste 10, 155
mold-free manufacturing 155
monolithic and complex shapes 298
monolithic ceramics 17, 201–203, 281
multi-jet powder delivery systems 196
multi-material additive manufacturing
  VPP 275–276
multi-material asymmetric micro supercapacitor (MSC) 229
multi-material ceramic components 12, 217–219
multi-material jetting (MMJ) 87–90, 92, 93, 95, 96, 100, 226, 227, 229–232
multi material jetting (CerAM MMJ) 226
multi-material printing 12, 86, 97, 139, 218, 220, 277–286, 289, 299, 318, 336

n

Nano-Particle Jetting (NPJ)™ 87–91, 94, 95
niobium carbide (NbC) 161, 204
non-oxide ceramics 1, 4, 5, 18, 76, 138, 160, 186, 196, 201

o

Objet Geometries Inc. 86
Ohnesorge number 59–61
open porosity 98, 133, 134, 258, 259, 305
Optical Coherence Tomography (OCT) 259, 287
optical scanning and image analysis techniques 257
ordinary Portland cement (OPC) 65, 67–69
oscillatory measurements 251

p

particle size distribution (PSD) 112, 166, 172, 173, 177, 219, 226, 246–248, 250, 260, 269, 270, 325
paste coextrusion 223
photopolymer-based additive manufacturing 41, 46
photo-polymerization 25–34, 39, 41, 42, 111, 114, 212, 250
piezoelectric ceramics 5, 254, 330
piezoelectric DOD inkjet printing (PIJ) 54–58
plasma spray deposition 202
plasma spraying 202
Plateau-Rayleigh-instability 52, 53
platform lowering 154
PolyJet machines 86
polymer-derived ceramics (PDCs) 213, 215, 333
polymer infiltration and pyrolysis (PIP) 8, 100
polypropylene feedstock 184

pore size distribution 247, 259
porosity, density and 257–259
porous/dense alumina 278, 282–288
post-consolidation mechanical strength 253
post-processing 8, 17, 25, 38, 46, 51, 65, 70, 77, 82, 83, 87, 88, 98, 109, 112, 124, 131–135, 153, 157, 159, 164, 165, 181–187, 195, 198, 206, 224, 243, 247, 252, 256, 270, 276, 286, 302, 304, 306, 312, 322, 330, 337
powder-based AM 245–250
powder bed fusion (PBF) 11
  defined 153
  direct PBSLP 161–171
  features 171–174
  hardware parameters 174–182
  heat treatment and post-processing 182–186
  indirect PBSLP 157–161
  terminology 155–157
powder bed selective laser processing (PBSLP) 164–171
powder bed spreading 154
powder flow and spreadability 246, 249–250
powder flowability 97, 172, 173, 248, 249
powder morphology 246
powder packing density 172, 248
powder properties 153, 166, 172–174, 248
powder removal 154
powder rheometry 250
powder rheometry tests 250
powder spreading process 172–174
Power-to-Gas (P2G) process 306
precision modeling 302
print head systems 124–125
process category
  material extrusion 221–226
  material jetting 226–227
  vat photopolymerization 220–221
process repeatability 257, 317

q
qualitative classification, PSD 247
quality control 255, 269, 286, 287, 302, 316
quality management 17, 271–273, 289

r
radiofrequency (RF) components 227
Raman spectroscopy 255
rapid prototyping (RP) 155, 198, 199, 244, 303
rapid response and sovereignty 298
reaction-bonded silicon carbide (RBSiC) 100, 160
recoverability 119, 120
relative permittivity 227, 228
Reynolds number (Re) 59–61, 122
rheology tests 251
ring recoater 174
ripple effects 315
robocasting 109, 224, 229, 256, 307, 324
rotating drum method 250
roughness-induced defects 256

s
scanning electron microscopy (SEM) 121, 255
scanning speed 161, 171, 176–178, 180, 181
scanning strategy 164, 178–180, 182
second approach, LCM multi-material printing 279
sedimentation and instability 272
selected area electron diffraction (SAED) 255
selective binder activation (SBA) 66
selective paste intrusion (SPI) technique 69, 70
selective powder deposition (SPD) recoater 174
semiconductor industry 59, 96, 97, 301
shear cells 250
sheet lamination (SHL) 12, 275
shorter process chain 297
shrinkage and warpage 256

sieve analysis 247
silica-rich feedstocks 324
silicate ceramics 72
silicon carbide (SiC) 1, 4, 5, 71, 73, 76, 80, 88, 96, 97, 100, 112, 136, 160, 168, 169, 180, 184–186, 196, 224, 301
silicon-infiltrated silicon carbide (SiSiC) material 71, 100
siliconization 185
single edge-notched beam (SENB) method 253
sintering 8, 9, 12, 14, 17, 18, 25–27, 31–34, 38, 44, 63, 70, 72, 73, 76, 77, 88, 96, 98–100, 129–133, 135, 153, 154, 156, 157, 159, 160, 164, 165, 168, 176, 178, 182–184, 186, 195, 213–215, 218–220, 222–224, 227–229, 232, 243, 245, 247, 252, 253, 255–259, 270–275, 277, 279, 282, 300, 302, 304, 305, 312, 323–326, 328–331, 333, 334, 336, 337
sintering step 133, 157
size and complexity 296
slurry composition 220
slurry-based deposition 75
solid oxide cells (SOCs) 17, 229, 295, 306–310
Solidscape Inc. 86
solid-state sintering 156
static image analysis (SIA) 246
stereolithography (SL) 10, 14, 17, 25, 27, 33, 215, 245, 255, 295, 307, 324–328, 335
strain-controlled oscillatory shear experiments 117
Stratasys 86
structural ceramics 1, 3, 4, 12, 71, 286
surface roughness 44, 164, 166, 172, 173, 180, 256, 272, 300, 303

t

Taguchi method 82
technology transfer 297, 317
tensile strength tests 253
tensile test 251
theory of ink printability 59–62
thermally induced phase separation (TIPS) 159
thermal post-processing 70, 87, 88, 270, 286, 330
thermal shock tests 281
thermal stability 2, 3, 184, 186, 255, 296
thermal treatment 27, 32–33, 42, 44, 76, 96, 127, 131, 134, 183, 187, 218, 229, 331
thermogravimetric analysis (TGA) 204, 252
thermomechanical analysis (TMA) 254
thermomechanical characterization 254
thermomechanical elasticity 254
thermomechanical stability 17, 252, 254
thermoplastic 3D printing (T3DP) 87, 226
time of discharge method 249
titanium carbide (TiC) 161, 197
topology optimization 17, 244, 259
traditional ceramics 1, 10, 62, 70, 112, 153, 198, 199, 301, 307, 315, 324
transmission electron microscopy (TEM) 255
triple act system 81
tungsten carbide (WC) 4, 161, 164, 168, 171
two-photon polymerization (TPP) 10, 41–43, 215, 245, 305

u

ultra-high temperature ceramics (UHTCs) 18, 136, 197
UV-assisted direct ink writing (UV-DIW) 212
UV-assisted MEX 212–213
UV-sensitive ceramic particles 27

v

valve DOD inkjet printing (VIJ) method 58, 59
vanadium pentoxide (V_2O_5) 229

Vat photopolymerization (VPP) 10
  advantages and limits of 37–39
  AM hybridization opportunities 39–41
  biomedical engineering 273–274
  chemical and process engineering 274
  compliance and 271–273
  context 34–35
  digital light processing (DLP) 27–28
  economic and regulatory 273
  electronics and micro-systems 274
  energy and aerospace engineering 274
  feedstock formulations 30–32
  LCM multi-material printing 277–286
  lithography-based ceramic manufacturing (LCM) 276–277
  market and selection guidance 274–275
  materials and rheology 270
  mechanical modeling 35–37
  multi-material additive manufacturing 275–276
  and performance metrics 271–273
  process category 220–221
  processing 270–271
  quality management 271–273
  reproducibility 271
  stereolithography process (SLA) 27, 33
  thermal treatment 32–33
  top-down and bottom-up approaches 28–30
  two-photon polymerization 42–43
volumetric additive manufacturing (VAM) process 10, 41–45
volumetric porosity 259
VP/2PP 214, 215
VP/MEX 214–215

w

water-based suspension 72, 75
warm isostatic pressing (WIP) 183
water quench tests 284
wear-resistant coatings 202
welding, downholders for 299–300
welding process 300
within-layer approach 279, 283

x

X-ray diffraction (XRD) 255

y

yield optimization 317
yield stress fluids (YSF) 114, 224
yttria-stabilized zirconia (YSZ) 5, 112, 164, 182, 184, 307, 325

z

zirconia (ZrO_2) 1, 4, 5, 17, 70, 87, 129, 164–167, 169, 175, 196, 200, 203, 206, 213, 221–223, 231, 256, 270, 279–282, 297, 301, 303–305, 307, 309, 325, 328, 330, 336
zirconia-toughened alumina (ZTA) 175, 279–282
Zr-coatings, 203